AF595554

Selected Papers from the 2018 IEEE International Workshop on Metrology for the Sea

Selected Papers from the 2018 IEEE International Workshop on Metrology for the Sea

Special Issue Editors

Attilio Di Nisio

Francesco Picariello

MDPI • Basel • Beijing • Wuhan • Barcelona • Belgrade • Manchester • Tokyo • Cluj • Tianjin

Special Issue Editors
Attilio Di Nisio
Polytechnic University of Bari
Italy

Francesco Picariello
University of Sannio, Benevento
Italy

Editorial Office
MDPI
St. Alban-Anlage 66
4052 Basel, Switzerland

This is a reprint of articles from the Special Issue published online in the open access journal *Sensors* (ISSN 1424-8220) (available at: https://www.mdpi.com/journal/sensors/special_issues/MetroSea2018).

For citation purposes, cite each article independently as indicated on the article page online and as indicated below:

LastName, A.A.; LastName, B.B.; LastName, C.C. Article Title. *Journal Name* **Year**, *Article Number*, Page Range.

ISBN 978-3-03928-406-1 (Hbk)
ISBN 978-3-03928-407-8 (PDF)

Contents

About the Special Issue Editors vii

Preface to "Selected Papers from the 2018 IEEE International Workshop on Metrology for the Sea" ix

Maria Francesca Bruno, Matteo Gianluca Molfetta, Luigi Pratola, Michele Mossa, Raffaele Nutricato, Alberto Morea, Davide Oscar Nitti and Maria Teresa Chiaradia
A Combined Approach of Field Data and Earth Observation for Coastal Risk Assessment
Reprinted from: *Sensors* **2018**, *19*, 1399, doi:10.3390/s19061399 1

Luca Parlagreco, Lorenzo Melito, Saverio Devoti, Eleonora Perugini, Luciano Soldini, Gianluca Zitti and Maurizio Brocchini
Monitoring for Coastal Resilience: Preliminary Data from Five Italian Sandy Beaches
Reprinted from: *Sensors* **2019**, *19*, 1854, doi:10.3390/s19081854 21

Elvira Armenio, Mouldi Ben Meftah, Diana De Padova, Francesca De Serio and Michele Mossa
Monitoring Systems and Numerical Models to Study Coastal Sites
Reprinted from: *Sensors* **2019**, *19*, 1552, doi:10.3390/s19071552 39

Davide Tognin, Paolo Peruzzo, Francesca De Serio, Mouldi Ben Meftah, Luca Carniello, Andrea Defina and Michele Mossa
Experimental Setup and Measuring System to Study Solitary Wave Interaction with Rigid Emergent Vegetation
Reprinted from: *Sensors* **2019**, *19*, 1787, doi:10.3390/s19081787 55

Boleslaw Dudojc and Janusz Mindykowski
New Approach to Analysis of Selected Measurement and Monitoring Systems Solutions in Ship Technology
Reprinted from: *Sensors* **2019**, *19*, 1775, doi:10.3390/s19081775 67

Fabiana Di Ciaccio, Paolo Menegazzo and Salvatore Troisi
Optimization of the Maritime Signaling System in the Lagoon of Venice
Reprinted from: *Sensors* **2019**, *19*, 1216, doi:10.3390/s19051216 87

Umberto Robustelli, Guido Benassai and Giovanni Pugliano
Signal in Space Error and Ephemeris Validity Time Evaluation of Milena and Doresa Galileo Satellites
Reprinted from: *Sensors* **2019**, *19*, 1786, doi:10.3390/s19081786 107

Enrico Petritoli, Fabio Leccese and Marco Cagnetti
High Accuracy Buoyancy for Underwater Gliders: The Uncertainty in the Depth Control
Reprinted from: *Sensors* **2019**, *19*, 1831, doi:10.3390/s19081831 123

Paolo De Girolamo, Mattia Crespi, Alessandro Romano, Augusto Mazzoni, Marcello Di Risio, Davide Pasquali, Giorgio Bellotti, Myrta Castellino and Paolo Sammarco
Estimation of Wave Characteristics Based on Global Navigation Satellite System Data Installed on Board Sailboats
Reprinted from: *Sensors* **2019**, *19*, 2295, doi:10.3390/s19102295 141

Federico Spagnoli, Pierluigi Penna, Giordano Giuliani, Luca Masini and Valter Martinotti
The AMERIGO Lander and the Automatic Benthic Chamber (CBA): Two New Instruments to Measure Benthic Fluxes of Dissolved Chemical Species
Reprinted from: *Sensors* **2019**, *19*, 2632, doi:10.3390/s19112632 . . . . **153**

Giovanni Chimienti, Attilio Di Nisio, Anna M. L. Lanzolla, Gregorio Andria, Angelo Tursi and Francesco Mastrototaro
Towards Non-Invasive Methods to Assess Population Structure and Biomass in Vulnerable Sea Pen Fields
Reprinted from: *Sensors* **2019**, *19*, 2255, doi:10.3390/s19102255 . . . . **183**

Federica Foglini, Valentina Grande, Fabio Marchese, Valentina A. Bracchi, Mariacristina Prampolini, Lorenzo Angeletti, Giorgio Castellan, Giovanni Chimienti, Ingrid M. Hansen, Magne Gudmundsen, Agostino N. Meroni, Alessandra Mercorella, Agostina Vertino, Fabio Badalamenti, Cesare Corselli, Ivar Erdal, Eleonora Martorelli, Alessandra Savini and Marco Taviani
Application of Hyperspectral Imaging to Underwater Habitat Mapping, Southern Adriatic Sea
Reprinted from: *Sensors* **2019**, *19*, 2261, doi:10.3390/s19102261 . . . . **195**

About the Special Issue Editors

Attilio Di Nisio was born in Bari, Italy, in 1980. He received his M.S. (Hons.) degree and Ph.D. degree in electronic engineering from the Polytechnic of Bari, in 2005 and 2009, respectively, where he is currently Assistant Professor (RTDb) in Electrical and Electronic Measurements. Since 2005, his research has spanned the fields of analog-to-digital and digital-to-analog converter modeling and testing, estimation theory, software for automatic test equipment, sensors, image processing for quality control applications and document understanding, medical imaging, photovoltaic panel modeling and testing, environmental monitoring systems, and soil mechanics testing equipment. His current research interests include DSP-based systems for power quality analysis, UAV performance measurement, wireless sensor networks, and electromagnetic tracking systems for surgical navigation. Dr. Di Nisio is a member of the IEEE I&M Society and of the Italian Association "Electrical and Electronic Measurements Group" (GMEE).

Francesco Picariello received his B.Sc. (2009) and M.Sc. (2012) (cum laude) degree in electronic engineering, from the Faculty of Engineering, University of Salerno, Fisciano, Italy. He received his Ph.D. degree in Information Engineering from the University of Sannio, Benevento, Italy in 2016. He is currently working in the Laboratory L.E.S.I.M., Department of Engineering of the University of Sannio, where he is a postdoctoral researcher in the field of electrical and electronic measurements. His research interests include electrical and electronic circuit and system modeling, applied electronics, embedded measurement systems, microelectronics, power electronics, wireless sensor networks, and road safety. Francesco Picariello has published around 50 papers in international journals and in national and international conference proceedings on the following subjects: embedded systems, intelligent sensors, wireless sensor networks, distributed measurement systems, mobile devices, power consumption analysis for workstation computers, unmanned aerial vehicles, and aerial photogrammetry.

Preface to "Selected Papers from the 2018 IEEE International Workshop on Metrology for the Sea"

This Special Issue offers extended versions of selected papers from the 2018 IEEE International Workshop on Metrology for the Sea. It is devoted to recent developments in instrumentation and measurement techniques applied to the marine field.

The sea encompasses a wide range of different environments and human activities, and that richness emerged clearly from the scope of the proposed research papers. That variety is reflected, for example, by the different applications of measurement systems to satellites, airborne devices, boats, buoys, and submarine drones.

Furthermore, the selected papers delineate important application areas of measurement related to the sea: monitoring and defense of coasts; technologies for navigation above and below sea surface; analysis of waters' quality and their pollution; and monitoring and safeguarding marine species. Main themes illustrated in this Special Issue are outlined below.

Developing coastal monitoring systems and measuring the interaction with waters are pivotal for implementing correct and timely defense and remedial actions. Several methodologies are reported in this issue, from remote sensing to fixed installations and buoys, to laboratory-scale models.

Improving navigation and operational safety and efficiency of ships is an area of continuous research. The correct operation of measurement and monitoring systems in addition to their high quality should be preserved in the case of harsh and hazardous working conditions, as well as for buoyancy and depth control of underwater drones. Virtual Automatic Identification System Aid to Navigation represents a valid alternative to traditional maritime signaling in the case of restricted visibility conditions and changing bathymetry restrictions. Greater positioning accuracy of Galileo GNSS may be achieved by properly taking into account the orbit anomalies of the Milena and Doresa satellites. Interestingly, GNSS can be used not only for reconstructing the movements of a sailboat, but also for getting a reliable estimate of wave parameters.

Several papers are devoted to measurement systems for studying the underwater habitat. Benthic fluxes of dissolved chemical species can be measured with the use of benthic landers, to provide useful data on the biogeochemical cycle and pollution. Hyperspectral imaging paves the way to automated mapping and monitoring of the benthic habitat, while integrity and characteristics of habitats constituted by coral forests, which can be severely affected by anthropic activities, can be evaluated by non-invasive image-processing-based methods.

The papers collected in this issue face challenges related to the previously mentioned themes, offering insights into new solutions to key measurement problems for marine applications, and shedding light on unique opportunities. The presented research activities are very challenging, and the adopted methodologies may be useful not only for people working in the specific field of metrology for the sea, but also for researchers interested in advanced measurement topics.

Attilio Di Nisio, Francesco Picariello
Special Issue Editors

Article

A Combined Approach of Field Data and Earth Observation for Coastal Risk Assessment

Maria Francesca Bruno [1,*], Matteo Gianluca Molfetta [1], Luigi Pratola [1], Michele Mossa [1], Raffaele Nutricato [2], Alberto Morea [2], Davide Oscar Nitti [2] and Maria Teresa Chiaradia [3]

1 Department of Civil, Environmental, Building Engineering and Chemistry, Technical University of Bari—Via E. Orabona, 4, 70125 Bari, Italy; matteogianluca.molfetta@poliba.it (M.G.M.); luigi.pratola@poliba.it (L.P.); michele.mossa@poliba.it (M.M.)

2 Geophysical Applications Processing (GAP) s.r.l., 70126 Bari, Italy; raffaele.nutricato@gapsrl.eu (R.N.); alberto.morea@uniba.it (A.M.); davide.nitti@gapsrl.eu (D.O.N.)

3 Department of Physics "M. Merlin", Technical University of Bari, 70126 Bari, Italy; mariateresa.chiaradia@ba.infn.it

* Correspondence: mariafrancesca.bruno@poliba.it; Tel.: +39-080-4605207

Received: 15 February 2019; Accepted: 15 March 2019; Published: 21 March 2019

Abstract: The traditional approach for coastal monitoring consists in ground investigations that are burdensome both in terms of logistics and costs, on a national or even regional scale. Earth Observation (EO) techniques can represent a cost-effective alternative for a wide scale coastal monitoring. Thanks to the all-weather day/night radar imaging capability and to the nationwide acquisition plan named MapItaly, devised by the Italian Space Agency and active since 2010, COSMO-SkyMed (CSK) constellation is able to provide X-band images covering the Italian territory. However, any remote sensing approach must be accurately calibrated and corrected taking into account the marine conditions. Therefore, in situ data are essential for proper EO data selection, geocoding, tidal corrections and validation of EO products. A combined semi-automatic technique for coastal risk assessment and monitoring, named COSMO-Beach, is presented here, integrating ground truths with EO data, as well as its application on two different test sites in Apulia Region (South Italy). The research has shown that CSK data for coastal monitoring ensure a shoreline detection accuracy lower than image pixel resolution, and also providing several advantages: low-cost data, a short revisit period, operational continuity and a low computational time.

Keywords: integrated coastal zone management; coastal risk; shoreline erosion; Earth Observation; COSMO-SkyMed; ground truths

1. Introduction

Sea and the coastal areas constitute a driving force for the economic activity based on maritime resources, the so-called "blue economy" [1]. Actually, maritime and coastal tourism are the leading sectors of the European maritime economy. The priority objective of the coastal policy should hence be to enhance ecological and landscape resources in order to further increase their attractiveness and economic and social growth. Any enhancing measure strongly requires proper governance of coastal areas through integrated management plans and actions targeted to mitigate and counteract coastal risks [2]. Therefore, a detailed outline of the dynamics of coastal erosion phenomena over different time-scales is necessary for an effective coastal risk assessment and for land-use planning purposes [2,3].

Growing urbanization, human activities and climate change are exacerbating the impact of shoreline erosion phenomena [4]. Hence, shoreline position and its variability over time are an essential element for coastal management [5,6], coastal protection structure evaluation [7], validation and calibration of numerical models [8–10] and climate change studies [11,12]. The coastline can ideally be defined as the land-sea interface [13], and its position is constantly changing over time, mainly due to the dynamic nature of the sea level that is influenced by astronomical tides, waves, storm surge, set-up, run-up and sediment transport (cross-shore and long-shore currents) [14,15]. Therefore, the shoreline position must be defined in the most appropriate temporal scale; in the case of a long-term study, a yearly coastline can be adequate [12], whereas in the case of run-up analysis, it may be necessary to sample positions with much greater frequency [9,16].

In situ measurement techniques have been successfully adopted for shoreline observation obtaining accurate results, such as the ones that involve Global Navigation Satellite System (GNSS) campaigns [5,17,18], but a continuous GPS monitoring turns out to be impractical in terms of costs and logistics over a wide scale.

Innovative monitoring approaches, based on remote sensing techniques, could be an economically sustainable alternative providing information on shoreline evolution trend and promptly indicating also the occurrence of risk situations [19]. Different remote sensing measurement techniques can be successfully adopted for shoreline observation: optical sensors on both aerial and space platforms [19], LiDAR campaigns [20,21], Unmanned Aerial Vehicle (UAV) [22–25], videomonitoring systems [26–29]. The exploitation of very-high resolution aerial optical images and LIDAR surveys allows very accurate positioning of the shoreline over a wide scale, but their high acquisition costs strongly limit the frequency of monitoring, whereas small UAVs and videomonitoring systems offer a low cost solution but they are able to provide information only about limited areas.

Over the years, Synthetic Aperture Radar (SAR) has proved to have a positive cost-benefit ratio for territorial analysis, from small/medium to wide scale and the all-weather day/night imaging capability has made this approach very popular, especially for mapping remote areas or under extreme climatic conditions [19]. The use of satellite SAR sensors has been limited in the past by the low spatial resolution of the images [30,31] . Since 2007, the new constellations of satellites equipped with high resolution X-band SAR sensors, such as the COSMO-SkyMed (CSK) and TerraSAR-X (TSX) missions, potentially allowed to identify and monitor shoreline evolution with meter or sub-meter level accuracy (according to the selected acquisition mode), thus fostering the exploitation of SAR data for coastline extraction. Moreover, the nationwide acquisition plan named MapItaly has been devised by the Italian Space Agency (ASI), to provide CSK high-resolution satellite SAR images covering the whole Italian territory with a 16-day revisit period [32] .

The existing approaches for shoreline extraction from SAR images are classified into two types: radiometric approach, certainly the most used, and interferometric approach. The first approach attempts to locate the discontinuities present in significant intensity values of the images [33–36] whereas the second exploits the loss of coherence typical of water-covered areas [37,38]. To overcome the issues related to the presence of noise in SAR images, numerous other approaches, based on complex mathematical models such as Active Contour Model [39,40], Level Set Functions [41–44], Snake models [45] have been tested with good results over large areas.

In the last few years, the very fine spatial and radiometric resolution of COSMO-SkyMed SAR images has encouraged the development of several procedures aimed at shoreline extraction. CSK amplitude SAR image processing [46,47] has provided very satisfactory results in coastline mapping achieving an accuracy equivalent to the spatial resolution, proved by comparing results with in situ shoreline measures acquired during satellite observation. A Bayesian estimation theory able to detect sea boundaries at full resolution and low error rate in a totally unsupervised way has also been proposed [48].

An interferometric technique for the coastline extraction from CSK interferometric pairs, using the coherence loss of pixels belonging to the sea surface, has been developed [38]. Dual polarimetric CSK data for coastline extraction have also been experimented [49–51], exploiting the correlation between co- and cross-polarized amplitude channels. Furthermore, an innovative coastline extraction procedure from Full-Polarized SAR imagery has been developed to integrate an Autoassociative Neural Network (AANN) and a Pulse-Coupled Neural Network (PCNN) [52].

Field data are always essential to improve the robustness, accuracy and reliability of any approach for coastal risk assessment based on Earth Observation (EO) data. In particular, remotely sensed data must be accurately geocoded, validated and corrected on the basis of the actual marine weather conditions.

This paper presents a semi-automatic integrated approach for shoreline detection and monitoring, as well as its application on two different test sites in Apulia Region (South Italy). An innovative model combining backscatter intensity data with the interferometric analysis has been tested on CSK Stripmap HIMAGE mode data providing an accurate shoreline detection and a coastal type classification. Thanks to the integration of satellite SAR images and ground truths (that are used for proper EO data selection, tidal corrections and validation of EO products), the developed processing chain provides a complete SAR image analysis system for shoreline monitoring, allowing to overcome some issues in the processing of SAR data and ensures the best possible results in coastal monitoring, as remarked in the following sections. The paper is structured as follows: Section 2 provides details on the developed integrated EO system and its outlines; Section 3 illustrates the features of the case study and available data and in Section 4 the obtained results are presented. Section 5 contains the discussion on the proposed scheme and in Section 6, conclusions are provided.

This paper is an extended version of "Remote Sensed and In Situ Data: an integrated approach for Coastal Risk Assessment" published in "2018 IEEE International Workshop on Metrology for the Sea; Learning to Measure Sea Health Parameters (MetroSea) Conference" Proceedings [53].

2. Methodology

The principle behind all applications of SAR intensity data to shoreline extraction is based on the active nature of the sensor: smooth surfaces reflect incident radiation mainly in the specular direction, while rougher surfaces have a more diffuse scattering pattern. Therefore, backscattering level is negligible on calm water bodies with respect to land areas. This suggests immediately the use of a threshold approach to discriminate land from water within a SAR image. However, precise shoreline extraction from SAR images is a non trivial task since several factors influence the actual backscattering level, including surface roughness and soil moisture (for land areas), and the presence of capillary waves (for water bodies). Moreover, SAR images are affected by speckle noise that causes statistical fluctuations in the backscattering levels, hindering stable detection of threshold values.

Many methods have been developed to deal with the shoreline extraction from SAR data. They generally consist of two steps: (i) despeckling; (ii) segmentation of the SAR image through the identification of the land-sea interface.

Concerning the first step, it can be performed in different ways, i.e., through spatial or temporal filtering operators, or by exploiting polarization diversity. Since the aim of our research has been to design an operational EO system for the continuous monitoring of coastal areas, full polarized data have not been considered because the MapItaly program routinely collects only Stripmap HIMAGE single-polarization acquisitions. The adaptive despeckling algorithm proposed by [54] has been adopted in this study, since it has been demonstrated to provide a satisfactory trade-off between noise reduction and edge sharpness.

The traditional techniques for land-sea image partitioning exploit edge-detection operators [55,56], generally gradient-based, that, although effective, are not very robust, since they are not based on explicit mathematical models of speckle noise and edge extraction. The "classic" edge-detection algorithms are very sensitive to noise and, consequently, their effectiveness depends on the filters used to reduce noise to acceptable levels. However, too powerful filters can blur the edges; so, in order to obtain satisfactory results it is necessary to find a fair compromise between the noise reduction and edge sharpness. In addition, the choice of threshold values for use in edge-detection algorithms represents a crucial point, due to the risk of including some spurious elements, not belonging to the land-sea interface.

In this paper different segmentation techniques have been experimented for land-sea interface extraction from SAR images and an innovative algorithm, based on active contour analysis, has been tested [57]. A performance evaluation of the innovative algorithm has been carried out by comparing the extracted shorelines to GPS shorelines measured simultaneously with the SAR acquisition.

Moreover, as part of the experimental investigation, a data processing chain, known as COSMO-Beach [58,59], has been implemented to perform shoreline monitoring using both SAR amplitude and interferometric coherence information [60]. The developed system exploits remote sensed images from CSK constellation combined with detailed topographic and meteorological data collected in situ (Figure 1). The implemented procedure consists in (i) the selection of satellite images to be included in the processing, based on automatic checks of the marine weather conditions over the target area, (ii) the morphological classification of the coastline, (iii) the segmentation of SAR data for land-sea interface identification and (iv) the final tidal correction. Following these steps, a diachronic analysis of the extracted shorelines can be achieved. Hereafter, the main processing steps are outlined.

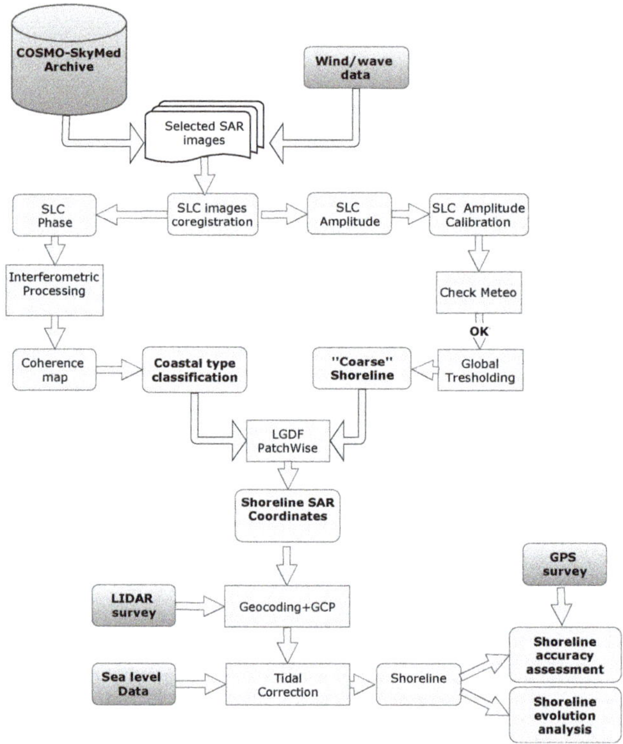

Figure 1. Diagram flow of the COSMO-Beach coastal monitoring system.

2.1. Marine Weather Conditions

Wind and wave conditions during the SAR acquisition greatly influence the backscattering on the sea surface. In the presence of waves, the values of the backscattering coefficient (σ_0) are much higher than under calm conditions. In particular, X-band SAR data seems to be more sensitive to the waves compared to L-band SAR [61].

Strong wave motion can lead to erroneous definition of the coastline, which would be landward by several pixels compared to the real position [39]. Therefore, sea conditions are an essential element to avoid selecting "noisy" images, whose processing could produce unreliable results and large errors in shoreline extraction.

The exploitation of wind and wave data allows a preliminary selection of the SAR images archived in the catalogue, and to discard *a priori* all the images acquired during storms or immediately following them. Post storm images, indeed, could be useful for the quantification of damage, but they could lead to a wrong assessment of coastal dynamics. Since wave buoys are often too sparsely distributed along the coastal zone and very few coastal areas are provided with local wind and wave measures, an analysis has been carried out aimed at extracting information on sea conditions directly from SAR data amplitudes, to be exploited in the case of unavailability of in situ measurements. Wind and wave fields can be directly inferred from SAR amplitude through a number of complex algorithms [62–64]. Actually, we are not interested in the estimation of the wave field, but rather in selecting only the SAR images acquired under calm weather conditions. Therefore, a simplified procedure has been proposed and implemented in the COSMO-Beach processing chain that consists in a first-order statistical analysis, based on the computation of mean and standard deviation of backscattered amplitude data in open sea areas located offshore the target regions.

2.2. Coastal Type Classification

A further key element of the MapItaly program is to provide SAR images taken from slightly different observation directions, thus allowing SAR interferometric (InSAR) analyses, based on the coherent nature of SAR imaging.

InSAR analysis has been applied in this study, since the estimation of the interferometric coherence through ergodic estimators can provide an additional information layer to be used for remote classification of the coastal morphology. This allows in turn: (i) the automatic identification of rocky and sandy coastal areas, (ii) the adaptive and optimal configuration of the segmentation algorithm, thus ensuring satisfactory performances regardless of the coastal type of the investigated areas.

A further preliminary step consists in the coastal type classification, which is recommended since it can enhance the land-sea boundary extraction. The next processing step, consisting in the segmentation of the SAR images, must be indeed properly and differently configured for sandy and rocky segments of the coast.

The COSMO-Beach processing chain automatically detects the shoreline type, by taking into account both amplitude and interferometric coherence information: high coherence and signal amplitude values allow to classify image pixels as "rock" or "coastal structure", while the pixels with lower values are recognized as "sand". The selected SAR dataset is first co-registered with sub-pixel accuracy [65] and an average coherence map is hence computed from multiple interferometric pairs. Optimal co-registration results are obtained by properly selecting a reference image named Master. More specifically, the perpendicular baselines and dates of the available acquisitions are analyzed in order to identify the acquisition closer to the "mass center" of the point cloud representing the various images in the space of temporal and perpendicular baselines. This selection is performed automatically although it can be changed by the user in those cases in which the automatic result is not adequate due to the presence of strong artifacts (e.g., atmospheric or focusing artifacts).

Furthermore, SAR amplitude images are radiometrically calibrated and despeckled. In order to mask out ambiguities and artifacts, apodization filters are also recommended [66].

Available ortophotos are, hence, used for the tuning of the amplitude and coherence thresholds to be configured in the coastal classification process.

2.3. Land-Sea Interface Extraction

A radiometric approach has been used for the extraction of the shoreline: it exploits the amplitude of SAR images by assuming that, in the absence of wave and/or wind, surface water exhibits very low backscattering values with respect to onshore areas.

Three different segmentation approaches have been implemented and tested on real data: (i) edge-detection methods, (ii) region-growing algorithms, (iii) active contour analyses. The first two approaches have been widely experimented for shoreline extraction from SAR images, with satisfactory results. The last one, known as Local Gaussian Distribution Fitting (LGDF) [57], is an innovative approach, originally developed for bio-medical applications, and never experimented before for the segmentation of SAR images. The LGDF approach has been tested because SAR images have a similar behavior as medical diagnostic images that are also affected by many noise sources and classical edge- or region-based segmentation approaches often fail.

In order to assess the potentials of shoreline extraction algorithms based on high resolution SAR data acquired by the CSK constellation, tests have been performed on both sandy and rocky (or urbanized) shores. The LGDF algorithm has been proved to work well in all the examined situations and it has been implemented in a two-step shoreline extraction process. The co-registered dataset is initially segmented using a global thresholding algorithm [55], in order to extract a preliminary "coarse" shoreline. The land-sea interface detection is hence refined by exploiting the "first-guess" land-sea boundary as well as the coastal type information, that are both provided as input to the LGDF segmentation code.

After the LGDF segmentation step, for noisy images, the end result might not be what is expected and a refinement step must be performed. In such cases, COSMO-Beach enables the operator to re-analyze SAR images and manually change input parameters of the segmentation algorithm. This change can be applied globally (i.e., to the whole image) or locally, i.e., only on those stretches that exhibit critical results, thus allowing a faster and more accurate refinement. In any case, a visual inspection of segmentation results by a skilled operator is recommended, although time-consuming, since land-sea interface from SAR images is not clear as shoreline from optical images. The visual inspection and the refinement step require operator interaction, whereas all the other processes are fully automated.

2.4. Precise Geocoding

All the shorelines generated by processing the input dataset are available in the Master SAR geometry and are all co-registered at sub-pixel level. Additionally, the results can be projected in geographic coordinates. Interestingly, the conversion from SAR coordinates to geographic coordinates is performed only for the Master geometry since all the shorelines are precisely co-registered on it. In this way, apart from reducing the computational cost of the geocoding (which is performed only for one image), we also avoid the random perturbation of geocoding errors that would result if each single image of the dataset was geocoded independently. The geolocation of the Master image is performed by taking into account the atmospheric propagation delay correction which is estimated through Numerical Weather Modeling (NWM) and is integrated into the coastal monitoring system in order to ensure a geolocation accuracy close to or better than the spatial resolution of the SAR images [67,68]. Finally, the extracted land-sea interface is further precisely geolocated using Ground Control Points (GCP).

2.5. Tidal Correction

After precise geocoding, land-sea boundaries extracted from SAR images have to be corrected for the effect of tidal levels and a datum-based shoreline can be defined. Neglecting tidal effects along low slope sandy beaches could lead to significant errors in evolution assessment, since high and low tide shoreline variation can be remarkable. Therefore, land-sea interface extracted from the images needs a final refinement, considering tidal oscillations, in order to be referred at 0 m above sea level. This refinement step requires a high resolution Digital Elevation Model (DEM) in the intertidal zone in order to evaluate slope along shore-perpendicular transects and tide values recorded by a tidal station in a near location. HRTI-5 is the DEM product level recommended for this correction [69].

3. Case Studies

The proposed approach has been tested in two target areas located along the Apulian coastline (Figure 2): the first one, Torre Canne, is located along the Adriatic coast whereas the second one, Porto Cesareo, lies along the Ionian coast.

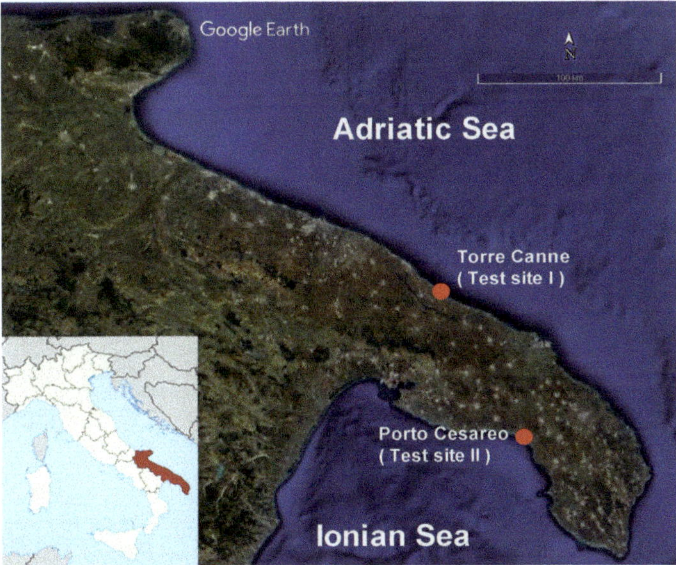

Figure 2. Selected test sites: Torre Canne and Porto Cesareo, both located in Southern Italy. The optical image is from GoogleEarthTM.

Both areas show a low-slope sandy beach affected by a remarkable erosive process that has significantly reduced the average emerged beach width. In both cases, human activities, started in the 1960s during Italian economic development, were significantly increased in recent decades and the increased land exploitation has brought a lack of sediment in the littoral cell, triggering a massive erosive trend. Due to the shoreline regression phenomena and their popularity, the selected areas have been constantly monitored for many years, so that high quality topographic and meteorological data are available in both locations.

Since 2004 several wide coastal monitoring programmes supported by the European Union (POR 2000–2006 and FESR 2007–2013 funds) have been carried out in Apulia region, allowing to obtain a detailed picture of the coastal trends and a huge database for scientific research. Those programmes included many actions: bathymetric, topographic and sedimentologic field campaigns, aerial and LIDAR surveys, real-time shoreline monitoring using webcams [14,16,28,29] and a deployment of a meteomarine network (Apulia Region Meteomarine Network, hereinafter referred to as "SIMOP") [70–72]. The availability of marine weather measures turns out to be extremely helpful for accurate selection of SAR acquisitions available in the CSK archive, based on a preliminary check on the sea weather conditions at the acquisition time. All the available field data collected in the target sites are listed in Table 1.

The examined SAR images are all HH (horizontal transmitting, horizontal receiving) polarized and captured by the SAR1 satellite of the CSK constellation in Stripmap HIMAGE mode (spatial resolution: 3×3 m^2) acquired in the framework of the MapItaly plan. The horizontal polarization is generally preferred to the vertical one since it ensures the largest contrast of the backscattering coefficient between the beach and the sea of the polarization modes [61].

Table 1. Remote sensed and field data selected for the experimental analysis over the two test sites.

	Test Site I	**Test Site II**
SAR dataset	19 CSK Stripmap HIMAGE	38 CSK Stripmap HIMAGE
DEM	LIDAR DEM 2008 1 m (POR PUGLIA 2000–2006)	LIDAR DEM 2008 1 m (POR PUGLIA 2000–2006)
Aerial photos	Maritime State Office (Apulia Region)	Maritime State Office (Apulia Region)
Waves	RON Wave Buoy (Monopoli)	SIMOP Wave Buoy (Taranto)
Tides	RMN Tidal station (Bari)	SIMOP Tidal station (Porto Cesareo)
GPS survey	60 GPS transects spaced 10 m	50 GPS transects spaced 10 m

A 14-km long coastal stretch has been selected as test site I, between towns of Torre Canne and Savelletri. The nearest tide gauge is located in the harbour of Bari, belonging to the Italian National Tide gauge Network (hereinafter referred to as "RMN") and wave data are collected by a wave buoy moored 15 km north of the site which is part of the Italian Data Buoy Network (hereinafter referred to as "RON") [73]. The H4-05 dataset of CSK acquisitions, spanning from March 2010 to May 2013, contains images acquired at around 16:52 (UTC time), along right-descending passes (Table 2).

Table 2. COSMO-SkyMed (CSK) dataset characteristics.

	CSK_H4-05_HH_RD_009	**CSK_H4-01_HH_RA_009**	**CSK_H4-02_HH_RA_009**
Instrument Mode	STR_HIMAGE	STR_HIMAGE	STR_HIMAGE
Polarization	HH	HH	HH
Look Side	right	right	right
Pass Direction	D	A	A
Track	9	209	209
Beam ID	H4-05	H4-01	H4-02
Off Nadir Angle	30620	24130	24670
Satellite ID	SAR1	SAR1	SAR1

The second experimental case study (Test site II) is located along a very popular beach (Bacino Grande) in the southern part of the Porto Cesareo municipality. It is a 3000 m long sandy beach, backed by a sub-parallel coastline dune ridge and oriented W-E. Since 2006, the area of interest has been constantly monitored by the SIMOP network, that includes: (i) an anemometric station, (ii) a tide gauge located a few hundred meters towards the South, and (iii) a wave buoy moored offshore in the Gulf of Taranto, about 60 km north of Porto Cesareo. The examined SAR images are part of two distinct CSK data takes, with an off-nadir angle of 24.13° and 24.67° (beams H4-01 and H4-02, respectively) and along ascending orbits (Table 2). The H4-01 dataset spans the period from 2009 to 2011, whereas H4-02 dataset was acquired between 2010 and 2015.

4. Results

From the CSK catalogue, two Stripmap HIMAGE acquisitions datasets, covering the target regions, have been selected according to the wind and wave conditions provided by the nearest marine observing stations from SIMOP, RON and RMN. When observed data were unavailable, that preliminary selection step has been performed exploiting ERA-Interim Re-Analysis dataset from the European Centre for Medium-Range Weather Forecasts (ECMWF) [74].

The first selected dataset (Test Site I) contains 19 images, spanning from 2010 to 2013, whereas the second one (Test Site II) contains 38 CSK images acquired between 2009 and 2015. Remotely sensed and in situ data, provided as input to the COSMO-Beach processing chain used for validating the experimental results, are listed in Table 1.

Taking into account that, in case of calm sea conditions, the surface of the sea appears as a very homogeneous region characterized by rather low amplitude values, mean and standard deviation of the amplitude values have been calculated in a sub-region, located offshore the examined beach, and extracted from the filtered data, and their relationship with recorded wave heights has been investigated. The calm sea is characterized, as expected, by low backscattering coefficient values, whereas in rough sea conditions σ_0 values grow considerably (Figure 3).

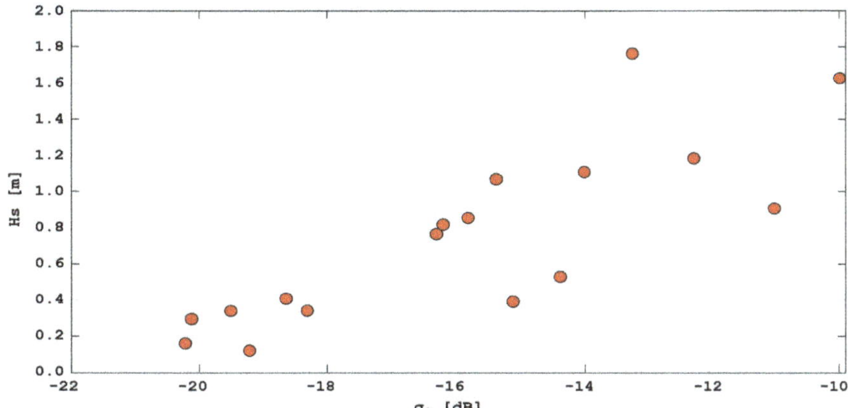

Figure 3. Scatter plot showing the correlation between the mean backscattering coefficient (computed offshore) and the significant wave height measured by a wave buoy (Test Site I).

Sea conditions can be likely distinguished using the average amplitude values: it was observed that when the backscattering coefficient exceeded a threshold value equal to −19 dB, images could certainly be discarded, since acquired in conditions of rough seas with significant wave heights (H_s), typically greater than 0.5 m. These results are in good agreement with results published in the recent literature for TSX constellation, i.e., the other X-band satellite SAR mission, which confirmed that the reflectivity of water surfaces in light wind conditions are less than −19 dB [75].

This step would allow to select part of the SAR datasets for further analysis when field data or high resolution wave model data are not available.

After filtering out noisy images, the selected SAR images were preprocessed and precisely geocoded thanks to the availability of high quality LIDAR surveys covering both areas of interest (AOI). Concerning the use of the interferometric coherence for coastal type classification, multiple acquisition pairs have been selected, with small temporal and geometric baseline in order to minimize temporal and spatial decorrelation noise sources. An average coherence map has been hence computed for minimizing time-uncorrelated fluctuations, allowing shoreline classification (Figure 4).

Very good results have been achieved in coastal type discrimination (with a 90% success rate for Test site I and close to 100% for Test site II), as proved by aerial photos provided by the Maritime State Office of Apulia Region. The coastline classification has also been fruitfully exploited for the application of the segmentation algorithm in the next step.

Three different segmentation algorithms have been tested on three 165 × 130 pixel matrices cut out of the original image, each of them representing a different coastal type (rock and sand) and one of them containing artifacts. In this step the performance of the tested segmentation algorithms has been visually inspected, overlaying the extracted shoreline on the original SAR image. In Figure 5, the land-sea interfaces extracted in the three sub-sites using three different segmentation approaches are displayed with different colors (red line for thresholding and region based, green line for LGDF). It can be seen from Figure 5 that SAR images have been correctly segmented in all the sub-sites, even on the sandy coast, by the LGDF algorithm, which is able to extract a "visually correct" land-sea interface. The results of all tests performed have been classified into two categories (Table 3): (i) Good if the SAR land-sea interface has no anomalies and is close to the visual boundary, and (ii) Poor if the SAR land-sea interface presents anomalies or significantly departs from the visual boundary. As reported in

Table 3, LGDF has been the only method capable of providing accurate coastline for both the scenarios and to be also robust to handle SAR intensity artefacts. Since it is an adaptive method capable of exploiting local variability of the SAR images, it has been proven to be reliable and efficient even in the case of SAR images with high intensity dispersion and power noise. Hence, the LGDF approach has been clearly preferred over the other two and it has been integrated in the processing chain for land-sea boundary extraction.

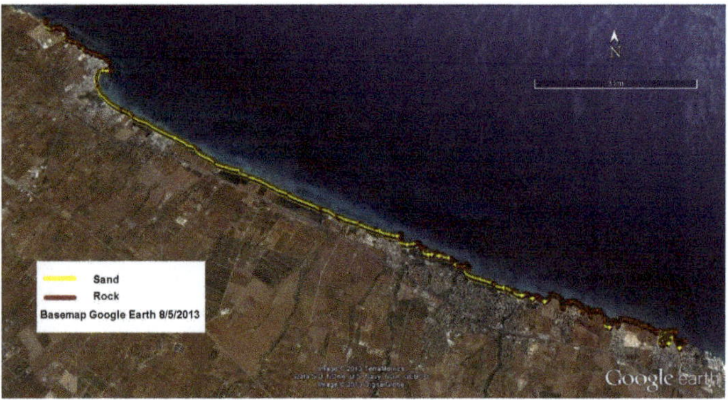

Figure 4. Coastal type classification over Test site I: rocky and sandy stretches are marked in brown and yellow, respectively.

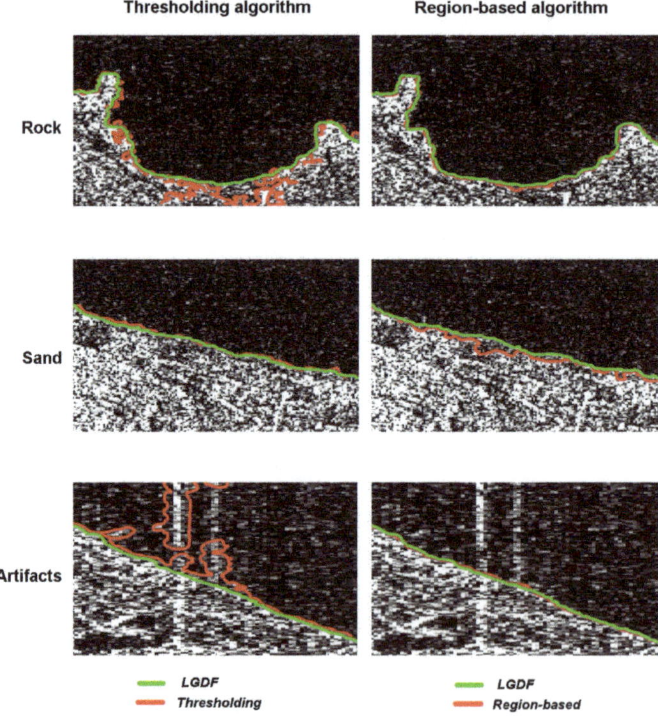

Figure 5. Segmentation results using thresholding, region-based algorithm and LGDF in the three examined sub-sites (rock coast, sandy coast and artifacts).

Table 3. Performances of different segmentation algorithms on the selected test site.

Algorithm	Rocky Coast	Sandy Coast	Artifact
Thresholding	GOOD	GOOD	POOR
Region-based	GOOD	POOR	GOOD
LGDF	GOOD	GOOD	GOOD

The LGDF algorithm accuracy for land-sea interface extraction in the sandy areas have been tested on two wide sandy beaches, because in the case of narrow beaches, where beach width is smaller or equal to spatial resolution, the backscattering of the sand portion is indistinguishable from SAR echo signal of structures delimiting the beach. In selected SAR images, both the sandy beaches are represented by a good number of pixels (about 10 pixels in the widest part of the beach) even during the winter season (Figure 6). For each test site, a SAR acquisition has been taken for the segmentation assessment, selecting optimal weather and sea conditions (calm sea conditions and light winds) in the off-season in order to avoid the crowded coasts.

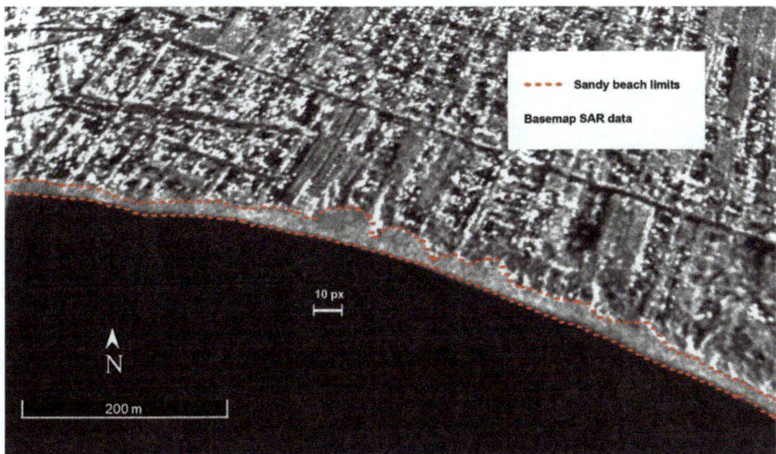

Figure 6. Synthetic Aperture Radar (SAR) Pixels belonging to sandy beach region.

Each SAR image has been segmented using an LGDF approach in order to extract the instantaneous shoreline. The LGDF algorithm, initialized in the first approximation by a "coarse" line, extracted applying a global thresholding, iteratively calculates, for subsequent steps, the dry/wet separation line using different segmentation parameters according to the coastal type. Extracted shorelines were affected by tidal fluctuations; they have been hence corrected by taking into account sea level values recorded by nearest tide stations and the local slopes derived from the available high resolution DEMs, and then referred to 0 m above sea level.

During both experiments, an accuracy assessment of shorelines extracted by the COSMO-Beach system has been performed by comparing the outcomes of the processing chain to ground truths collected through GPS survey campaigns carried out simultaneously with the SAR acquisition. The surveys have been conducted using two high accuracy GPS receivers (Leica GX1230) operating in Real Time Kinematic (RTK) mode and real-time corrected through the permanent GNSS network of the Apulia Region. Within the in situ campaigns, regularly spaced shore-perpendicular transects have been surveyed and the 0 m contour has been extrapolated.

The comparison between the corrected SAR shorelines and the observed 0 m GPS contour highlights a good performance in coastline detection using the proposed procedure (Figure 7a,b). For the statistical error analysis, 50 transects in test site I and 160 transects in test site II, located at a fixed distance of 10 m between each other, have been considered.

(**a**)

(**b**)

Figure 7. Accuracy assessment of SAR extracted shoreline: (**a**) Test site I; (**b**) Test site II.

The overall performances of the LGDF algorithm have been estimated using a series of statistical indicators. Mean error, standard deviation and Root Mean Square Error (RMSE) have been calculated for the sample of distances between SAR and GPS shorelines for each transect. Results have shown an overall mapping accuracy (RMSE) of 1.35 m for Test Site I and 2.2 m for Test Site II, i.e., lower than image pixel resolution (Table 4).

Table 4. Shoreline accuracy assessment.

Test Site	Mean (m)	Standard Deviation (m)	RMSE (m)
Test Site I	1.12	0.78	1.35
Test Site II	1.36	1.47	2.2

The algorithm performance has also been assessed using a criterion based on neighbourhood pixels [40], estimating the pixel distance between the SAR shoreline and the GPS shoreline (Table 5)

The accuracy analysis carried out on Test Site I shows that 87% of extracted shoreline segments lie within 1 pixel distance from the surveyed shoreline, and, with a tolerance of 2 pixels, the accuracy rises to 100%. Test Site II experiment, even with a greater number of transects, has achieved similar results: 85% of SAR coastline lies within 1 pixel distance from the GPS coastline, 98% of SAR coastline has a tolerance of 2 pixels, and the accuracy rises to 100% with a tolerance of 3 pixels.

Table 5. Algorithm performance for different neighborhood sizes.

	Accuracy		
Test Site	**<1 Pixel Distance**	**1–2 Pixel Distance**	**2–3 Pixel Distance**
Test Site I	87%	100%	100%
Test Site II	85%	98%	100%

It has to be highlighted that these results have been achieved by excluding, from the statistical analysis, transects where organic deposits of *Posidonia oceanica* were detected during on-site inspections. Larger errors have been detected only over those areas, since the intensity of the radar signal backscattered by the stockpiles hinders the reliable estimation of the shoreline through any segmentation algorithm (Figure 8). In such cases, as expected, the extracted land-sea interface lies along the algal deposit on the shore in the zone extending seaward from the "real" shoreline.

Figure 8. Organic deposits of *Posidonia oceanica* along the coastline during SAR acquisition.

After the accuracy assessment of shoreline detection, the coastal evolution trend has been analyzed in both target areas. Shorelines have been extracted from each available SAR image and have been corrected taking into account the tidal oscillations. Shoreline movements have been measured along user-specified orthogonal transects and rates-of-change and associated statistics have been estimated. During the investigated time periods, both test sites were erosion-prone areas: the diachronic shoreline analysis revealed a maximum retreat of 24 m from 2011 to 2013 in Test Site I (Figure 9a), and a shoreline erosion up to 20 m from 2009 to 2015 in Test Site II (Figure 9b).

A comparison between SAR extracted shorelines and 2018 satellite maps (Figure 9a,b) reveals that beach erosion in both examined areas still occurs, attesting the need for a continuous coastal monitoring system for coastal risk assessment.

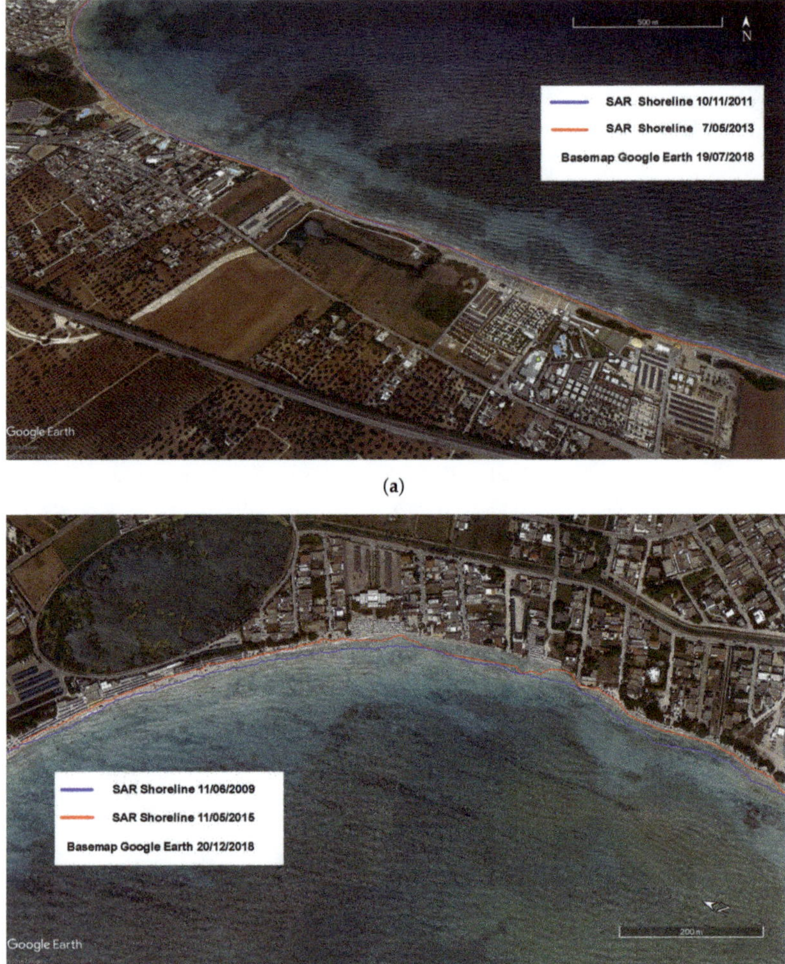

(a)

(b)

Figure 9. Shoreline changes detected from SAR images (blue line for the oldest SAR shoreline, red line for the most recent): (**a**) Test site I (Porto Cesareo); (**b**) Test site II (Torre Canne).

5. Discussion

A coastal monitoring system requires some key characteristics such as cost effectiveness, high-speed data acquisition and processing and the accuracy of the collected information. Traditional methods for coastal monitoring, generally ensuring higher accuracy than EO systems, determine a significant financial burden due to the extension of the areas, the survey frequency and the long-term continuity.

Since the aim of our research has been to set up an operational tool for coastal observation ensuring high accuracy, the experimental analysis has been carried out on CSK SAR data due to their high spatial resolution, low-cost, short revisit period and operational continuity. Moreover, the CSK satellite constellation has been operating since 2007, providing a large archive of SAR images that can be exploited for short and long-term analysis. The fast revisit/response time makes it possible to employ these data even for risk situation assessment. Besides, COSMO-SkyMed Second Generation is planned to become operational in 2019–2020, thus ensuring continuity of service for many more years. Currently, there are very few SAR satellite missions that combine all these features.

The direct comparison of accuracy results in shoreline extraction with previous studies is not straightforward since they depend on several factors including sensor characteristics, spatial resolution, sensor acquisition geometry, shoreline morphology, sea state, and processing methodology [76]. Moreover, it has to be underlined that in most studies the algorithm performance is estimated by an overall eye inspection [42,43,50] or a comparison with a manually traced coastline [45,48,77]. Few studies comply with ground truth data for accuracy assessment [46,47,51,78]. Focusing attention only on shoreline identification performance from CSK data estimated by comparing the extracted coastline with ground truths, our experiments are consistent with previous results [47], that found accuracies of 1 pixel for shorelines extracted by the CSK Stripmap HIMAGE data using a procedure based on image segmentation using locally adaptive thresholding method. CSK polarimetric data produce less accurate results in shoreline detection than those that can be achieved using CSK Stripmap images, providing a 7 pixel accuracy [51].

Interesting results have also been reported in the recent literature with high-resolution X-band SAR images acquired by the TSX mission. In particular, Vandebroek et al. [78] monitored a beach renourishment in the Netherlands. Thanks to the high accuracy of the timing information and orbital state vectors, TSX allows a sub-decimetric ranging accuracy and a high-quality geocoding [79,80], thus potentially ensuring a very precise georeferencing of the shoreline. Nevertheless, the authors have stated in [78] that the errors in the shoreline estimate may amount to some tens of meters, which is well above the TSX spatial resolution ($\sim$3 m).

Although the spatial resolutions of CSK and TSX stripmap data are similar, the results detailed in [78] are not directly comparable with the outcomes of the present study, because of the very different settings of the selected test sites. In particular, the significant discrepancies between the performances achieved in [78] and in the present study, respectively, can be explained in terms of different environmental conditions, such as meteorological, tidal and topographic factors among the areas of interest.

In particular, results reported in [78] are heavily affected by the local exposure of the test site, a mega scale beach renourishment (known as "Sand Motor") on the Dutch coast facing the North Sea, and exposed to severe winds, wave and tide conditions. Instead, tidal oscillations, wind and wave climate conditions are significantly milder in the areas selected for the present study (i.e., Torre Canne and Porto Cesareo), thus reducing the uncertainty in the still sea level. By selecting only images acquired in the most favorable weather conditions, wind and wave setup and also wave run-up have been neglected in the present study.

Moreover, even by assuming a comparable uncertainty in the measurement of the water level, the errors in the shoreline detection are expected to be significantly affected by the local slope of the shoreline: the higher the local slope, the better the accuracy of the tidal correction. The slope of the Dutch test site is much lower than the slope of the two Apulian coastal areas. Finally, SAR backscattering should be further investigated both in the intertidal zone and in areas covered by only a few centimeters of water. This further highlights the influence that environmental parameters play in the coastline extraction.

The main limitation of the presented integrated approach is related to the nature of the radar data. Unlike optical images, SAR images are not straightforward to be visually interpreted. As an example, the algorithm cannot recognize objects along the coast, as in the case of *Posidonia oceanica* blocks, because as external elements they could not be masked out in the segmentation process. Since they can affect the backscattering values in SAR images, their occurrence can hinder the accurate detection of the shoreline through SAR data segmentation analysis.

COSMO-Beach could be successfully applied to other coastal areas because it does not depend on the location of coastal areas, but it has to be highlighted that the performance in shoreline detection is strongly linked to the quality and quantity of available field and remotely sensed data, so the "one-pixel" potential accuracy can be achieved as long as all the image quality checks have been satisfied and high accuracy in situ observed data are available.

The proposed procedure for beach monitoring could be also effectively extended to SAR data provided by other European Space Agency missions released in the framework of the Copernicus Earth Observation programme. Interestingly, Sentinel-1 interferometric data can reach a revisit time of one image every six days, thus allowing a quasi real-time monitoring but their use is limited by spatial resolution (5×25 m^2 in Interferometric Wide Swath mode). The use of these low-resolution images would likely allow the monitoring of beaches presenting relevant shoreline evolution rates (sand spits, river deltas) rather than beach environments having a small dynamic range.

6. Conclusions

An integrated approach of remotely sensed and in-situ data has been presented, aimed at coastal observation on both the local and regional scale for short and long-term analysis. A semi-automatic coastal monitoring system, known as COSMO-Beach, has been implemented to detect and analyze the coastline from Cosmo-SkyMed Stripmap HIMAGE data exploiting a mixed radiometric/interferometric approach. The research conducted on the COSMO-SkyMed SAR images has shown that this data satisfactorily meets coastal monitoring requirements: (i) a large archive of low-cost SAR images with a short revisit period and operational continuity; (ii) a segmentation processing can be quite easily implemented with a low computational time and (iii) a good performance in shoreline detection ensuring an accuracy lower than image pixel resolution.

The segmentation stage adopts a two-step processing procedure: a coarse shoreline, detected using a thresholding method, feeds an Active Contour Model (LGDF) that extracts a fine resolution land-sea interface. A good performance in shoreline detection has been achieved ensuring a high sub-pixel. The whole processing chain is fully automated except the fine segmentation step that requires an operator intervention in cases of anomalies in the extracted features. The integration of remote sensed and in situ data tested in this research has proven to be effective for an accurate estimation of the shoreline and for the study of its dynamics, allowing: (i) selection of high quality SAR images, (ii) precise geocoding, (iii) tidal correction, and (iv) shoreline accuracy assessment.

Author Contributions: Conceptualization, M.F.B., M.G.M., R.N. and D.O.N.; Data curation, M.F.B., M.G.M., R.N. and D.O.N.; Formal analysis, M.F.B., M.G.M., R.N. and D.O.N.; Funding acquisition, M.M.; Investigation, M.F.B., M.G.M., L.P., R.N. and D.O.N.; Methodology, M.F.B., M.G.M., L.P., M.M., R.N., D.O.N., A.M. and M.T.C.; Resources, M.F.B., M.G.M., L.P., R.N., D.O.N. and A.M.; Software, M.F.B., M.G.M., R.N., D.O.N. and A.M.; Supervision, M.M. and M.T.C.; Validation, M.F.B., M.G.M., L.P., R.N., D.O.N. and A.M.; Visualization, M.F.B., M.G.M., R.N. and D.O.N.; Writing—original draft, M.F.B., R.N. and D.O.N.; Writing—review and editing, M.F.B., M.G.M., L.P., M.M., R.N., D.O.N., A.M. and M.T.C..

Funding: This research was partially funded by the FP7 EU projects COCONET (http://www.coconet-fp7.eu).

Acknowledgments: CSK Products of the Italian Space Agency (ASI), delivered by ASI under a license to use.

Conflicts of Interest: The authors declare no conflict of interest.

References

1. Pauli, G. *The Blue Economy: 10 Years, 100 Innovations, 100 Million Jobs*; Paradigm Publications: London, UK, 2010.
2. Clark, J.R. *Coastal Zone Management Handbook*; CRC Press: Boca Raton, FL, USA, 2018.
3. Di Risio, M.; Bruschi, A.; Lisi, I.; Pesarino, V.; Pasquali, D. Comparative Analysis of Coastal Flooding Vulnerability and Hazard Assessment at National Scale. *J. Mar. Sci. Eng.* **2017**, *5*, 51. [CrossRef]
4. Nicholls, R.J.; Woodroffe, C.; Burkett, V. Chapter 20—Coastline Degradation as an Indicator of Global Change. In *Climate Change*, 2nd ed.; Letcher, T.M., Ed.; Elsevier: Boston, MA, USA, 2016; pp. 309–324.
5. Furmańczyk, K.; Andrzejewski, P.; Benedyczak, R.; Bugajny, N.; Cieszyński, Ł.; Dudzińska-Nowak, J.; Giza, A.; Paprotny, D.; Terefenko, P.; Zawiślak, T. Recording of selected effects and hazards caused by current and expected storm events in the Baltic Sea coastal zone. *J. Coast. Res.* **2014**, *70*, 338–342.

6. Baart, F.; van der Kaaij, T.; van Ormondt, M.; van Dongeren, A.; van Koningsveld, M.; Roelvink, J.A. Real-time forecasting of morphological storm impacts: A case study in the Netherlands. *J. Coast. Res.* **2009**, *SI 54*, 1617–1621.
7. Anastasiou, S.; Sylaios, G. Assessment of Shoreline Changes and Evaluation of Coastal Protection Methods to Mitigate Erosion. *Coast. Eng. J.* **2016**, *58*, 1650006. [CrossRef]
8. Deng, J.; Harff, J.; Zhang, W.; Schneider, R.; Dudzińska-Nowak, J.; Giza, A.; Terefenko, P.; Furmańczyk, K. The dynamic equilibrium shore model for the reconstruction and future projection of coastal morphodynamics. In *Coastline Changes of the Baltic Sea from South to East*; Springer: Berlin, Germany, 2017; pp. 87–106.
9. Paprotny, D.; Andrzejewski, P.; Terefenko, P.; Furmańczyk, K. Application of empirical wave run-up formulas to the Polish Baltic Sea coast. *PLoS ONE* **2014**, *9*, e105437. [CrossRef] [PubMed]
10. Roelvink, D.; Reniers, A.; Van Dongeren, A.; de Vries, J.v.T.; McCall, R.; Lescinski, J. Modelling storm impacts on beaches, dunes and barrier islands. *Coast. Eng.* **2009**, *56*, 1133–1152. [CrossRef]
11. Enríquez, A.R.; Marcos, M.; Álvarez-Ellacuría, A.; Orfila, A.; Gomis, D. Changes in beach shoreline due to sea level rise and waves under climate change scenarios: Application to the Balearic Islands (western Mediterranean). *Nat. Hazards Earth Syst. Sci.* **2017**, *17*, 1075–1089. [CrossRef]
12. Robinet, A.; Castelle, B.; Idier, D.; Le Cozannet, G.; Déqué, M.; Charles, E. Statistical modeling of interannual shoreline change driven by North Atlantic climate variability spanning 2000–2014 in the Bay of Biscay. *Geo-Mar. Lett.* **2016**, *36*, 479–490. [CrossRef]
13. Dolan, R.; Hayden, B.P.; May, P.; May, S. The reliability of shoreline change measurements from aerial photographs. *Shore Beach* **1980**, *48*, 22–29.
14. Lisi, I.; Molfetta, M.G.; Bruno, M.F.; Risio, M.D.; Damiani, L. Morphodynamic classification of sandy beaches in enclosed basins: The case study of Alimini (Italy). *J. Coast. Res.* **2011**, *SI 64*, 180–184.
15. Armenio, E.; Meftah, M.B.; Bruno, M.; De Padova, D.; De Pascalis, F.; De Serio, F.; Di Bernardino, A.; Mossa, M.; Leuzzi, G.; Monti, P. Semi enclosed basin monitoring and analysis of meteo, wave, tide and current data: Sea monitoring. In Proceedings of the 2016 IEEE Workshop on Environmental, Energy, and Structural Monitoring Systems (EESMS), Bari, Italy, 13–14 June 2016; pp. 1–6.
16. Damiani, L.; Saponieri, A.; Valentini, N. Validation of swash model for run-up prediction on a natural embayed beach. *Ital. J. Eng. Geol. Environ.* **2018**, *SI 1*, 27–37.
17. Morton, R.A.; Leach, M.P.; Paine, J.G.; Cardoza, M.A. Monitoring Beach Changes Using GPS Surveying Techniques. *J. Coast. Res.* **1993**, *9*, 702–720.
18. Kostrzewski, A.; Zwoliński, Z.; Winowski, M.; Tylkowski, J.; Samołyk, M. Cliff top recession rate and cliff hazards for the sea coast of Wolin Island (Southern Baltic). *Baltica* **2015**, *28*, 109–120. [CrossRef]
19. Gens, R. Remote sensing of coastlines: Detection, extraction and monitoring. *Int. J. Remote Sens.* **2010**, *31*, 1819–1836. [CrossRef]
20. Stockdon, H.F.; Sallenger, A.H., Jr.; List, J.H.; Holman, R.A. Estimation of Shoreline Position and Change Using Airborne Topographic Lidar Data. *J. Coast. Res.* **2002**, *18*, 502–513.
21. Terefenko, P.; Zelaya Wziątek, D.; Dalyot, S.; Boski, T.; Pinheiro Lima-Filho, F. A High-Precision LiDAR-Based Method for Surveying and Classifying Coastal Notches. *ISPRS Int. J. Geo-Inf.* **2018**, *7*, 295. [CrossRef]
22. Pereira, E.; Bencatel, R.; Correia, J.; Félix, L.; Gonçalves, G.; Morgado, J.; Sousa, J. Unmanned air vehicles for coastal and environmental research. *J. Coast. Res.* **2009**, *SI 54*, 1557–1561.
23. Mancini, F.; Dubbini, M.; Gattelli, M.; Stecchi, F.; Fabbri, S.; Gabbianelli, G. Using Unmanned Aerial Vehicles (UAV) for High-Resolution Reconstruction of Topography: The Structure from Motion Approach on Coastal Environments. *Remote Sens.* **2013**, *5*, 6880–6898. [CrossRef]
24. Gonçalves, J.; Henriques, R. UAV photogrammetry for topographic monitoring of coastal areas. *ISPRS J. Photogramm. Remote Sens.* **2015**, *104*, 101–111. [CrossRef]
25. Turner, I.L.; Harley, M.D.; Drummond, C.D. UAVs for coastal surveying. *Coast. Eng.* **2016**, *114*, 19–24. [CrossRef]
26. Holman, R.A.; Stanley, J. The history and technical capabilities of Argus. *Coast. Eng.* **2007**, *54*, 477–491. [CrossRef]
27. Archetti, R.; Paci, A.; Carniel, S.; Bonaldo, D. Optimal index related to the shoreline dynamics during a storm: the case of Jesolo beach. *Nat. Hazards Earth Syst. Sci.* **2016**, *16*, 1107–1122. [CrossRef]

28. Valentini, N.; Saponieri, A.; Molfetta, M.G.; Damiani, L. New algorithms for shoreline monitoring from coastal video systems. *Earth Sci. Informat.* **2017**, *10*, 495–506. [CrossRef]
29. Valentini, N.; Saponieri, A.; Damiani, L. A new video monitoring system in support of Coastal Zone Management at Apulia Region, Italy. *Ocean Coast. Manag.* **2017**, *142*, 122–135. [CrossRef]
30. Cracknell, A. Remote sensing techniques in estuaries and coastal zones an update. *Int. J. Remote Sens.* **1999**, *20*, 485–496. [CrossRef]
31. Richards, J.; Jia, X. *Remote Sensing Digital Image Analysis*; Springer: Berlin, Germany, 1999.
32. Virelli, M.; Coletta, A.; Battagliere, M.L. ASI COSMO-SkyMed: Mission Overview and Data Exploitation. *IEEE Geosci. Remote Sens. Mag.* **2014**, *2*, 64–66. [CrossRef]
33. Lee, J.S.; Jurkevich, I. Coastline Detection And Tracing In SAr Images. *IEEE Trans. Geosci. Remote Sens.* **1990**, *28*, 662–668.
34. Bijaoui, J.; Cauneau, F. Separation of sea and land in SAR images using texture classification. In Proceedings of the OCEANS '94. 'Oceans Engineering for Today's Technology and Tomorrow's Preservation, Brest, France, 13–16 September 1994; Volume 1, pp. I/522 –I/526.
35. Niedermeier, A.; Hoja, D.; Lehner, S. Topography and morphodynamics in the German Bight using SAR and optical remote sensing data. *Ocean Dyn.* **2005**, *55*, 100–109. [CrossRef]
36. Liu, H.; Jezek, K. A complete high-resolution coastline of Antarctica extracted from orthorectified Radarsat SAR imagery. *Photogramm. Eng. Remote Sens.* **2004**, *70*, 605–616. [CrossRef]
37. Wang, Z.; Yang, S.; Chen, T. Coastline extraction from tandem ERS 1/2 interferometric coherence map. In Proceedings of the 2012 International Workshop on Image Processing and Optical Engineering. International Society for Optics and Photonics, Harbin, China, 9–10 January 2012; p. 83351L.
38. Dellepiane, S.; De Laurentiis, R.; Giordano, F. Coastline extraction from SAR images and a method for the evaluation of the coastline precision. *Pattern Recogn. Lett.* **2004**, *25*, 1461–1470. [CrossRef]
39. Mason, D.; Davenport, I. Accurate and efficient determination of the shoreline in ERS-1 SAR images. *IEEE Trans. Geosci. Remote Sens.* **1996**, *34*, 1243 –1253. [CrossRef]
40. Modava, M.; Akbarizadeh, G. Coastline extraction from SAR images using spatial fuzzy clustering and the active contour method. *Int. J. Remote Sens.* **2016**, *38*, 355–370. [CrossRef]
41. Osher, S.; Sethian, J.A. Fronts propagating with curvature-dependent speed: Algorithms based on Hamilton-Jacobi formulations. *J. Comput. Phys.* **1988**, *79*, 12–49. [CrossRef]
42. Shu, Y.; Li, J.; Gomes, G. Shoreline extraction from RADARSAT-2 intensity imagery using a narrow band level set segmentation approach. *Mar. Geod.* **2010**, *33*, 187–203. [CrossRef]
43. Ouyang, Y.; Chong, J.; Wu, Y. Two coastline detection methods in Synthetic Aperture Radar imagery based on Level Set Algorithm. *Int. J. Remote Sens.* **2010**, *31*, 4957–4968. [CrossRef]
44. Zhao, L.; Fan, L.; Wang, C.; Tang, Y.; Zhang, B. A non-supervised method for shoreline extraction using high resolution SAR image. In Proceedings of the 2012 International Conference on Computer Vision in Remote Sensing (CVRS), Xiamen, China, 16–18 December 2012; pp. 317 –322.
45. Sheng, G.; Yang, W.; Deng, X.; He, C.; Cao, Y.; Sun, H. Coastline Detection in Synthetic Aperture Radar (SAR) Images by Integrating Watershed Transformation and Controllable Gradient Vector Flow (GVF) Snake Model. *IEEE J. Ocean. Eng.* **2012**, *37*, 375 –383. [CrossRef]
46. Palazzo, F.; Latini, D.; Baiocchi, V.; Del Frate, F.; Giannone, F.; Dominici, D.; Remondiere, S. An application of COSMO-SkyMed to coastal erosion studies. *Eur. J. Remote Sens.* **2012**, *45*, 361–370. [CrossRef]
47. Braga, F.; Tosi, L.; Prati, C.; Alberotanza, L. Shoreline detection: Capability of COSMO-SkyMed and high-resolution multispectral images. *Eur. J. Remote Sens.* **2013**, *46*, 837–853. [CrossRef]
48. Baselice, F.; Ferraioli, G.; Pascazio, V. Coastal line extraction from SAR multi-channel images. In Proceedings of the 2012 Tyrrhenian Workshop on Advances in Radar and Remote Sensing (TyWRRS), Naples, Italy, 12–14 September 2012; pp. 322 –325.
49. Nunziata, F.; Migliaccio, M.; Li, X. Dual-polarized COSMO-SkyMed SAR data for coastline detection. In Proceedings of the 2012 IEEE International on Geoscience and Remote Sensing Symposium (IGARSS), Munich, Germany, 22–27 July 2012; pp. 5109–5112.
50. Nunziata, F.; Migliaccio, M.; Li, X.; Ding, X. Coastline extraction using dual-polarimetric COSMO-SkyMed PingPong mode SAR data. *IEEE Geosci. Remote Sens. Lett.* **2014**, *11*, 104–108. [CrossRef]
51. Nunziata, F.; Buono, A.; Migliaccio, M.; Benassai, G. Dual-polarimetric C-and X-band SAR data for coastline extraction. *IEEE J. Sel. Top. Appl. Earth Observ. Remote Sens.* **2016**, *9*, 4921–4928. [CrossRef]

52. De Laurentiis, L.; Latini, D.; Del Frate, F.; Schiavon, G. Automatic procedure for coastline extraction exploiting full-pol SAR imagery and AANN-PCNN processing chain. *Proc. SPIE* **2018**, *10789*, 10789.
53. Bruno, M.F.; Molfetta, M.G.; Pratola, L.; Mossa, M.; Nutricato, R.; Morea, A.; Nitti, d.O.; Teresa, C.M. Remote Sensed and In Situ Data: An integrated approach for Coastal Risk Assessment. In Proceedings of the International Workshop on Metrology for the Sea (MetroSea 2018), Bari, Italy, 8–10 October 2018; pp. 13–17.
54. Lee, J.S. Speckle analysis and smoothing of synthetic aperture radar images. *Comput. Graph. Image Process.* **1981**, *17*, 24–32. [CrossRef]
55. Otsu, N. A threshold selection method from gray-level histograms. *Automatica* **1975**, *11*, 23–27. [CrossRef]
56. Canny, J. A computational approach to edge detection. *IEEE Trans. Pattern Anal. Mach. Intell.* **1986**, *PAMI-8*, 679–698. [CrossRef]
57. Wang, L.; He, L.; Mishra, A.; Li, C. Active contours driven by local Gaussian distribution fitting energy. *Signal Process.* **2009**, *89*, 2435–2447. [CrossRef]
58. Bruno, M.F.; Molfetta, M.G.; Mossa, M.; Morea, A.; Chiaradia, M.T.; Nutricato, R.; Nitti, D.O.; Guerriero, L.; Coletta, A. Integration of multitemporal SAR/InSAR techniques and NWM for coastal structures monitoring: Outline of the software system and of an operational service with COSMO-SkyMed data. In Proceedings of the 2016 IEEE Workshop on Environmental, Energy, and Structural Monitoring Systems (EESMS), Bari, Italy, 13–14 June 2016; pp. 1–6.
59. Bruno, M.F.; Molfetta, M.G.; Mossa, M.; Nutricato, R.; Morea, A.; Chiaradia, M.T. Coastal Observation through Cosmo-SkyMed High-Resolution SAR Images. *J. Coast. Res.* **2016**, *75*, 795–799. [CrossRef]
60. Bovenga, F.; Refice, A.; Nutricato, R.; Guerriero, L.; Chiaradia, M. SPINUA: A flexible processing chain for ERS/ENVISAT long term interferometry. In Proceedings of the ESA-ENVISAT Symposium, Salzburg, Austria, 6–10 September 2004; pp. 6–10.
61. Wu, L.; Tajima, Y.; Yamanaka, Y.; Shimozono, T.; Sato, S. Study on characteristics of synthetic aperture radar (SAR) imagery around the coast for shoreline detection. *Coast. Eng. J.* **2018**, 1–19. [CrossRef]
62. Hasselmann, S.; Brüning, C.; Hasselmann, K.; Heimbach, P. An improved algorithm for the retrieval of ocean wave spectra from synthetic aperture radar image spectra. *J. Geophys. Res.* **1996**, *101*, 16615–16. [CrossRef]
63. Rana, F.M.; Adamo, M.; Pasquariello, G.; De Carolis, G.; Morelli, S.; Bovenga, F. A simplified local gradient method for the retrieval of SARderived sea surface wind directions. In Proceedings of the EUSAR 2014; 10th European Conference on Synthetic Aperture Radar, Berlin, Germany, 3–5 June 2014; pp. 1–4.
64. Benassai, G.; Migliaccio, M.; Nunziata, F. The use of COSMO-SkyMed© SAR data for coastal management. *J. Mar. Sci. Technol.* **2015**, *20*, 542–550. [CrossRef]
65. Nitti, D.O.; Hanssen, R.F.; Refice, A.; Bovenga, F.; Nutricato, R. Impact of DEM-assisted coregistration on high-resolution SAR interferometry. *IEEE Trans. Geosci. Remote Sens.* **2011**, *49*, 1127–1143. [CrossRef]
66. Locorotondo, P.; Guerriero, A.; Chiaradia, M.T.; Nitti, D.O.; Nutricato, R.; Bovenga, F. Preliminary assessment of side-lobe mitigation techniques for proper coherent targets selection in MTInSAR applications. In Proceedings of the Active and Passive Microwave Remote Sensing for Environmental Monitoring II, Berlin, Germany, 12 September 2018; Volume 10788, p. 1078804.
67. Nitti, D.O.; Bovenga, F.; Nutricato, R.; Refice, A.; Bruno, M.F.; Petrillo, A.F.; Chiaradia, M.T. On the use of Numerical Weather Models for improving SAR geolocation accuracy. In Proceedings of the EUSAR 2014; 10th European Conference on Synthetic Aperture Radar, Berlin, Germany, 3–5 June 2014; pp. 1–4.
68. Nitti, D.O.; Nutricato, R.; Lorusso, R.; Lombardi, N.; Bovenga, F.; Bruno, M.F.; Chiaradia, M.T.; Milillo, G. On the geolocation accuracy of COSMO-SkyMed products. In Proceedings of the SAR Image Analysis, Modeling, and Techniques XV, Toulouse, France, 21-24 September 2015; Volume 9642, p. 96420D.
69. National Imagery and Mapping Agency . *High Resolution Terrain Information (HRTI): Performance Specification*; Technical Report; NIMA: Springfield, VA, USA, 2000.
70. Damiani, L.; Bruno, M.F.; Molfetta, M.G.; Nobile, B. Coastal zone monitoring in Apulia region: First analysis on meteomarine climate. In Proceedings of the Vth International Symposium on Environmental Hydraulics (ISEH 2007), Tempe, AZ USA, 4–7 December 2007.
71. Bruno, M.F.; Molfetta, M.G.; Petrillo, A.F. The influence of interannual variability of mean sea level in the Adriatic Sea on extreme values. *J. Coast. Res.* **2014**, 241–246, doi:10.2112/SI70-041.1. [CrossRef]
72. Pasquali, D.; Bruno, M.; Celli, D.; Damiani, L.; Risio, M.D. A simplified hindcast method for the estimation of extreme storm surge events in semi-enclosed basins. *Appl. Ocean Res.* **2019**, *85*, 45–52. [CrossRef]

73. Bencivenga, M.; Nardone, G.; Ruggiero, F.; Calore, D. The Italian data buoy network (RON). *Adv. Fluid Mech. IX* **2012**, *74*, 321.
74. Dee, D.P.; Uppala, S.; Simmons, A.; Berrisford, P.; Poli, P.; Kobayashi, S.; Andrae, U.; Balmaseda, M.; Balsamo, G.; Bauer, P.; et al. The ERA-Interim reanalysis: Configuration and performance of the data assimilation system. *Q. J. R. Meteorol. Soc.* **2011**, *137*, 553–597. [CrossRef]
75. Infoterra. *Radiometric Calibration of TerraSAR-X Data*; Technical Report; Infoterra: Friedrichshafen, Germany, 2008.
76. Mitra, S.S.; Mitra, D.; Santra, A. Performance testing of selected automated coastline detection techniques applied on multispectral satellite imageries. *Earth Sci. Informat.* **2017**, *10*, 321–330. [CrossRef]
77. Modava, M.; Akbarizadeh, G.; Soroosh, M. Integration of spectral histogram and level set for coastline detection in SAR images. *IEEE Trans. Aerosp. Electron. Syst.* **2018**. [CrossRef]
78. Vandebroek, E.; Lindenbergh, R.; van Leijen, F.; de Schipper, M.; de Vries, S.; Hanssen, R. Semi-automated monitoring of a mega-scale beach nourishment using high-resolution terrasar-x satellite data. *Remote Sens.* **2017**, *9*, 653. [CrossRef]
79. Yoon, Y.T.; Eineder, M.; Yague-Martinez, N.; Montenbruck, O. TerraSAR-X precise trajectory estimation and quality assessment. *IEEE Trans. Geosci. Remote Sens.* **2009**, *47*, 1859–1868. [CrossRef]
80. Eineder, M.; Minet, C.; Steigenberger, P.; Cong, X.; Fritz, T. Imaging geodesy—Toward centimeter-level ranging accuracy with TerraSAR-X. *IEEE Trans. Geosci. Remote Sens.* **2011**, *49*, 661–671. [CrossRef]

Article

Monitoring for Coastal Resilience: Preliminary Data from Five Italian Sandy Beaches †

Luca Parlagreco [1,*], Lorenzo Melito [2], Saverio Devoti [1], Eleonora Perugini [2], Luciano Soldini [2], Gianluca Zitti [2] and Maurizio Brocchini [2]

1 Istituto Superiore per la Protezione e la Ricerca Ambientale, 00144 Rome, Italy; saverio.devoti@isprambiente.it

2 Department of DICEA, Università Politecnica delle Marche, Ancona 60131, Italy; l.melito@pm.univpm.it (L.M.); e.perugini@pm.univpm.it (E.P.); l.soldini@univpm.it (L.S.); g.zitti@univpm.it (G.Z.); m.brocchini@univpm.it (M.B.)

* Correspondence: luca.parlagreco@isprambiente.it

† This paper is an extended version of our paper published in Melito, L., Parlagreco, L., Perugini, E., Postacchini, M., Zitti, G., Brocchini, M. Monitoring for coastal resilience: A project for five Italian beaches. In Proceedings of the 2018 IEEE International Workshop on Metrology for the Sea, Bari, Italy, 8–10 October 2018; pp. 76–81.

Received: 15 March 2019; Accepted: 15 April 2019; Published: 18 April 2019

Abstract: Video-monitoring can be exploited as a valuable tool to acquire continuous, high-quality information on the evolution of beach morphology at a low cost and, on such basis, perform beach resilience analyses. This manuscript presents preliminary results of an ongoing, long-term monitoring programme of five sandy Italian beaches along the Adriatic and Tyrrhenian sea. The project aims at analyzing nearshore morphologic variabilities on a time period of several years, to link them to resilience indicators. The observations indicate that most of the beach width variations can be linked to discrete variations of sandbar systems, and most of all to an offshore migration and decay of the outermost bars. Further, the largest net shoreline displacements across the observation period are experienced by beaches with a clear NOM (Net Offshore Migration)-type evolution of the seabed.

Keywords: video-monitoring; beach morphological evolution; beach resilience; sand bars

---

1. Introduction

A natural system is resilient when it is able to restore perturbations without losing its functionality. In the same way, the resilience of a beach could be defined as the capability to preserve its functionality under changing hydro-morphological conditions. Among the main and most relevant functions of wave-dominated beach systems is its capability to dissipate and absorb the energy of waves coming from the offshore.

The morphologic impact of waves on coasts, focus of a growing volume of literature (e.g., [1–3]), has been proven a crucial matter for the subsistence of coastal communities. The thrive of coastal ecosystems and the related benefits for society are closely linked to beach integrity and well-functioning. Thus, efficient management policies require evidence-based strategies to protect coasts without affecting the beach system capability to adapt to the hazards of climate change [4]. A clear example of beach resilience are the fluctuations in shoreline position in response to a storm [5]. The shoreline recovery acts as a long-term process of maintenance of the subaerial beach width and affects the net impact of storms in time [6–10].

Nearshore sand bars are a crucial component of sandy coastal systems, since they act as sediment storage and wave energy preferential dissipation points, thus playing a key role in the overall morphologic stability of the shoreline. Generally, during large wave events sand is shifted offshore and deposited in submerged sandbars [11]. During the following calm periods, sandbars slowly migrate back onshore and possibly weld to the shoreline to restore the pre-storm subaerial beach width and morphology [12].

Relationships between the shoreline and the sandbar morphology were first described by the "morphodynamic model" of Wright and Short [13], in which beaches were classified into specific morphological states, each characterized by different sandbar configurations and connections to the shoreline. Transitions between morphological states are described in the model as a response to wave and tide forcings [14].

Among the most commonly recognized sandbar evolutionary processes is the Net Offshore Migration (NOM) pattern [15], characterized by a long-term cyclic evolution in three stages [16]. At first, a bar is generated close to the shoreline by interaction of undertow, wave velocity asymmetries and long waves effects [17]. The bar then experiences a net offshore migration, a byproduct of an alternation of gradual onshore movements during calms and strong offshore motions during storms [18]. Finally, the bar moves offshore of the breaking line and degenerates [19]. NOM cycles have been identified on several barred beaches worldwide. In the Mediterranean Sea this phenomenon was observed along the Gulf of Lions and described both by bathymetric [20] and remote-sensed data [21]. Relations between the extent and typology of beach erosional phenomena and bar stage are observed [15] since a decaying outer bar offers a reduced protection to the shoreline and inner bar, leading to enhanced erosion of the shoreline [22–24]. Recently, the analysis of shoreline recovery processes has provided further insight in the close relation between shoreline recovery rates, the bar location and the tidal regime [10,25].

In the past decades, the increase of remote sensing techniques for the monitoring of coasts have promoted a deeper understanding of the multi-year response of the beach system, allowing for a much greater spatial and temporal resolution than traditional survey methods. Moreover, they ensure that all significant events involving coastline alteration can be analyzed in a retrospective way [26]. In particular, the evaluation of barlines and shoreline positioning, as well as their medium-to-long term evolution, via remotely sensed images has been proven a rather affordable and reliable approach that allows for the collection of a large volume of data. Several metrics for a thorough analysis of beach response to hydrodynamic forcing can then be extracted by video-monitoring products. Among these are the wave run-up [27,28] which can be used as a proxy for the residual wave energy reaching the shore after wave breaking, the swash zone properties (e.g., see [29]), and the mean shoreline net changes after storm events (e.g., see [30]). A number of video-monitoring stations have been productively deployed along Italian coasts in recent years (e.g., see [30–34]) and their informational value can be proficiently used to satisfy the increasing need for informed coastal management decision planning by local authorities [35].

In this paper, we present preliminary results of nearshore morphological evolution derived from multi-year beach monitoring. The study sites are located along five Italian beaches characterized by different coastal protection demands and environmental forcing. This work is part of an ongoing, long-term monitoring programme of sandy beaches belonging to environmentally-protected areas, with the aim of providing the information on beach resilience needed to ensure the survival and good functioning of such environmental sanctuaries [36].

Section 2 provides an overview of the study sites, the local climates and video-monitoring system used for the analyses. The main results are detailed in Section 3. A discussion of the results and some concluding remarks are finally proposed in Section 4.

2. Materials and Methods

2.1. Study Sites

The study sites are located in central Italy, three along the Adriatic Sea and two along the central Tyrrhenian Sea (Figure 1). The sites present sandy beaches free from coastal defense structures and characterized by different wave exposures and seabed slopes. All the sites are micro-tidal beaches with the major tidal excursions experienced by the Adriatic beaches, in all cases the tidal excursion never exceeding 0.6 m (Rete Mareografica Nazionale—ISPRA). The main differences between the Adriatic and Tyrrhenian sites are due to inherited geological and physiographic constraints, resulting in different local wave climates, sedimentary input densities and degrees of coastal fragmentation by geological forcing (rocky headlands), as well as different amplitudes of storm surges and wave set-up, both larger in the semi-enclosed Adriatic basin.

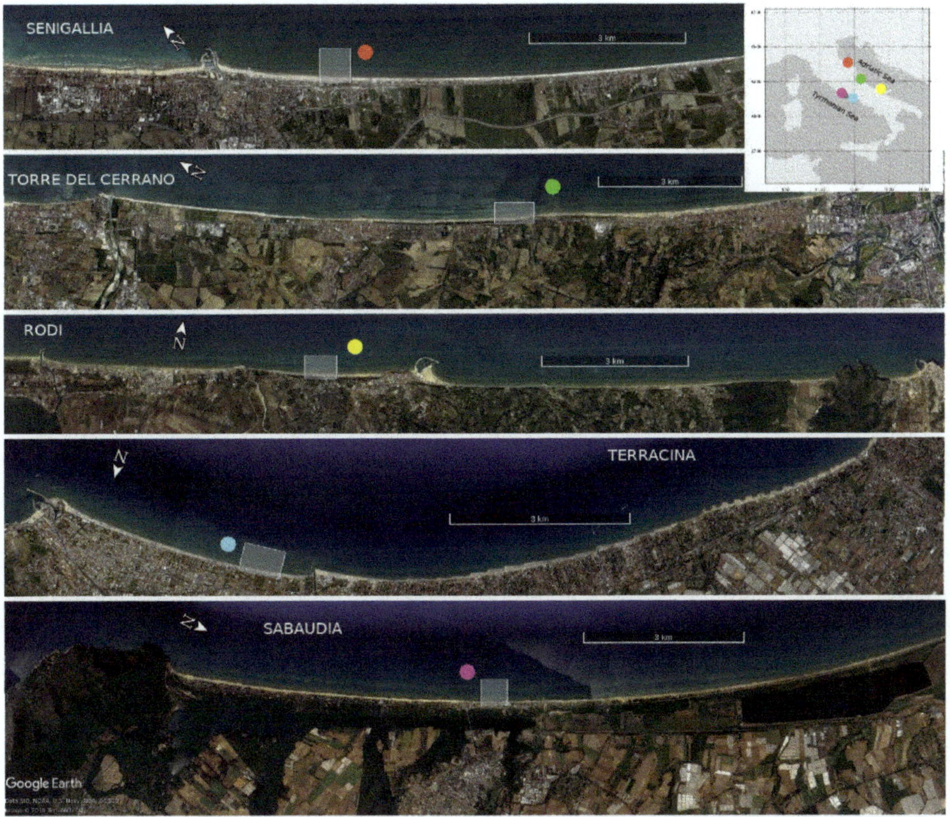

Figure 1. The five investigated locations (pictures taken from Google Earth). From top to bottom: Senigallia, Torre del Cerrano, Rodi Garganico, Terracina and Sabaudia. The shaded areas indicate the video-monitored portions of beach.

The Senigallia beach is located along the northern portion of the Central Adriatic Sea, 1 km south of the Misa river estuary, in a region of touristic attractiveness that is included in the longest stretch of unprotected beach of the Marche region [37]. The coastline is oriented approximately NW–SE and the beach is characterized by fine and medium sands ($d_{50} \approx 0.25$ mm), with a nearshore slope of about 0.5% computed, as for all the sites, as the average slope between the shoreline and the -10 m depth isoline.

The second Adriatic site is located along the Central Adriatic margin inside the marine protected area of Torre del Cerrano. The shoreline is NW–SE oriented and the beach is composed of fine/medium sand ($d_{50} \approx 0.3$ mm) with nearshore slope of about 0.5–0.6%.

The Rodi Garganico site, part of the Gargano National Park, is the latest observing systems installed and is located along the northern side of the Gargano promontory in the southern Adriatic

Sea. The coastline orientation is about E–W facing the north, with medium grained beach sediments ($d_{50} \approx 0.3$ mm) and nearshore slope of about 0.8%. The harbour located 1 km eastward of the monitored beach may have an important influence on the morphological evolution of the nearby coastal areas, since it may alter the longshore sediment flow patterns significantly.

Two adjacent beaches were analyzed along the central Tyrrhenian side: Sabaudia and Terracina. They are separated by the rocky Circeo Cape and characterized by very different touristic demand. The Sabaudia sandy beach-dune system, spanning for 24 km northward of this rocky headland, is entirely included in the Circeo National Park since 1934, representing one of the largest and most natural littoral zones of the Tyrrhenian coast. The video-monitored portion is located about 5 km north of the Circeo Cape and is composed of fine to medium sand ($d_{50} \approx 0.3$ mm) with a nearshore slope of about 1.3%. The site orientation is NNW–SSE.

The Terracina beach, located eastward of the Circeo Cape, is enclosed into a 15 km long embayment with an average WSW–ENE orientation and a nearshore slope of about 1.7%, the maximum nearshore slope value of all the five sites here studied. Two major rivers are present inside the embayment and the analyzed sector is bounded at one end by one of these river jetties. Native beach sediments consist of fine to medium-fine grained sand ($d_{50} \approx 0.3$ mm) evolved into mixed beach sediments after a beach face nourishment was deployed in 2007, with a mean beach infill of 270 m^3/m [31]. Because of this, the evolution of the Terracina beach has to be regarded as a long-term self-adjustment of a "not in equilibrium" beach. This is a situation different from those of the other sites, where no relevant coastal interventions were undertaken before and during the monitoring period, and thus the natural beach adaptation processes in response to local wave forcing is not artificially altered.

2.2. The Video-Monitoring System and Data Processing

All the analyzed areas are provided with double-camera video-monitoring stations working with the same setting since 2016, except the Rodi Garganico site, operative since the end of 2018 and, for this reason only, partly described in this manuscript. The Terracina system, previously made of a single-camera system, was upgraded to the actual double-camera setting in 2015.

The monitoring stations are all located on the roof of beach-front buildings, at a height larger than 20 m above the mean sea level (Table 1) and at a distance smaller than 150 m from the shoreline. The acquiring systems routinely collect images of the nearshore zone at 2 Hz during 10 min every daylight hour, through a double digital video-camera equipped with CMOS 1/1.8″ sensor and fitted with fixed-focal lens. The alongshore extent of the monitored coast is site-specific, but always ensures visibility of the active nearshore zone along a sector extending 500–1000 m in the longshore direction. Information on camera systems are given in Table 1.

In order to achieve quantitative morphologic information from remote images, standard photogrammetric procedures [38] were used to project the oblique video image onto a planar surface, coincident with the local sea level [39]. The geometrical solutions of the projection procedure were constrained with several Ground Control Points (GCP), clearly visible points in the image with known coordinates both in the image-space and in ground-space. The geometries were computed for each selected image, corrected for optical distortions following the method used by Bouguet [40] and finally transformed into geometrically-correct plan views on a 1 m grid. A local metric reference system was set for each station, with the origin fixed at the video-station position and rotated in order to ensure that the mean shoreline direction is parallel to the x-axis and the y-axis is positive toward the offshore (Figure 2). The cross-shore and alongshore limits of the analyzed portion of each rectified image were fixed at distances from the stations where the pixel footprint was not higher than 5 m and 20 m, respectively.

Quantitative information on the morphological features of the nearshore zone has been extracted from optical signatures visible on the optical images and used to investigate the dynamics of the analyzed area [39,41]. Shoreline and sandbar locations have been indirectly estimated using composite images resulting from time averaging of the pixel intensity (Timex images). In this way moving features, e.g., propagating waves and bores, are not visible but more stable characteristics, e.g., regions of wave breaking, are highlighted as white bands [42].

Table 1. Specifics of the deployed video-monitoring stations.

Location	Sea	Coordinates	Height (m)	Cameras Azimuth (°N)	Focal Length (mm)	Monitored Beach (m)
Senigallia	Adriatic	43° 42′ 24″ N 13° 14′ 14″ E	28.0	60° 115°	8 12	800
Torre del Cerrano	Adriatic	42° 35′ 06″ N 14° 05′ 21″ E	37.5	40° 330°	12 25	1000
Rodi Garganico	Adriatic	41° 55′ 39″ N 15° 52′ 34″ E	54.0	310° 345°	8 8	500
Sabaudia	Tyrrhenian	41° 18′ 00″ N 13° 00′ 33″ E	21.1	165° 240°	12 6	800
Terracina	Tyrrhenian	41° 17′ 03″ N 13° 12′ 27″ E	39.0	95° 165°	25 8	1000

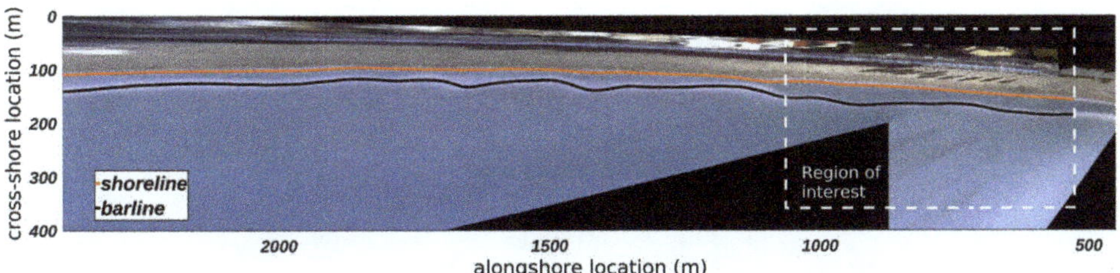

Figure 2. An example of rectified and georeferenced Timex image from the video-monitoring station in Terracina. The remotely sensed locations of the shoreline and the bar are shown with the orange and the black line, respectively. The dashed line box shows the region of interest along the Terracina coast.

Once Timex images have been rectified and georeferenced, many approaches to sample patterns of high intensity from plan view images can be used. In the present work the Barline Intensity Mapper (BLIM) by van Enckevort and Ruessink [18] and Pape et al. [43] has been used. In this algorithm, pixels within a region of interest are scanned along the image in order to detect the position lines of maximum pixel intensity values (Figure 2). These lines have been taken as proxies for wave breaking regions and, therefore, for the positions of sandbar crests and shorelines with a suitable accuracy.

One single image per day was chosen close to the lowest tide level, when environmental conditions allowed to clearly analyze the image. For each sampled feature (bar line or shoreline), the cross-shore position was alongshore averaged and used as a proxy for the overall cross-shore position of the remotely sensed morphology. Alongshore standard deviation could also be assumed as representative for the alongshore unevenness of the morphology. Other morphological parameters, like the bar height and the existing bathymetry, are also deemed to be relevant to the beach morphodynamic processes; unfortunately, no information on this regard could be extracted by video-monitoring products in our study, although there are some undergoing studies on the matter of estimating the bathymetry from remotely sensed images [44].

Since the location of barlines and shorelines could be affected by uncertainties related to sea level existing during the Timex construction [18,45], a conservative approach was used to filter out sea level-induced errors on features positions. Cut-off resolutions were fixed at ±5 m for the shoreline, ±10 m for the inner bars and ±20 m for the outermost bar, so that all consecutive feature motions lower than such thresholds were not regarded as real cross-shore displacements. The present analyses focused on the long-term behaviour rather than on the single event response, allowing, however, to identify all beach feature variations whose amplitude were larger than the above-mentioned thresholds.

Further, in order to analyze the long-term behaviour of selected beaches, a down-sampling of time series was operated by averaging values over different temporal windows (7 days, 15 days and 30 days). Some periods, however, remained unsampled because of system failure, dirty lenses due to persisting bad weather, or calm weather giving no visible wave breaking patterns.

Throughout the manuscript the shoreline is labelled as "sl" and bars as "b*n*", where *n* is an identifying number that increases with the temporal occurrence and the offshore position of the given feature. Only for the beach in Sabaudia, the presence of transient sandy features between the bars and the shoreline (labelled as SPAW, Shoreward Propagating Accretionary Waves [46]) has been detected.

Table 2 gives basic information on the time extension and a summary of the sampled features in each image dataset, with the number of days each feature appears, and its percentage of occurrence over the whole observation period.

Table 2. The image datasets used for the study. For each location, the observational time intervals and the total number of observed days are reported. For each feature detected at a specific location, the number of days that the feature has been detected and its percentage of occurrence over the total number of days are also given.

Location	**Dataset Specifics & Days of Feature Occurrence**					
Senigallia	Time interval: 16 December 2016–9 February 2019 Number of days: 786					
feature	sl	b1	b2	b3	b4	b5
n° of days	370	447	282	89	9	307
% occurrence	47	57	36	11	19	39
Torre del Cerrano	Time interval: 15 May 2016–14 January 2019 Number of days: 985					
feature	sl	b1	b2	b3		
n° of days	515	458	214	15		
% occurrence	52	47	22	2		
Sabaudia	Time interval: 13 October 2015–22 December 2018 Number of days: 1167					
feature	sl	spaw	b1	b2		
n° of days	812	155	604	157		
% occurrence	70	13	52	13		
Terracina	Time interval: 28 September 2007–28 December 2018 Number of days: 4110					
feature	sl	b1	b2	b3	b4	
n° of days	1307	54	285	484	218	
% occurrence	32	1	7	12	5	

2.3. Wave Climate

2006–2016 hindcast and 2017–2018 forecast wave data delivered by the Copernicus Marine Environment Monitoring Service (EU Copernicus programme, http://marine.copernicus.eu/) have been used to characterize wave forcings typical of the study sites. For the mid/long term analysis described here, a preliminary wave height threshold equivalent to a 97th percentile has been adopted [5,47]. In other words, the threshold for the definition of a storm event is assumed so that the recorded time series of wave heights stays above it for 3% of the whole analyzed period. The proposed method identified about 150 events per region as classified storm events throughout the period 2006–2018. As suggested by Boccotti [48], two consecutive storm records must be regarded as distinct if they are separated by a continuous interval greater than 12 h. The wave data and characteristics are here primarily used to compute morphodynamic parameters (Section 3.1), and no specific analysis was performed to characterize storm statistics. Statistical information on the storm wave forcings at the investigated sites are summarized in Table 3.

Table 3. Wave climate statistics for each study site: wave height threshold for storm classification $H_{s,0.97}$, mean significant wave height at classified storm peaks $\langle H_{s,peak} \rangle$, and mean peak period at classified storm peaks $\langle T_{p,peak} \rangle$.

	Senigallia	Torre del Cerrano	Rodi Garganico	Terracina	Sabaudia
$H_{s,0.97}$ (m)	2.04	1.91	1.99	2.26	2.41
$\langle H_{s,peak} \rangle$ (m)	2.81	2.67	2.67	3.18	3.33
$\langle T_{p,peak} \rangle$ (s)	7.30	7.42	7.41	8.71	8.59

Since the variability of nearshore patterns is highly related to the balance between cross- and alongshore components of the wave-induced currents, a crucial parameter to be considered across all sites is the wave approach angle with respect to the coast. As a consequence of the physiographic differences between the Adriatic and Tyrrhenian basins (elongated and enclosed the former, more open the latter), wave obliquity in the Adriatic sites is clearly bimodal (due to the action of the local Scirocco and Bora winds), while the incidence distribution for Tyrrhenian sites is mainly unimodal (Figure 3).

Along the Adriatic sites, the prevailing NE and E waves approach the coasts at different angles: at Senigallia the waves coming from N approach the coast at very large angles with respect to the shore normal, whereas at Torre del Cerrano waves approach the beach from two symmetric directions with large angles. At Rodi Garganico waves coming mainly from N–NW approach the coast almost normally (Figure 3; the panel pertaining to Rodi Garganico is not shown).

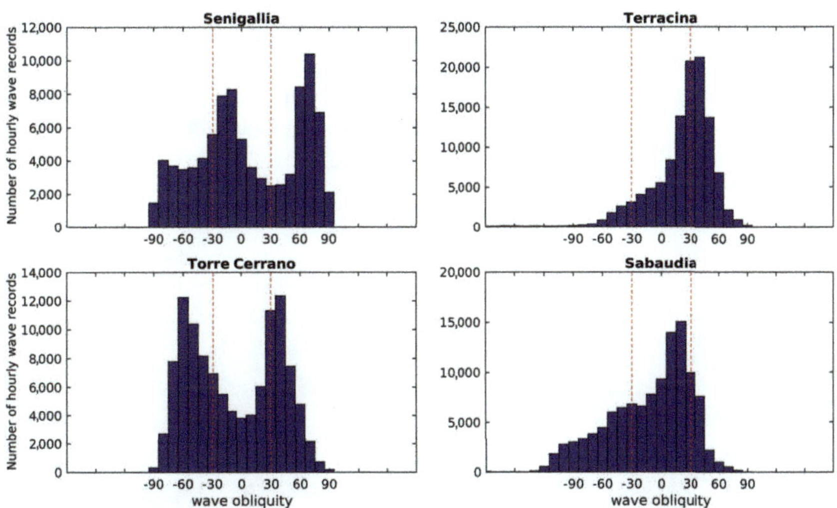

Figure 3. Distribution of wave obliquity at Senigallia, Terracina, Torre del Cerrano and Sabaudia. The red dashed lines demarcate the ±30° interval of wave incidence with respect to the shore normal. In the *y*-axis the number of hourly wave records coming from a given direction interval throughout the total observation period is given.

Concerning the Tyrrhenian sites, although both Sabaudia and Terracina are subjected to roughly the same wave climate, the dominant waves coming from W, SW and S induce different inshore hydrodynamics since the respective coastlines have strongly different orientation. This results in a dominance of waves approaching at low angles for Sabaudia, and at high angles for Terracina.

2.4. Beach Classification Parameters

A simple and general description of beach morphology at the investigated sites can be achieved by means of synthetic parameters, such as the dimensionless fall velocity Ω [13], the bar parameter B^* [49], the relative tide range RTR [14] and the Iribarren parameter Ir [50]:

$$\Omega = \frac{H_b}{w_s T_p}, \tag{1}$$

$$B^* = \frac{x_s}{g T_p^2 \tan\beta}, \tag{2}$$

$$\mathrm{RTR} = \frac{\mathrm{TR}}{H_b}, \tag{3}$$

$$\mathrm{Ir} = \frac{\tan\beta}{\sqrt{H_b/L_0}}, \tag{4}$$

where H_b is the breaking wave height, w_s is the settling sediment velocity, T_p is the peak wave period, L_0 is the estimated offshore mean wave length, x_s is an offshore distance corresponding to a specific depth at which the beach slope becomes very small (beach "closure"), g is the gravitational acceleration, $\tan\beta$ is the bottom slope and TR is the tide range. The settling sediment velocity is evaluated with the Zanke formulation [51,52], valid for sediment diameters in the range 0.1–1 mm:

$$w_s = 10\frac{\nu}{d_{50}}\left(\sqrt{1+\frac{0.01 g d_{50}^3 \left(\rho_s - \rho_w\right)}{\nu^2 \rho_w}} - 1\right), \tag{5}$$

where ν is the kinematic viscosity of water, d_{50} is the sediment median-diameter, ρ_s is the particle density, ρ_w is the water density.

The breaking wave height H_b appearing in Equations (1) and (3) has been computed starting from the mean value of the (offshore) peak wave heights of the identified storms, $\langle H_{s,peak}\rangle$. The offshore datum has been transported to the breaking line by taking into account the effect of wave shoaling over an equilibrium beach profile:

$$H_b = \langle H_{s,peak}\rangle\, k_s \tag{6}$$

where k_s is the shoaling coefficient. The wave period T_p used in Equations (1) and (2) has been calculated as the mean of the peak wave periods at the peaks of identified storms $\langle T_{p,peak}\rangle$, and the Iribarren parameter as the mean of the Iribarren numbers computed for each classified storm event.

3. Results

3.1. Beach Classification

The classification parameters estimated for each site are summarized in Table 4. The dimensionless fall velocity Ω takes into account both wave and sediment characteristics; values of Ω larger than 6 have been found for all sites, allowing us to classify all the beaches as dissipative. In this kind of environment submerged bars may be present and rips are usually absent [13,14]. Low values of the surf similarity parameter, Ir (see Table 4), also display a dissipative beach behaviour, characterized by spilling type breaking waves ($\mathrm{Ir} < 0.5$). Following the conceptual beach model of Masselink and Short [14] that makes use of the relative tide range RTR, all beaches can be classified as barred dissipative beaches, further confirming the presence of submerged bars.

Table 4. Environmental and classification parameters used to characterize the main morphological behaviour of the study sites.

	Slope (%)	TR (m)	d_{50} (mm)	Ω [13]	B^* [49]	RTR [14]	Ir [50]
Senigallia	0.5	0.6	0.25	11.3	497	0.2	0.03
Torre del Cerrano	0.5	0.5	0.30	8.6	447	0.2	0.03
Rodi Garganico	0.8	0.4	0.30	8.6	209	0.1	0.05
Sabaudia	1.3	0.4	0.30	9.3	55	0.1	0.09
Terracina	1.7	0.4	0.30	8.9	67	0.1	0.08

To predict the possible number of sandbars, use is made of the dimensionless bar parameter B^* [49], which considers both the nearshore geometry and the wave characteristics (Table 4). According to the bar parameter, the presence of three or more bars has been estimated for Senigallia, Torre del Cerrano and Rodi Garganico ($B^* > 100$), while only two bars are predicted for Terracina and Sabaudia ($50 < B^* < 100$). As it is shown in the following, these estimations are in fairly good agreement with the bar systems actually observed from the deployed video-monitoring stations (see Figure 4).

The method by Wright and Short [13], based on the parameter Ω and developed for energetic wave climates, usually does not provide a correct estimation of the morphodynamic states for enclosed basins (e.g., [53,54]). Nonetheless, this classification method still identifies all the analyzed beaches as dissipative, similarly to the findings of Lisi et al. [55] for the south Adriatic coast.

The above shows that, for inter-site comparison purposes, environmental parameters like the wave obliquity (Figure 3) and the nearshore slope (Table 4) are both important, as they significantly affect the nearshore hydrodynamics and the way in which sites are classified.

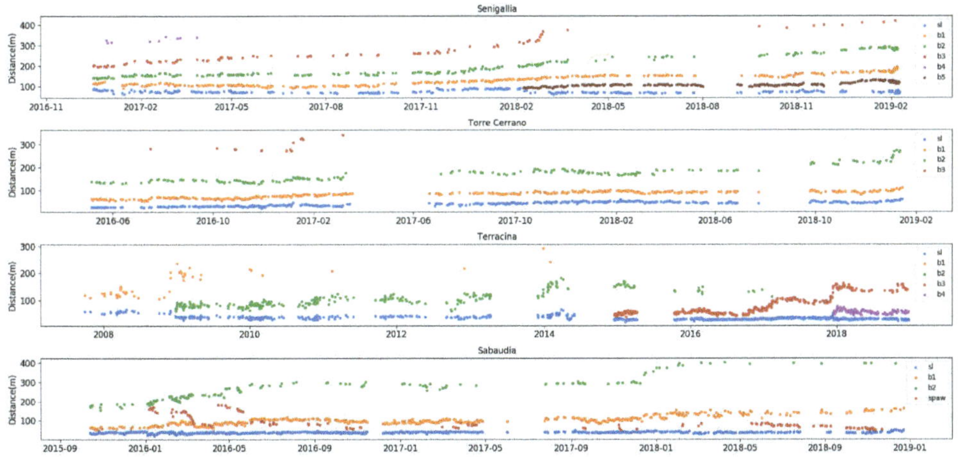

Figure 4. Time evolution of locations of shoreline and barlines in Senigallia, Torre del Cerrano, Terracina and Sabaudia. Each dot represents the daily-averaged location of the relative beach feature.

3.2. Nearshore Morphological Settings in Time

The seabed morphologies of the study sites are characterized by a number of nearshore bars variable from one to four. Only in Sabaudia the number of bars remained constant throughout the whole observation period (Figure 4 and Table 5).

The morphological setting in Senigallia (evaluated in an observational period spanning 2.1 years) is characterized by a 4-3-4 bars transition, with the 4-bar setting persisting for about 60% of the total observation time. During the 2.7 years of observations at Torre del Cerrano the nearshore setting experienced a transition from a 3- to a 2-bars system, which lasted for 69% of the total time.

Recordings from Terracina, spanning 11.3 years, evidenced a complex variability of the bar structure, with continuous changes from a single- to a double-barred system, with a dominance of the latter configuration (74% of the total time). The beach in Sabaudia constantly exhibits two bars and a SPAW.

Table 5. Persistence of sand bars settings at the study sites. The left section of the table shows the percentages of the observation period in which a specific bar number has been observed. The right section of the table shows the number of consecutive days in which a specific bar configuration (given under each number) has been observed.

Location	Number of Bars				Consecutive Days of Bar Setting					
	Single	Double	Three	Four						
Senigallia			40%	60%	102 four	316 three	368 four			
Torre del Cerrano		69%	31%		307 three	678 double				
Terracina	26%	74%			455 single	1863 double	317 single	795 double	303 single	377 double
Sabaudia		100%			all double					

The change in the number of bars also shows site-specific characteristics. In Senigallia (first panel in Figure 4), the outer bar (b4) lost its visibility for almost a year, in fact changing the seabed bar configuration into a 3-bar system. The restoration of the 4-bar setting is due to the generation of new inner bar (b5) in February 2018. At Torre del Cerrano (second panel in Figure 4) the decrease of the bar number is coeval with the Senigallia stage transition (after February 2017) but was not followed by the generation of a new inner bar at the shoreline. In Terracina (third panel in Figure 4) the change in the bar number is always linked to the generation of a new bar near the shoreline (b2 in 2009, b3 in 2015, b4 in 2018).

The temporal distribution of distances between shoreline and barlines (Figure 5) highlights the presence of broader or thinner multiple peaks, connected to a more or less wide breaking region in the cross-shore direction. In the Adriatic sites (Senigallia and Torre del Cerrano; top panels in Figure 5), the distances of the inner and middle bars from the shoreline show a reduced dispersion around the respective peaks, thus, giving an indication of their linearity and relative stability in time. The distribution of inner bar distances for the Tyrrhenian sites (Sabaudia and Terracina; bottom panels in Figure 5), conversely, shows high dispersion around the peaks, suggesting more complex inner bar planar shapes and increased mobility of the bars in response to wave forcings.

Notably, at the Senigallia beach the distance from the shoreline of the inner and intermediate bars remain approximately constant in time, although an established pattern of generation and offshore migration of the bars occurs, in accordance with the NOM model [15,19]. The inner bars b1 (before February 2018) and b5 (after February 2018) show a position peak at around 40 m from the shoreline; in the same way, the first intermediate bars b2 (before February 2018) and b1 (after February 2018) have a position peak of about 80 m from the shore.

The seabed morphology observed at Terracina is representative of a 10-years nearshore morphological evolution after the artificial nourishment of the beach. The reshaping of this area is characterized by a variable "equilibrium" condition of the sandbars that follows a decrease of the surf zone size. Since the distance reached by most offshore bar is never reached by the other barlines afterwards, the distribution of the shoreline-bar distances well shows a decrease of the distance of inner and outer bars in time (Figure 5).

The distribution of shoreline-barlines distances at Sabaudia differs from those observed in Senigallia and Terracina since, even if the outer bar (b2) experiences the largest range in positioning (from 150 m to 350 m from the shoreline), no new bar is generated at the shoreline in response;

transitional forms (SPAW) evolve instead, at first between the outer and inner bar, and then between the inner bar and the shoreline (Figure 5), suggesting a very complex morphology evolution both of the foreshore and in the outer surf zone. Similar behaviours are recorded for crescentic bars located along other Mediterranean sites [56].

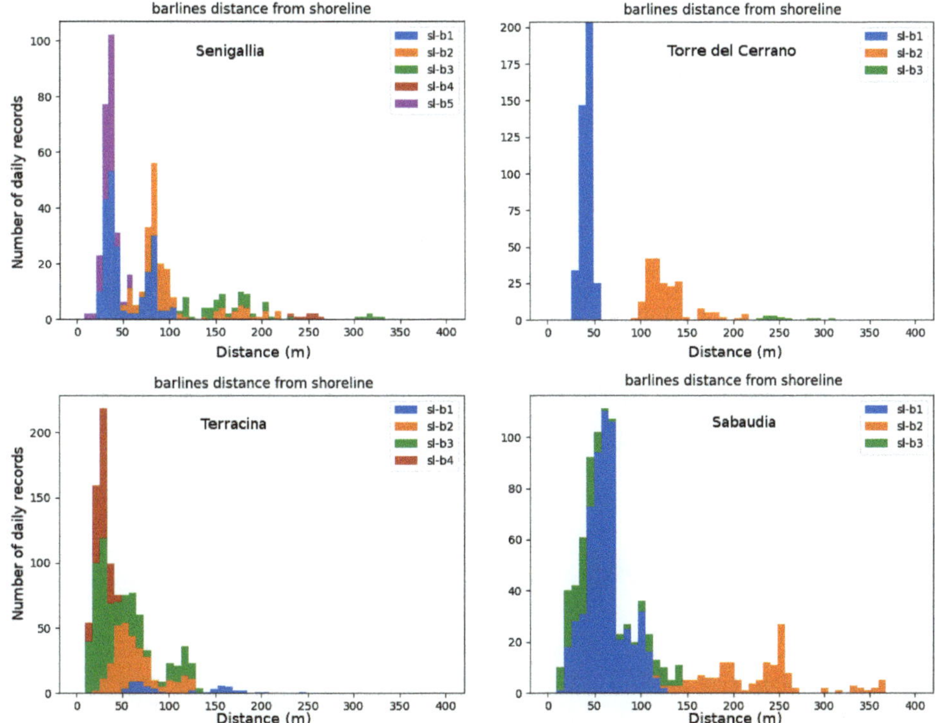

Figure 5. Distribution of distances between shoreline and bars at Senigallia, Torre del Cerrano, Terracina and Sabaudia. In the y-axis the number of daily occurrences in which the distance between a bar and the shoreline could be evaluated is reported.

The Torre del Cerrano bar setting, finally, exhibits a rather constant position of the inner and intermediate bars, located respectively at around 45–50 m and 100–150 m from the shoreline (Figure 5).

3.3. Rates of Variability

The net shoreline displacement rates (end point rates, EPR) do not display specific trends, apart from the case of Terracina, where a 40 m shoreline retreat is registered over 11 years (Table 6).

Table 6. Net shoreline displacement rates (as end point rates) at Senigallia, Torre del Cerrano, Terracina and Sabaudia.

	Senigallia	Torre del Cerrano	Terracina	Sabaudia
EPR (m/month)	−0.5	0.9	−0.2	0.3

The monthly rates of shoreline displacements (Figure 6) show that the highest rates of displacement were experienced at the change in bar setting, reaching their maximum values in Terracina and Senigallia. On the opposite, the results for Sabaudia and Torre del Cerrano are always within the range of uncertainty.

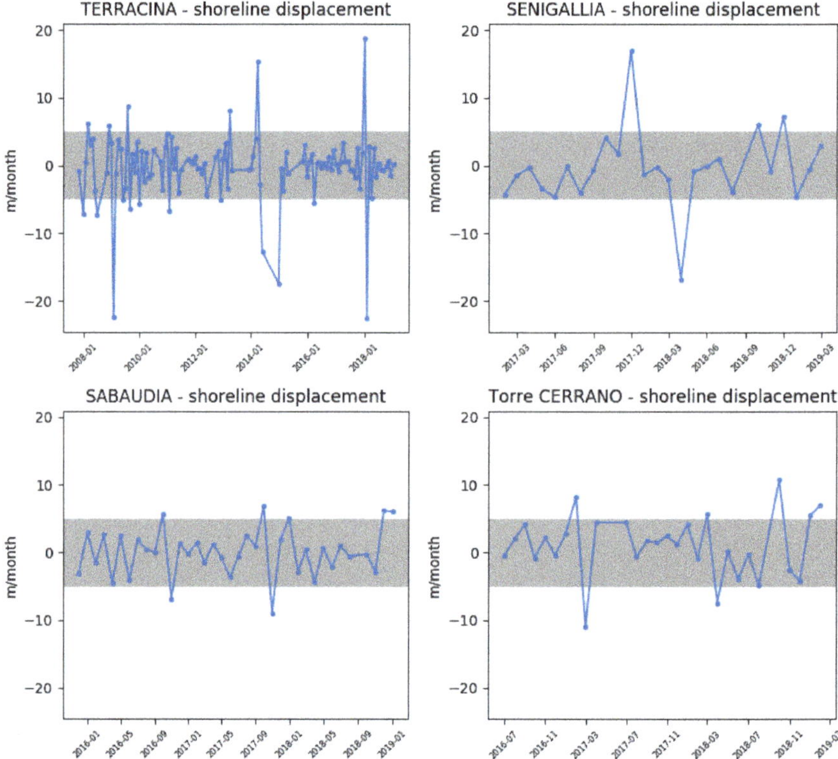

Figure 6. Maximum shoreline displacement rates at Senigallia, Torre del Cerrano, Terracina and Sabaudia, calculated from monthly-resampled time series of shoreline displacement. The shaded area indicates a displacement rate range from −5 to +5 m/month. Negative values correspond to shoreline retreat, while positive values show shoreline advancement.

Finally, at all sites the maximum rate of cross-shore variability reached its highest value at the outermost bars (b3 and b4 for Senigallia, b3 for Torre del Cerrano, b1 for Terracina, and b2 for Sabaudia; Figure 7) independently of the temporal windows used for the mean rate calculation (7 days, 15 days or 1 month). Furthermore, the rates are almost twice as large for the beaches along the Tyrrhenian sea than for those along the Adriatic sea.

The highest rate of cross-shore movement was recorded by the SPAW in Sabaudia, also regardless of the temporal windows used for the rate computation.

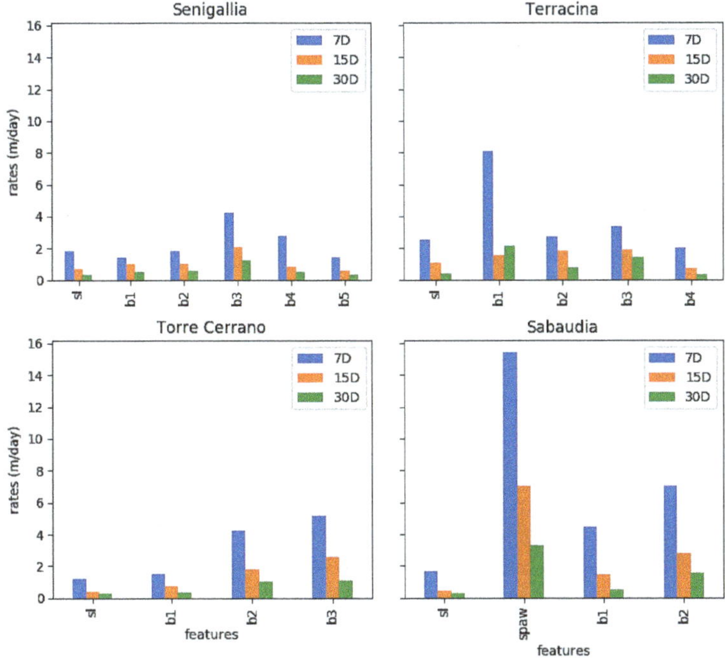

Figure 7. Maximum rates of variability for sampled morphology features at the investigated sites. The rates are computed with different temporal windows: 7 days (blue bars), 15 days (orange bars), and 1 month (green bars).

4. Discussion and Conclusions

The above findings provide a preliminary inter-site comparison of rates, amplitudes and modes of morphologic variability for the analyzed beach systems, with a focus on the long-term evolution. For all the sites, the rates of variability of sandbar cross-shore positions are much larger than for the shoreline, providing further indication on the large relative importance of the nearshore morphologic variability on the overall beach evolution (e.g., see [57]).

The beaches of Terracina and Senigallia experienced a similar net offshore bar migration process, but with some differences in rates, likely related to differences in the Iribarren number (much larger for the Tyrrhenian sites than for the Adriatic sites). During the observation period in Terracina (11 years), the generation of three bars was related to the fast seaward migration of the outermost bar. The width of the emerged beach sensibly decreased at the times in which the new inner bars are generated (this is particularly evident in 2009; see Figure 4), but remained stable afterwards. The evidence of outer bar decay was always related to waves breaking on a wider area (indicating a flatter bar) until the intermediate bar was forced to weld to the decaying outer bar. In Senigallia the nearshore setting experienced a more regular evolution, with lower but progressive offshore movements of the whole bar system, even if the two most offshore bars experienced higher migration rates. Shoreline displacement reached its maximum values during the near-shoreline generation of the inner bar, after which it remained fairly stable.

During the 2.9 years of observation of the Torre del Cerrano beach, no specific trends of sandbar migration were recorded, even though during the latest period of observation a fast offshore movement of the outer bar was recorded. In this site the shoreline is progressively advancing. The environmental parameters are similar to those of Senigallia, where a clear NOM-type bar evolution pattern occurs. Such difference in behaviour seems to be related to the largest wave obliquity observed in

Senigallia, this leading to larger longshore currents and associated larger seabed friction, giving larger sediment mobility.

The Sabaudia site is mainly characterized by complex foreshore morphologies and very high rates of outer bars seaward motion. No erosional trends of the shoreline are observed, but alongshore moving transient features indicate a three-dimensional pattern of variability along this analyzed portion of beach, where the most of high waves approached the coast with small angles.

This preliminary inter-site comparison suggests some findings that seem to agree with similar studies on the matter:

1. the sandbar morphologic variability is much larger than the shoreline displacement. This is very likely related to the exposure of sandbars to more intense forcing flows than those experienced by the near-shoreline sediments;
2. for the first time, evidence of NOM-type sand bar evolution patterns along Italian coasts was recorded;
3. migration rates of the sandbars during NOM seem to be significantly influenced by the wave approach angle, as well as the nearshore slope (see also [58–64]), which in turn translates into different values of the Iribarren number;
4. the major erosional phenomena along the analyzed beaches coincide with discrete changes in the nearshore morphological patterns that occur in relation to the rapid seaward migration of the outer bars or to the decay of the offshore bar. In view of the results of our analysis, the reciprocal roles of wave obliquity and beach slope can be summarized as per Table 7;
5. the larger shoreline displacements have been observed along those beaches that are characterized by a NOM-type evolution of the entire sandbar system. This suggests that quasi-steady sandbar systems are related to more stable shorelines. However, more detailed analyses are needed to verify if such behaviour also leads to more resilient coastlines (i.e., with a lesser retreat in the long term);
6. high-frequency remote sensing technology can provide fundamental insight about such long-term shoreline evolution. Moreover, it can provide important information for coastal management purposes, especially along natural coasts were dynamics could be very fast and where no other (historic) data are available.

Table 7. Schematic of the roles of beach slope and wave obliquity on the bars-shoreline system evolution.

	Low Wave Obliquity	High Wave Obliquity
Low Ir	No trend–stable bars and shoreline	Cyclic behaviour (NOM) with low rates of change
High Ir	3D patterns with high rates of change	Cyclic behaviour (NOM) with high rates of change

Author Contributions: Conceptualization: L.P. and M.B.; data curation: L.M. and S.D.; formal analysis: L.P., M.L. and E.P.; supervision: L.S., G.Z. and M.B.; visualization: L.P.; writing—original draft: L.P.; writing—review & editing: L.M., G.Z. and M.B.

Funding: This study is funded by Circeo National Park, Gargano National Park and the "Torre del Cerrano" Marine Protected Area in the context of the National Biodiversity Strategy promoted by the Ministero dell'Ambiente e della Tutela del Territorio e del Mare (MATTM) (GAB0024444). M.B. gratefully acknowledges financial support from the ONR Global (UK) through the MORSE Research Grant (N62909-17-1-2148).

Conflicts of Interest: The authors declare no conflict of interest.

Abbreviations

The following abbreviations are used in this manuscript:

BLIM	BarLine Intensity Mapper
GCP	Ground Control Point
NOM	Net Offshore Migration (pattern)
SPAW	Shoreward Propagating Accretionary Wave

References

1. Shi, F.; Kirby, J.T.; Harris, J.C.; Geiman, J.D.; Grilli, S.T. A high-order adaptive time-stepping TVD solver for Boussinesq modeling of breaking waves and coastal inundation. *Ocean Model.* **2012**, *43–44*, 36–51. [CrossRef]
2. Soldini, L.; Antuono, M.; Brocchini, M. Numerical modeling of the influence of the beach profile on wave run-up. *J. Waterw. Port Coast. Ocean Eng.* **2013**, *139*, 61–71. [CrossRef]
3. Postacchini, M.; Lalli, F.; Memmola, F.; Bruschi, A.; Bellafiore, D.; Lisi, I.; Zitti, G.; Brocchini, M. A model chain approach for coastal inundation: Application to the bay of Alghero. *Estuar. Coast. Shelf Sci.* **2019**, *219*, 56–70. [CrossRef]
4. Klein, R.J.; Smit, M.J.; Goosen, H.; Hulsbergen, C.H. Resilience and vulnerability: Coastal dynamics or Dutch dikes? *Geogr. J.* **1998**, *164*, 259–268. [CrossRef]
5. Masselink, G.; van Heteren, S. Response of wave-dominated and mixed-energy barriers to storms. *Mar. Geol.* **2014**, *352*, 321–347. [CrossRef]
6. Zhang, K.; Douglas, B.; Leatherman, S. Do storms cause long-term beach erosion along the U.S. east barrier coast? *J. Geol.* **2002**, *110*, 493–502. [CrossRef]
7. Ferreira, Ó. The role of storm groups in the erosion of sandy coasts. *Earth Surf. Process. Landf.* **2006**, *31*, 1058–1060. [CrossRef]
8. Coco, G.; Senechal, N.; Rejas, A.; Bryan, K.; Capo, S.; Parisot, J.; Brown, J.; MacMahan, J. Beach response to a sequence of extreme storms. *Geomorphology* **2014**, *204*, 493–501. [CrossRef]
9. Karunarathna, H.; Pender, D.; Ranasinghe, R.; Short, A.D.; Reeve, D.E. The effects of storm clustering on beach profile variability. *Mar. Geol.* **2014**, *348*, 103–112. [CrossRef]
10. Phillips, M.S.; Harley, M.D.; Turner, I.L.; Splinter, K.D.; Cox, R.J. Shoreline recovery on wave-dominated sandy coastlines: The role of sandbar morphodynamics and nearshore wave parameters. *Mar. Geol.* **2017**, *385*, 146–159. [CrossRef]
11. Gallagher, E.L.; Elgar, S.; Guza, R.T. Observations of sand bar evolution on a natural beach. *J. Geophys. Res. Oceans* **1998**, *103*, 3203–3215. [CrossRef]
12. Hoefel, F.; Elgar, S. Wave-induced sediment transport and sandbar migration. *Science* **2003**, *299*, 1885–1887. [CrossRef]
13. Wright, L.; Short, A. Morphodynamic variability of surf zones and beaches: A synthesis. *Mar. Geol.* **1984**, *56*, 93–118. [CrossRef]
14. Masselink, G.; Short, A.D. The effect of tide range on beach morphodynamics and morphology: A conceptual beach model. *J. Coast. Res.* **1993**, *9*, 785–800.
15. Shand, R.; Bailey, D.; Shepherd, M. An inter-site comparison of net offshore bar migration characteristics and environmental conditions. *J. Coast. Res.* **1999**, *15*, 750–765.
16. Ruessink, B.; Kroon, A. The behaviour of a multiple bar system in the nearshore zone of Terschelling, The Netherlands: 1965–1993. *Mar. Geol.* **1994**, *121*, 187–197. [CrossRef]
17. Roelvink, J.A.; Stive, M.J.F. Bar-generating cross-shore flow mechanisms on a beach. *J. Coast. Res.* **1989**, *94*, 4785–4800. [CrossRef]
18. Van Enckevort, I.; Ruessink, B. Video observations of nearshore bar behaviour. Part 1: Alongshore uniform variability. *Cont. Shelf Res.* **2003**, *23*, 501–512. [CrossRef]
19. Wijnberg, K.M.; Terwindt, J.H. Extracting decadal morphological behaviour from high-resolution, long-term bathymetric surveys along the Holland coast using eigenfunction analysis. *Mar. Geol.* **1995**, *126*, 301–330. [CrossRef]
20. Aleman, N.; Certain, R.; Robin, N.; Barusseau, J.P. Morphodynamics of slightly oblique nearshore bars and their relationship with the cycle of net offshore migration. *Mar. Geol.* **2017**, *392*, 41–52. [CrossRef]
21. Bouvier, C.; Balouin, Y.; Castelle, B. Video monitoring of sandbar-shoreline response to an offshore submerged structure at a microtidal beach. *Geomorphology* **2017**, *295*, 297–305. [CrossRef]
22. Castelle, B.; Turner, I.; Ruessink, B.; Tomlinson, R. Impact of storms on beach erosion: Broadbeach (Gold Coast, Australia). *J. Coast. Res.* **2007**, 534–539. Available online: https://www.jstor.org/stable/26481646 (accessed on 10 April 2019).
23. Price, T.; Ruessink, B. State dynamics of a double sandbar system. *Cont. Shelf Res.* **2011**, *31*, 659–674. [CrossRef]

24. Splinter, K.D.; Turner, I.L.; Reinhardt, M.; Ruessink, G. Rapid adjustment of shoreline behavior to changing seasonality of storms: Observations and modelling at an open-coast beach. *Earth Surf. Process. Landf.* **2017**, *42*, 1186–1194. [CrossRef]
25. Angnuureng, D.B.; Almar, R.; Senechal, N.; Castelle, B.; Addo, K.A.; Marieu, V.; Ranasinghe, R. Shoreline resilience to individual storms and storm clusters on a meso-macrotidal barred beach. *Geomorphology* **2017**, *290*, 265–276. [CrossRef]
26. Splinter, K.; Harley, M.; Turner, I. Remote sensing is changing our view of the coast: Insights from 40 years of monitoring at Narrabeen-Collaroy, Australia. *Remote Sens.* **2018**, *10*, 1744. [CrossRef]
27. Aagaard, T.; Holm, J. Digitization of wave run-up using video records. *J. Coast. Res.* **1989**, *5*, 547–551.
28. Vousdoukas, M.I.; Wziatek, D.; Almeida, L.P. Coastal vulnerability assessment based on video wave run-up observations at a mesotidal, steep-sloped beach. *Ocean Dyn.* **2011**, *62*, 123–137. [CrossRef]
29. Ibaceta, R.; Almar, R.; Catalán, P.; Blenkinsopp, C.; Almeida, L.; Cienfuegos, R. Assessing the performance of a low-cost method for video-monitoring the water surface and bed level in the swash zone of natural beaches. *Remote Sens.* **2018**, *10*, 49. [CrossRef]
30. Archetti, R.; Paci, A.; Carniel, S.; Bonaldo, D. Optimal index related to the shoreline dynamics during a storm: The case of Jesolo beach. *Nat. Hazard. Earth Syst. Sci.* **2016**, *16*, 1107–1122. [CrossRef]
31. Parlagreco, L.; Archetti, R.; Simeoni, U.; Devoti, S.; Valentini, A.; Silenzi, S. Video-monitoring of a barred nourished beach (Latium, Central Italy). *J. Coast. Res.* **2011**, *SI 64*, 110–114.
32. Armaroli, C.; Ciavola, P. Dynamics of a nearshore bar system in the northern Adriatic: A video-based morphological classification. *Geomorphology* **2011**, *126*, 201–216. [CrossRef]
33. Brignone, M.; Schiaffino, C.F.; Isla, F.I.; Ferrari, M. A system for beach video-monitoring: Beachkeeper plus. *Comput. Geosci.* **2012**, *49*, 53–61. [CrossRef]
34. Valentini, N.; Saponieri, A.; Damiani, L. A new video monitoring system in support of coastal zone management at Apulia region, Italy. *Ocean Coast. Manag.* **2017**, *142*, 122–135. [CrossRef]
35. Bonaldo, D.; Antonioli, F.; Archetti, R.; Bezzi, A.; Correggiari, A.; Davolio, S.; Falco, G.D.; Fantini, M.; Fontolan, G.; Furlani, S.; et al. Integrating multidisciplinary instruments for assessing coastal vulnerability to erosion and sea level rise: Lessons and challenges from the Adriatic Sea, Italy. *J. Coast. Conserv.* **2018**, *23*, 19–37. [CrossRef]
36. Melito, L.; Parlagreco, L.; Perugini, E.; Postacchini, M.; Zitti, G.; Brocchini, M. Monitoring for coastal resilience: A project for five Italian beaches. In Proceedings of the 2018 IEEE International Workshop on Metrology for the Sea, Bari, Italy, 8–10 October 2018; pp. 76–81, [CrossRef]
37. Postacchini, M.; Soldini, L.; Lorenzoni, C.; Mancinelli, A. Medium-term dynamics of a middle Adriatic barred beach. *Ocean Sci.* **2017**, *13*, 719–734. [CrossRef]
38. Hartley, R.; Zisserman, A. *Multiple View Geometry in Computer Vision*; Cambridge University Press: Cambridge, UK, 2004.
39. Holland, K.; Holman, R.; Lippmann, T.; Stanley, J.; Plant, N. Practical use of video imagery in nearshore oceanographic field studies. *IEEE J. Ocean. Eng.* **1997**, *22*, 81–92. [CrossRef]
40. Bouguet, J.Y. Camera Calibration Toolbox for Matlab. Available online: http://www.vision.caltech.edu/bouguetj/calib_doc/ (accessed on 8 March 2019).
41. Holman, R.; Sallenger, A.; Lippmann, T.; Haines, J. The application of video image processing to the study of nearshore processes. *Oceanography* **1993**, *6*, 78–85. [CrossRef]
42. Lippmann, T.C.; Holman, R.A. The spatial and temporal variability of sand bar morphology. *J. Geophys. Res.* **1990**, *95*, 11575–11590. [CrossRef]
43. Pape, L.; Plant, N.G.; Ruessink, B.G. On cross-shore migration and equilibrium states of nearshore sandbars. *J. Geophys. Res.* **2010**, *115*. [CrossRef]
44. Perugini, E. The Application of Video-Monitoring Data to Understand Coastal And Estuarine Processes. PhD Thesis, Università Politecnica delle Marche, Ancona, Italy, 2019.
45. Ribas, F.; Ojeda, E.; Price, T.D.; Guillen, J. Assessing the suitability of video imaging for studying the dynamics of nearshore sandbars in tideless beaches. *IEEE Trans. Geosci. Remote Sens.* **2010**, *48*, 2482–2497. [CrossRef]
46. Wijnberg, K.M.; Holman, R. Video-observations of shoreward propagating accretionary waves. In Proceedings of the RCEM 2007, Enschede, NL, USA, 7–21 September 2007; pp. 737–743.

47. Castelle, B.; Marieu, V.; Bujan, S.; Splinter, K.D.; Robinet, A.; Sénéchal, N.; Ferreira, S. Impact of the winter 2013–2014 series of severe Western Europe storms on a double-barred sandy coast: Beach and dune erosion and megacusp embayments. *Geomorphology* **2015**, *238*, 135–148. [CrossRef]
48. Boccotti, P. *Wave Mechanics for Ocean Engineering*; Elsevier: Amsterdam, The Netherlands, 2000.
49. Short, A.D.; Aagaard, T. Single and multi-bar beach change models. *J. Coast. Res.* **1993**, *SI 15*, 141–157.
50. Battjes, J. Surf similarity. In Proceedings of the 14th International Conference on Coastal Engineering. American Society of Civil Engineers, Copenhagen, Denmark, 24–28 June 1974.
51. Zanke, U. *Berechnung der Sinkgeschwindigkeiten von Sedimenten*; University of Hannover, Hannover, Germany, 1977; Volume 46. (In German)
52. Van Rijn, L.C. Sediment transport, Part II: Suspended load transport. *J. Hydraul. Eng.* **1984**, *110*, 1613–1641. [CrossRef]
53. Gómez-Pujol, L.; Orfila, A.; Cañellas, B.; Alvarez-Ellacuria, A.; Méndez, F.; Medina, R.; Tintoré, J. Morphodynamic classification of sandy beaches in low energetic marine environment. *Mar. Geol.* **2007**, *242*, 235–246. [CrossRef]
54. Jiménez, J.A.; Guillén, J.; Falqués, A. Comment on the article "Morphodynamic classification of sandy beaches in low energetic marine environment". *Mar. Geol.* **2008**, *255*, 96–101. [CrossRef]
55. Lisi, I.; Molfetta, M.; Bruno, M.; Risio, M.D.; Damiani, L. Morphodynamic classification of sandy beaches in enclosed basins: The case study of Alimini (Italy). *J. Coast. Res.* **2011**, *SI 64*, 180–184.
56. Swart, R.L.D.; Ribas, F.; Ruessink, G.; Simarro, G.; Guillén Aranda, J. Characteristics and dynamics of crescentic bar events in an open, tideless beach. In Proceedings of the Coastal Dynamics 2017, Helsingør, Denmark, 12–16 June 2017; pp. 555–566.
57. Plant, N.G.; Holman, R.A.; Freilich, M.H.; Birkemeier, W.A. A simple model for interannual sandbar behavior. *J. Geophys. Res. Oceans* **1999**, *104*, 15755–15776. [CrossRef]
58. Ruessink, B.G.; Wijnberg, K.M.; Holman, R.A.; Kuriyama, Y.; van Enckevort, I.M.J. Intersite comparison of interannual nearshore bar behavior. *J. Geophys. Res. Oceans* **2003**, *108*. [CrossRef]
59. Smit, M.; Reniers, A.; Ruessink, B.; Roelvink, J. The morphological response of a nearshore double sandbar system to constant wave forcing. *Coast. Eng.* **2008**, *55*, 761–770. [CrossRef]
60. Ruessink, B.; Pape, L.; Turner, I. Daily to interannual cross-shore sandbar migration: Observations from a multiple sandbar system. *Cont. Shelf Res.* **2009**, *29*, 1663–1677. [CrossRef]
61. Aleman, N.; Robin, N.; Certain, R.; Barusseau, J.P.; Gervais, M. Net offshore bar migration variability at a regional scale: Inter-site comparison (Languedoc-Roussillon, France). *J. Coast. Res.* **2013**, *SI 65*, 1715–1720. [CrossRef]
62. Di Leonardo, D.; Ruggiero, P. Regional scale sandbar variability: Observations from the U.S. Pacific Northwest. *Cont. Shelf Res.* **2015**, *95*, 74–88. [CrossRef]
63. Tătui, F.; Vespremeanu-Stroe, A.; Ruessink, G.B. Alongshore variability of cross-shore bar behavior on a nontidal beach. *Earth Surf. Process. Landf.* **2016**, *41*, 2085–2097. [CrossRef]
64. Walstra, D.J.; Wesselman, D.; van der Deijl, E.; Ruessink, G. On the intersite variability in inter-annual nearshore sandbar cycles. *J. Mar. Sci. Eng.* **2016**, *4*, 15. [CrossRef]

Article

Monitoring Systems and Numerical Models to Study Coastal Sites †

Elvira Armenio [1,*], Mouldi Ben Meftah [1,2], Diana De Padova [1,2], Francesca De Serio [1,2] and Michele Mossa [1,2]

1 Department of Civil, Environmental, Land, Building Engineering and Chemistry (DICATECh), Polytechnic University of Bari (Italy), via Orabona 4, 70125 Bari, Italy; mouldi.benmeftah@poliba.it (M.B.M.); diana.depadova@poliba.it (D.D.P.); francesca.deserio@poliba.it (F.D.S.); michele.mossa@poliba.it (M.M.)

2 CoNISMa, National Interuniversity Consortium of Marine Sciences, Piazzale Flaminio 9, 00196 Roma, Italy

* Correspondence: elvira.armenio@poliba.it

† This is an extension version of our conference paper: Armenio, E.; de Padova, D.; de Serio, F.; Mossa, M. Monitoring System in Mar Grande Basin (Ionian Sea). In Proceedings of the 2018 IEEE International Workshop on Metrology for the Sea; Learning to Measure Sea Health Parameters (MetroSea), Bari, Italy, 8–10 October 2018.

Received: 26 February 2019; Accepted: 28 March 2019; Published: 30 March 2019

Abstract: The present work aims at illustrating how the joint use of monitoring data and numerical models can be beneficial in understanding coastal processes. In the first part, we show and discuss an annual dataset provided by a monitoring system installed in a vulnerable coastal basin located in Southern Italy, subjected to human and industrial pressures. The collected data have been processed and analysed to detect the temporal evolution of the most representative parameters of the inspected site and have been compared with recordings from previous years to investigate recursive trends. In the second part, to demonstrate to what extent such type of monitoring actions is necessary and useful, the same data have been used to calibrate and run a 3D hydrodynamic model. After this, a reliable circulation pattern in the basin has been reproduced. Successively, an oil pollution transport model has been added to the hydrodynamic model, with the aim to present the response of the basin to some hypothetical cases of oil spills, caused by a ship failure. It is evident that the profitable prediction of the hydrodynamic processes and the transport and dispersion of contaminants strictly depends on the quality and reliability of the input data as well as on the calibration made.

Keywords: monitoring station; numerical modelling; current circulation; oil spilling

1. Introduction

Coastal areas are highly vulnerable because they are exposed to both natural hazards, including flooding, storm impacts, sea-level rise, and coastal erosion, and to anthropic activities such as urbanization, industrialization and transportation on the other hand. Nearshore regions provide several important functions, with high monetary value, managed by different stakeholders, e.g., fishing, aquaculture, leisure and tourism, water supply, wastewater treatment, construction, harbours and energy production. Furthermore, they are often a valuable environmental heritage, thus requiring proper safeguarding and enhancement, to be achieved by means of sustainable development, education and environmental protection [1–3].

In recent years, considerable attention has been paid to the preservation of coastal basins and inlets, which are especially vulnerable to human interventions and climate change because of their semi-enclosed shapes and lagoon characteristics. They represent a noteworthy natural resource, characterized by dynamic ecosystems in which natural and anthropic processes interact. Thus, their massive exploitation due to dense population and industrial activities could alter their geomorphological, physical and biological features, causing heavy pollution phenomena [3]. Consequently, both politicians and coastal managers are strongly required to strive to understand the evolution of the coastal system and guaranteeing its preservation.

To achieve this goal and support the environmental policy, the monitoring activity is one of the most appropriate tools in controlling and preventing the response of coastal areas to human actions and natural hazards [4]. This is more necessary in sensitive coastal sites characterized by a plurality of pressure factors, such as urban and industrial discharges or intense naval traffic or large harbours and military arsenal settings, where accidental releases of crude oil, gas and chemical products could take place. Accidental and illegal oil pollution constitutes a major threat to the marine environment and the risk of coastal oil spills dramatically increases.

The most common nautical accidents occur due to sinking or foundering, grounding, structural failure, scuttling, by contact or collision, explosion or fire, or after disappearance or abandonment. The discharge from oil pipelines, oil platforms, and vessels also causes significant damage to the marine environment and coastal areas. As shown by [5] and [6], marine oil spills may lead to serious environmental disasters, often with significant long-term ecological impacts on the coastal environment and detrimental consequences on the socio-economic activities of the area.

Successful responses to oil spill events can be achieved, provided that detailed information on type and volume of the escaped oil is promptly available to field operators. With information on the oil slick location, extent, thickness, and expected drift direction, the response team can plan effective countermeasures to mitigate the effects of devastating pollution on the marine environment [7]. To facilitate the decision-making process, it is possible to obtain very positive support for oil spill monitoring thorough the joint use of oil drift models, remote-sensing observations and measurement stations. It is worth noting that remote sensing techniques, widely and effectively adopted for oil spill monitoring scopes [7] are based on both microwave observations given by synthetic aperture radar (SAR) and satellite optical imagery. However, these types of data are not always available for the target sites and for the desired time period. In such cases, at least the coupling of advanced monitoring technologies and numerical models is required [7]. Furthermore, for consistent and accurate results, it is essential to run numerical models previously implemented and calibrated with reliable field data.

Since the 1960s, numerous oil spill models have been developed by various organizations, companies, and researchers, to simulate weathering processes and forecast the fate of oil spilled, in terms of providing valuable support to both contingency planners and pollution response teams.

There are two categories according to the Industry Technical Advisory Committee (ITAC) for oil spill response. One category, known as oil weathering models, estimates how oil properties change over time under the influence of current and wind advection, but does not predict potential migration of the slick [8]. The second category includes trajectory or deterministic models, stochastic or probability models, hind cast and three-dimensional models. In addition to predicting weathering profiles, these models estimate the evolution of a slick over time [9,10].

Some of the oil spill models that are currently available are: General NOAA Operational Modeling Environment (GNOME) [11], MEDSLIK-II [12], SeaTrackWeb (STW) [13], Model Oceanique de Transport d' Hydrocarbures (MOTHY) [14], DieCAST-SSBOM (Shirshov-Stony Brook Oil spill transport Model) [15], COastal Zone OIL spill model (COZOIL) [16], POSEIDON Oil spill model [17].

Specifically, in the present study, we have addressed the DHI MIKE 3 FM Oil Spilling Model [18], which belongs to the second category of the models described above. This numerical model has effectively demonstrated its ability to accurately analyse oil spill events [19–26] provided that an adequate calibration for the hydrodynamic module is performed first. As shown by [27] for the Fu Shan Hai oil spill accident in the Danish waters, the DHI Oil Spilling model, forced by not calibrated currents, provided a poor performance.

In this paper, we have focused our interest on a very vulnerable coastal site in southern Italy, the Mar Grande basin. First, an extensive set of monitored data of winds, waves and currents is presented and discussed, recorded in this semi-enclosed coastal basin in the year 2015. These data have been then used to calibrate the DHI 3D hydrodynamic flow model [18]. After this, some scenarios of oil spreading arising from a continuous 72-h spill of crude oil from a ship failure, for selected winter and summer periods in 2015, have been hypothesized and examined. The MIKE 3 FM Oil Spilling Model [18] based on dispersive processes, has been added to the hydrodynamic module, to simulate the possible oil spreading.

2. Materials and Methods

2.1. Study Site and Monitoring Station

The Mar Grande is a coastal basin with a typical round shape, located in the inner northeastern area of the Ionian Sea, in Southern Italy (Figure 1). Its total surface area is about 35 km^2, while its maximum depth is about 35 m in its central part. The Mar Grande is connected to a smaller semi-enclosed basin, named Mar Piccolo, formed in turn by two embayments. The Mar Grande and the Mar Piccolo basins are joined by means of an artificial channel, i.e., the Navigable Channel, and a natural one, i.e., the Porta Napoli Channel. As shown in Figure 1, on its western side the Mar Grande is limited by two small isles (S. Pietro and S. Paolo) called Cheradi, connected each other by a long breakwater. The northwestern opening between the mainland and the Cheradi Isles is named Punta Rondinella and is about 100 m long, while the southwestern one between the Cheradi Isles and the mainland is about 1400 m long. All the port activities are located along the northern coast of the Mar Grande, while the Naval Arsenal is located along the southern coast. The urban centre is located along the eastern coast (Figure 1). Over the years, the basin has been subjected to a heavy polluting charge, affected by different outflows from sources of civil, military and industrial origins, which are authorized and monitored only in some cases.

To start collecting some information about the physical and hydrodynamic state of the basin, a meteo-oceanographic station was installed in its central area (here named for brevity MG station), in December 2013, at the geographical coordinates 40°27.6′ N and 17°12.9′ E, where the local depth is around 23.5 m. The devices were settled in the frame of the Italian Flagship Project RITMARE [28] and the entire system is managed by the research unit of the Polytechnic University of Bari–Laboratory of Coastal Engineering (LIC). The station is equipped with many instruments, including a bottom mounted Acoustic Doppler Current Profiler (ADCP), a multidirectional wave array (both by Teledyne RD) and a weather station (by Met Pack). Other scalar parameters, such as temperature, salinity, chlorophyll, are also measured at five meters below the sea surface [29].

Figure 1. Map of Mar Grande basin and location of the monitoring station MG.

2.2. Monitoring System

The monitoring system's configuration of the MG station is based on a telemetry data transmission technique, to acquire field data and transmit them in real time, ensuring high standards of efficiency and precision even over considerable distances. Its network architecture has been implemented to acquire measurements and communicate them to receivers, thus allowing a continuous exchange and update of information. The monitoring station is composed of various sensors and devices connected each other to implement and integrate the data management network. Through a client-server system and relying on a protocol architecture, it allows the sharing of the acquired measurements. In particular, the monitoring system configuration performs the following activities:

(1) continuous detection of the measured environmental parameters using suitable sensors and devices installed in the station;
(2) remote pre-processing of raw data for decoding from binary files in a user-friendly format to be easily managed by users;
(3) wireless data transmission via GSM/GPRS devices to connect the MG station to a Data Acquisition Center;
(4) data storage on a dedicated server on which authorized users can visualize and download data also by remote connection.

Figure 2 displays the conceptual configuration of the monitoring station. All data measured by the current profiler, the weather sensors and the multidirectional wave array are both stored onsite, available for remote access queries to check and display their status in real time, and transmitted in real time to the Data Acquisition Center, where it can be downloaded, archived and processed.

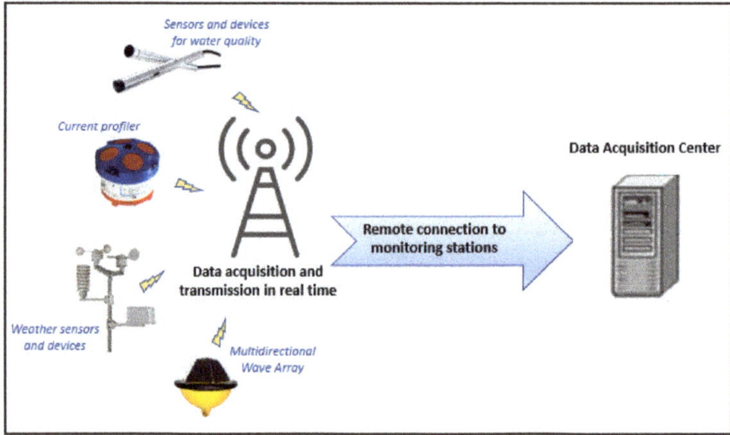

Figure 2. Conceptual scheme of the monitoring system MG.

More specifically, the sensors are all connected to an autonomous data acquisition unit, i.e., a datalogger (named LISC), which acquires data from 12 serial ports and 16 analog channels, which allows: (i) data processing onsite and in real time, (ii) remote control and (iii) data download. The datalogger processing, demanding in computing and energy-supply terms, is managed by a software program (MARLIN) installed in a suitable device, which communicates with the LISC at the end of each measurement for a short time window, just necessary to get the raw data and send back the processed ones. The datalogger can be remotely queried by means of a cellular modem 3G, connected to it through a serial port and equipped with a stack TCP/IP to send data on the web cloud. In this way, the communication from the remote system to the devices and vice versa is possible, managed by a proper software (ROCS). The ROCS software can call the LISC datalogger both automatically and manually, permitting to download the acquired data and to check or change the acquisition parameters. Configurations and data from the managed devices are all contained in a relational database, which is the ROCS core, and can be exported in binary raw data or ASCII files, according to user needs. This database is constantly updated. Depending on the type of data, a table is filled by specifying the date and time of recordings. Any anomalies or malfunctions are identified by a specific code.

2.3. Analysis of Monitored Data

The wind data for the whole year 2015 have been processed. To identify the predominant seasonal trends, the rose plots referring to the winter and summer period are shown respectively in Figure 3, based on their incoming directions. During the winter period (January–March), the most frequent winds come from NNW, NNE and WSW. Considering the location of the MG station, WSW winds, which are also the most intense ones (intensities >9 m/s), are significant, because they come from the open sea where they originate wind waves on longer fetches. Moderate (3–6 m/s) and high (6–9 m/s) wind intensities are observed along the other directions. During the summer period (July–September), the most frequent winds come mainly from NNW with peak velocities in the high range 6–9 m/s. Compared to these, all the winds coming from different directions are more occasional and almost weak. The wind distribution observed in 2015 replicates the records of the winds measured in both the years 2014 and 2016 [30], especially referring to the winter period, while, for the summer season, the 2015 data do not show significant winds from SE (Scirocco), which on the contrary are more evident in 2014 and 2016 [30].

Figure 4 displays the polar plots of the significant wave heights *Hs* monitored during the year 2015, for the winter and summer period respectively, where the directions of wave propagation are shown. The lowest waves are observed in the summer period as expected. In both seasons, a well-defined and evident path is recognized for high waves, which come from SW and propagate towards NE. This wave behaviour confirmed also by the observations of 2014 and 2016 [30], seems consistent with the presence of the wide opening located on the SW boundary of the Mar Grande basin, which allows external swell waves to enter the basin and spread towards the opposite border. The highest values of H_s along this direction are in the range 1.0–1.2 m for both winter and summer cases, although in summer the occurrence of these values is much rarer and generally low waves prevail along all directions (<0.3 m).

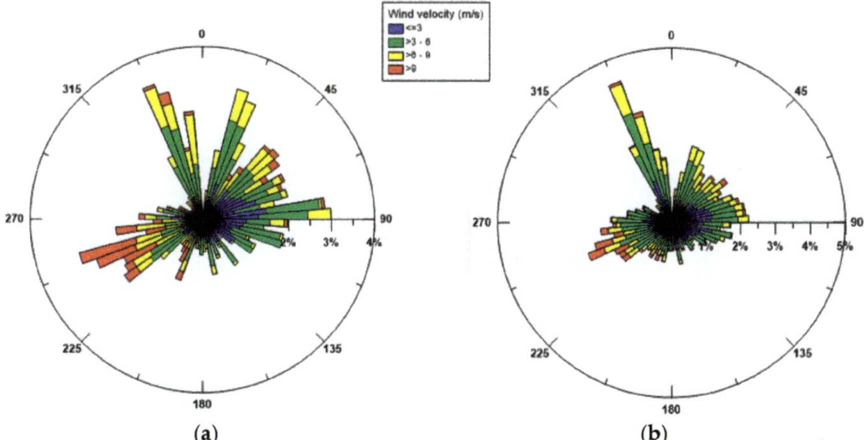

Figure 3. Wind polar diagrams for 2015: (**a**) winter; (**b**) summer. Incoming wind directions are shown.

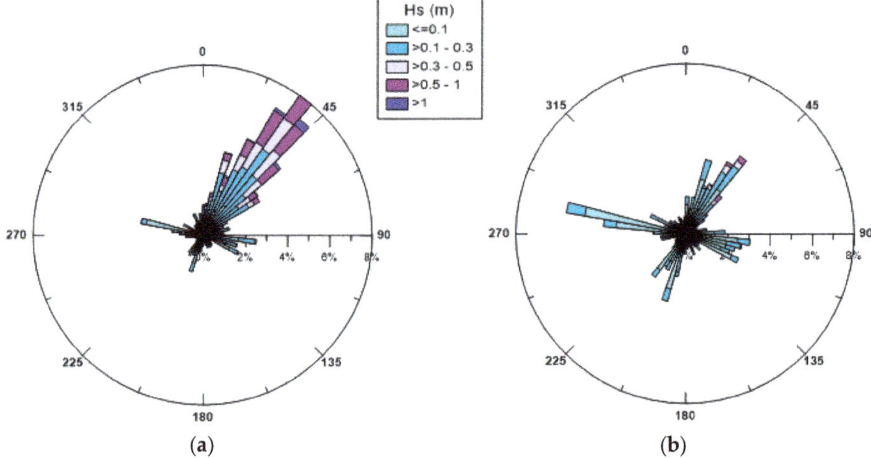

Figure 4. Seasonal trend of the significant wave heights (in m): (**a**) winter 2015; (**b**) summer 2015. Wave propagation directions are shown.

In Figure 5 the seasonal distribution of the superficial currents is shown, as measured at a depth of 2 m from the sea surface to disregard the possible influence of waves. Variability in the current direction is evident and especially reflects the variability of the blowing winds. The most frequent and intense surface currents are consistent with the most dominant winds, in fact both in winter and in summer they are mainly directed towards SE and SSE. This observation confirms what already pointed out by De Serio and Mossa [29,30], that is winds blowing from land do not have a direct effect on the origin of sea waves, but rather seem to drive the surface current.

Analogously, Figure 6 illustrates the polar diagrams of the currents measured near the bottom, for both seasons. The trend observed confirms what was also recorded in 2014 and 2016. In fact, the currents appear to have a preferred direction and tend to converge towards the SW opening of the basin which thus exerts a sort of topographical control [29,30]. By comparing the surface currents with the bottom ones, the highest values (>0.3 m/s) are always noted near the surface, rather than near the bed, where values in the range 0.05–0.1 m/s prevail. Thus, the effect of the wind shear stress, which is the principal driver of the surface circulation, is gradually lost along the water column, and rather at the bottom the currents are controlled by topography.

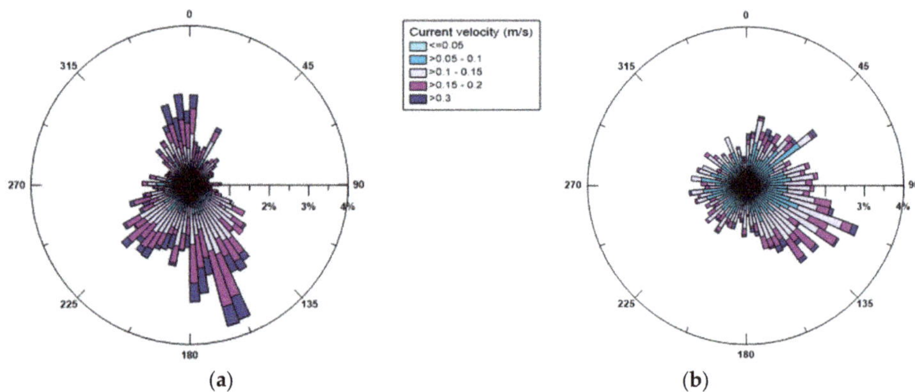

Figure 5. Seasonal currents (in m/s) near the sea surface in 2015 (**a**) winter and (**b**) summer. Direction of current propagation is shown.

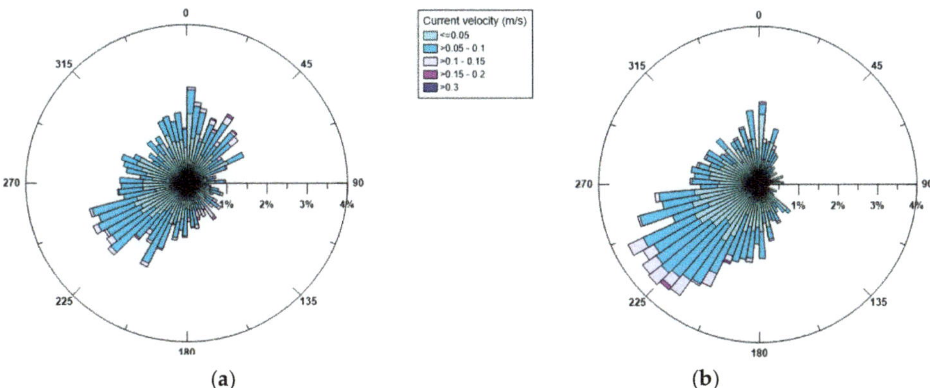

Figure 6. Seasonal currents (in m/s) near the bottom in 2015 (**a**) winter and (**b**) summer. Directions of current propagation is shown.

From the comparison with previous results [29,30], we can note that the behaviours of winds, waves and both superficial and near bottom currents are annually recursive, showing repeating features, remarkably typical for the two examined seasons.

3. Numerical Modelling

3.1. Calibration

The available field data described above have been used to implement and calibrate the 3D hydrodynamic numerical model MIKE 3 FM HD produced by the Danish Hydraulic Institute (DHI) [15]. A finite mesh of 7235 triangular elements with ten vertical layers has been used (Figure 7). To improve the numerical approach and model more realistic conditions, the hydrodynamic simulations have been carried out in baroclinic mode, with temperature and salinity vertical profiles extracted by the Mediterranean Sea Physics Reanalysis model, characterized by a horizontal grid resolution of $1/16°$ and by 72 unevenly spaced vertical levels. Moreover, the simulation has been forced at the sea open boundary by the time varying water levels measured at S. Eligio pier (Figure 1) by the National Institute for Environmental Protection and Research (ISPRA). The sea surface wind field has been deduced by the atmospheric model described in [2] and is variable in space and time

The turbulent closure model used within MIKE 3 FM HD model relies on the k-ε formulation for the vertical direction [31] and on the Smagorinsky formulation for the horizontal direction [32]. The Smagorinsky coefficient has been assumed uniform in space and temporally constant, equal to 0.6. According to the sensitivity analysis shown by De Padova et al. [28], the simulation has been performed by adopting a seabed roughness equal to 0.1 m.

Based on earlier studies [26,28], the wind drag coefficient C_d has been considered as the calibration parameter to which the model results are most sensitive. Therefore, it has been tuned, switching among different values. We have deduced that the most suitable value of C_d to get the best match between modelled and measured data is 0.02. The magnitude of this drag coefficient is one order higher than in most of the cases studied, but it is worth noting that lower magnitudes are typical of circulation studies involving oceanic sites or large seas, where the action of the wind is very strong. On the contrary, near coastal sites and in confined seas, where action of the wind is weaker, an increased C_d is needed to transmit effective wind stress to surface currents [29].

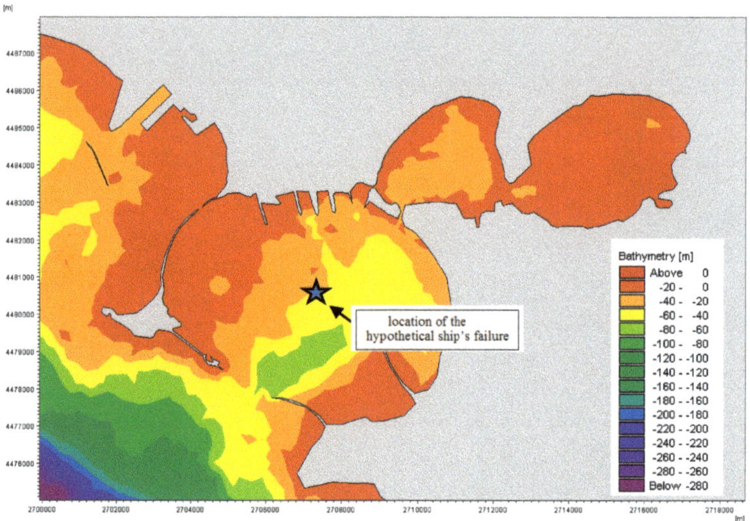

Figure 7. Computation domain used for the numerical simulations. The location of the hypothetical ship's failure is also shown. UTM coordinates used.

The comparison between modelled and real current velocities has been made for the point where the MG station is located, at a depth of 5 m from the surface, to disregard the possible influence of surface waves on current measurements. For the same reason, the month of July 2015 has been chosen for comparison, because it is characterized by the lowest recorded waves. Figure 8 shows the time series of the computed and observed current intensities. A good similarity between the modelled and measured currents is clearly shown. The degree of agreement has been estimated via the index I_w proposed by Wilmott [33], which has been computed. It takes into account the relative error among field and output values and can be equal to 1, at best. In our case it assumes the value 0.72, thus indicating a good reproduction of the measured data by the model in terms of magnitude. Analogously, I_w has been calculated also to evaluate the model performance in terms of velocity directions, providing, in this case, an average value around 0.6, which is still a satisfactory result, especially considering the complexity of the environment reproduced and the simplifications necessarily applied in the modelling.

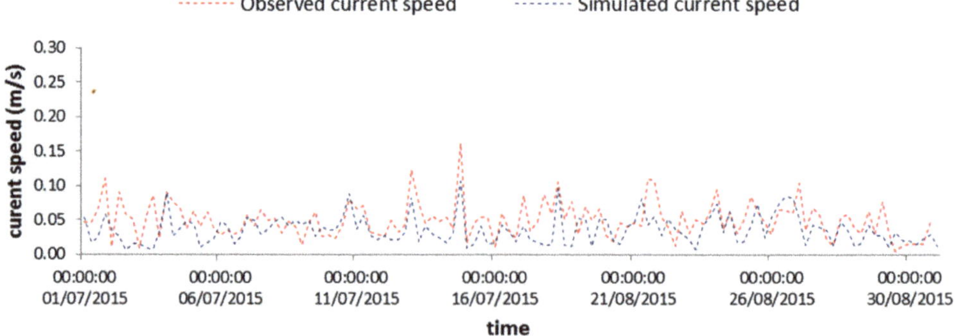

Figure 8. Comparison between time series of observed and simulated current speed at the selected point.

3.2. Oil Spilling Runs

Once verified the validity of the adopted hydrodynamic model, this has been combined with a dispersive module capable of reproducing the spreading of a contaminant, with the aim of evaluating the response of the model when an oil spilling occurs. Thus, the MIKE 3 FM Oil Spilling model by DHI has been implemented to simulate a hypothetical case of the oil spill caused by a ship failure in the central area of the Mar Grande basin (Figure 7). The oil spill model [18] solves the so-called Fokker–Planck equation for suspended oil substances in two dimensions, through the introduction of a consistent random walk particle method. It is solved by the Lagrangian discrete parcel method, while the weathering processes are solved by the Runga–Kutta fourth-order method. The pollutant is divided into discrete parcels and sets of spatial coordinates are assigned to each parcel. It is assumed that these parcels advect with the surrounding water body and diffuse as a result of random processes. The displacement of each Lagrangian particle is given by the sum of an advective deterministic and a stochastic component, the latter representing the chaotic nature of the flow field, the sub-grid turbulent dispersion. Further details on this module are in [18,24,25].

Two simulation runs, denoted as T1 and T2, have been carried out, with the aim of evaluating how the transport of contaminants in this area can be different during the winter and summer seasons respectively. The spill has been modelled as a continuous leakage of oil over a period of 72 h (3 days) starting on 12 December 2015 at 08:00 for test T1 and on 12 July 2015 at 08:00 for test T2. For each simulation the total mass of oil released is equal to 519 tons (rate of 2 kg/s) and the applied oil type is crude oil.

The sea surface wind field has been deduced by the atmospheric model described in [2] and is variable in space and time. There has been a considerable dispute among modellers about the best choice for the values of the wind drift factor and the wind-drift angle to be used. In fact, it depends on the physical process of the problem and the desired computational efficiency. Most of the models have used a value of around 3% for the former and a value between 0° and 25° for the latter [34–37]. In this study we have employed the well-established 3% wind factor and a wind drift angle equal to 20° also in accordance with what obtained from our previous study [26]. The oil spill model has been driven by the outputs of the calibrated hydrodynamic model MIKE 3FM, respectively for T1 and T2 cases.

4. Results and Discussion

In the winter case (T1), the selected period of the run is characterized by mainly NW and NE winds. At 18:00 on December 12, i.e., 10 h after the oil spilling started, NW winds are quite uniformly distributed on the whole domain (Figure 9a). At the same time, there is a superficial circulation mainly outflowing from the Mar Grande basin (Figure 9b) and the oil slick is transported from the centre of the basin towards the SW opening, elongated and directed towards the open sea. At 18:00 on December 13 (Figure 10), the oil slick has changed its orientation and seems more confined to the western boundary, driven by westwards surface currents and the wind from NE. At 18:00 on December 14, due to northerly winds and easterly currents, the oil slick widens along the western boundary and spreads outside the Mar Grande (Figure 11).

In the summer case (T2), the selected period of the run is characterized by very variable winds. After the first day (July 12) with currents directed outside the Mar Grande basin under the influence of NE winds, at 00:00 on July 13 (Figure 12) the wind direction changes and SE winds induce in the central part of the domain a clockwise trend, with increasing flow intensity. Consequently, the oil slick, is transported towards the NW border. At 12:00 on July 13, SW winds blow (Figure 13) and an intense northwards flow is observed along the coast of San Pietro Isle and along the northern coast of the basin. The oil slick is moved and spreads towards the northern border of the basin, in the port area. On July 14 at 00:00, the winds change direction again, becoming NE winds (Figure 14), inducing the formation of weak currents along the northern border, a cyclonic vortex in the central part of the domain and a weak outflow towards the open sea. Thus, in this case, the oil slick tends to return to the centre of the basin.

For both tests T1 and T2, the main results of the oil pollution transport model show that the oil slick is primarily moved by the action of surface currents, whose variability in turn depends strongly on winds, rather than on other forces. The intensity and direction of the surface currents drive the dispersion of the oil slick, determining its shape and size. In particular, in test T1, predominantly NW and NE winds induce in the basin an intense outward current or a clockwise one feeding an outflowing branch. They both allow the oil slick travelling throughout the basin, directed outside. In test T2, under the action of weaker winds with a varying direction (clockwise from South to North), a clockwise trend is observed in the surface circulation. As a result, the northern and western coasts of the Mar Grande are seriously exposed to an oil pollution load. This increases the hazard in such areas, which are already vulnerable, under the polluting pressure due to the presence of the port. Moreover, both in winter (T1) and in summer (T2), the eastern coast of the Cheradi Isles (Figure 1), where some important natural species such as *Poseidonia oceanica* have been observed [38–40], is reached and affected by oil pollution. Strong and negative impacts occurring on the marine environment, like this one, could be monitored and even predicted by a trustworthy numerical simulation. It is evident that the possible performed scenarios and the output of the oil pollution transport model can be considered reliable only when reliable input and boundary conditions are used. This further proves the need to adopt field data of high quality in the use of the model. Operating in this way permits a profitable coastal management and intervention in the event of accidents.

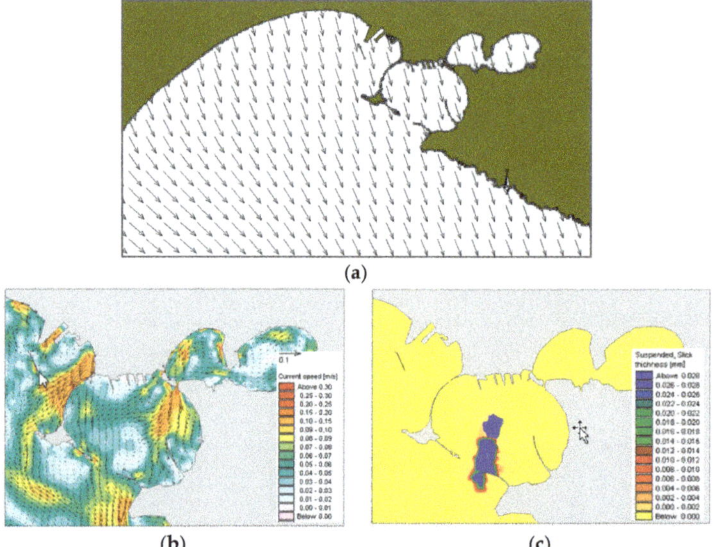

Figure 9. Test T1. Dec. 12 at 18:00. (**a**) Model wind fields; (**b**) Surface current field; (**c**) Oil slicks.

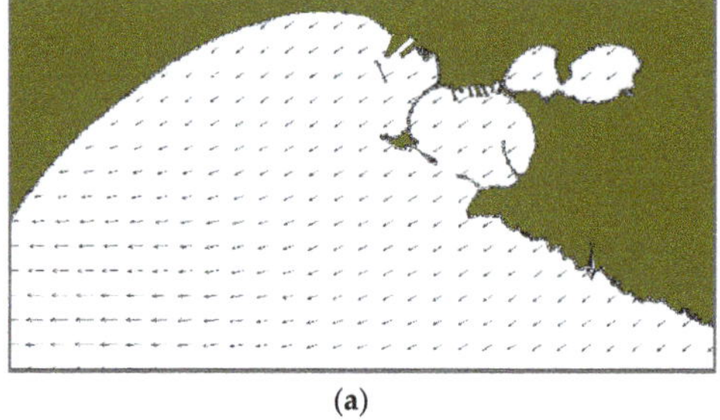

(**a**)

Figure 10. *Cont.*

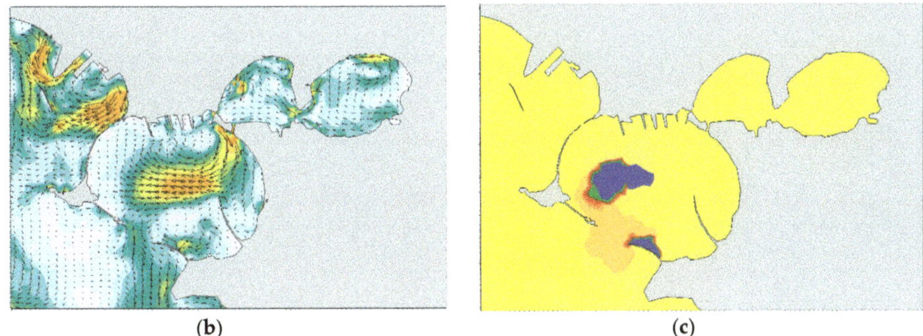

(b) (c)

Figure 10. Test T1. Dec. 13, at 18:00. (**a**) Model wind fields; (**b**) Surface currents; (**c**) Oil slicks. Legends as in Figure 9.

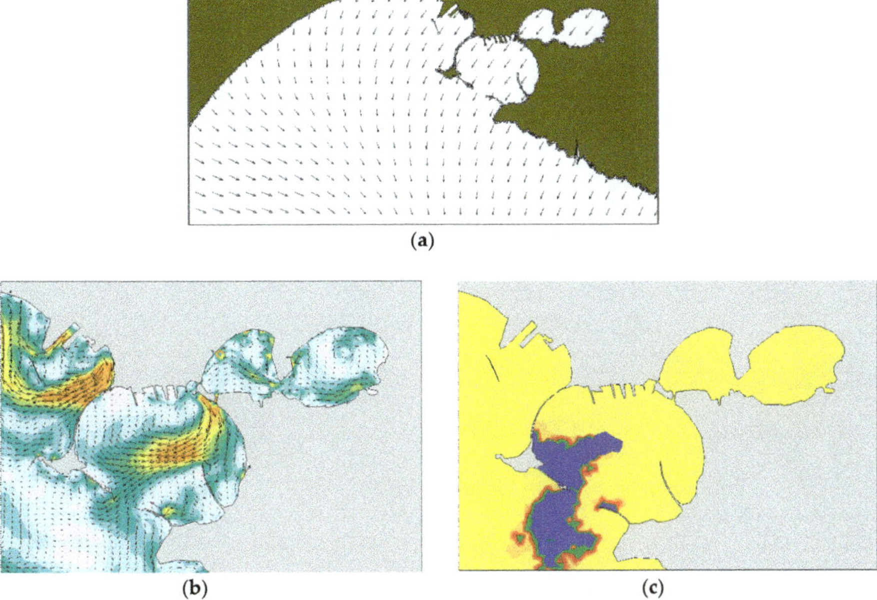

(a)

(b) (c)

Figure 11. Test T1. Dec. 14, at 18:00. (**a**) Model wind fields; (**b**) Surface currents; (**c**) Oil slicks. Legends as in Figure 9.

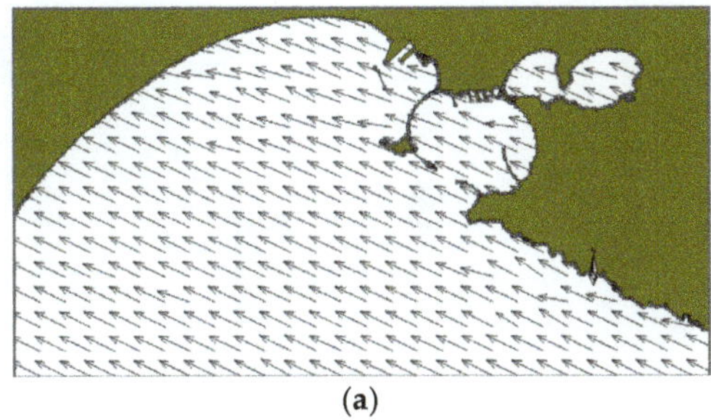

(a)

Figure 12. *Cont.*

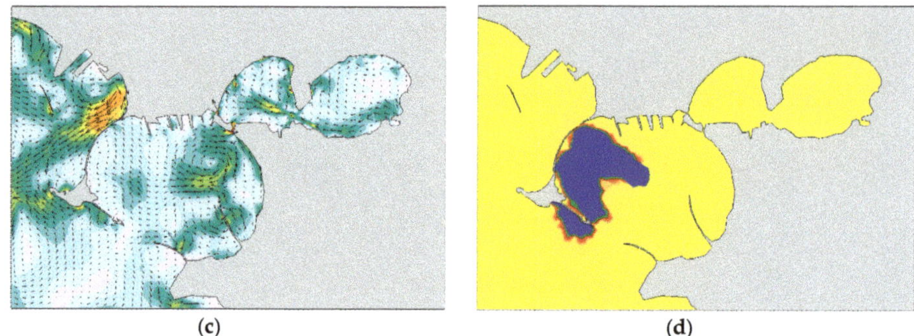

Figure 12. Test T2. July 13, at 00:00. (**a**) Model wind fields (**b**) Surface currents; (**c**) Oil slicks. Legends as in Figure 9.

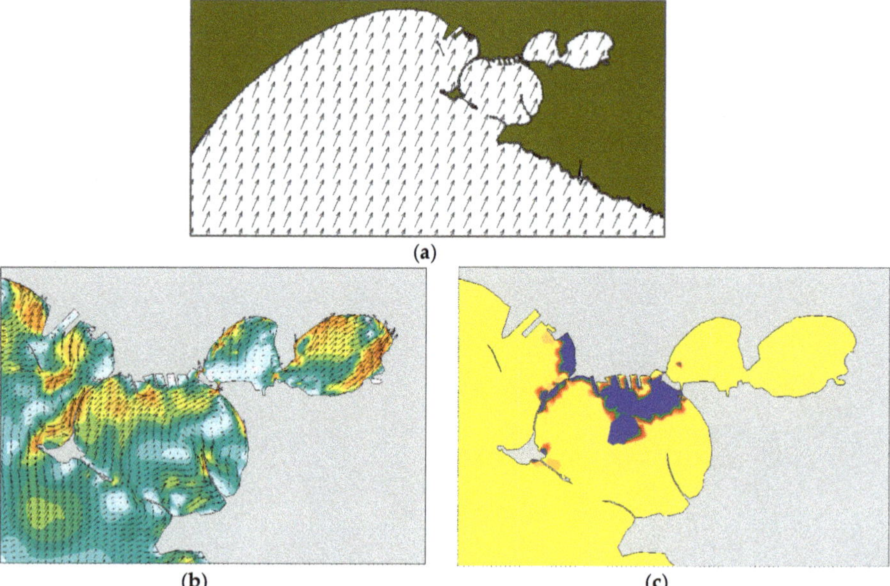

Figure 13. Test T2. July 13, at 12:00. (**a**) Model wind fields (**b**) Surface currents; (**c**) Oil slicks. Legends as in Figure 9.

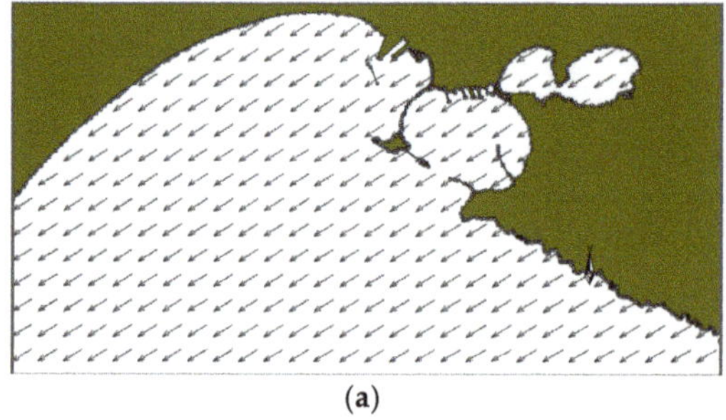

Figure 14. *Cont.*

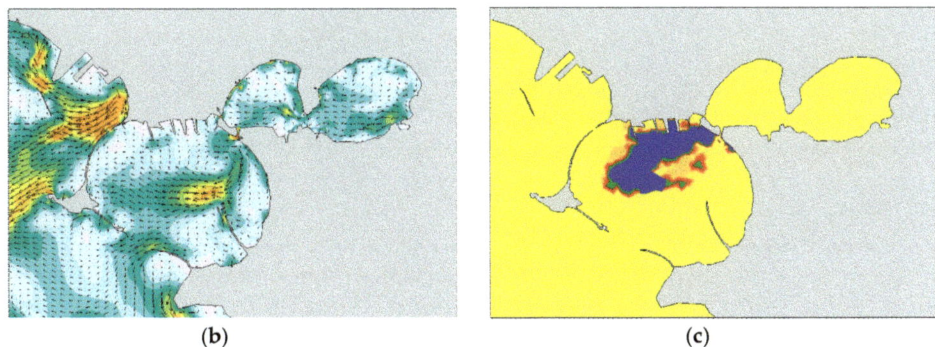

Figure 14. Test T2. July 14, at 00:00. (**a**) Model wind fields (**b**) Surface current field from; (**c**) Oil slicks. Legends as in Figure 9.

5. Conclusions

The paper has shown to what extent the joint use of data from monitoring stations and numerical models is necessary and useful in a very vulnerable coastal site. Specifically, we have proved that this synergy is possible if based on high quality available field data. In fact, data sets including many different parameters, which, assessed continuously and for a sufficiently long time, are needed to produce consistent hydrodynamics of the target area.

Firstly, the extensive set of monitored data of winds, waves and currents recorded in the Mar Grande basin, referring to the year 2015, has been examined. We have deduced that during the winter period, the most frequent winds are from NNW, NNE and WSW, with WSW winds being the most intense ones, coming from the open sea where they originate wind waves on longer fetches. During the summer period, the most frequent winds come mainly from NNW, with peak velocities in the high range 6–9 m/s. The polar plots of the significant wave heights at the monitoring station reveals that in both seasons there is an evident path for the highest waves, reaching the station from SW and propagating towards NE, due to the presence of the wide opening located on the SW boundary of the basin. The distribution of the superficial currents is consistent with the most dominant winds, in both winter and summer. Near the bottom, the measured currents have a preferred direction and tend to converge towards the SW opening, which thus exerts a sort of topographical control.

In the second part of the paper, the MIKE 3D FM model has been calibrated with the examined field data, by tuning the wind drag coefficient. These results highlight that the calibrated model can detect and reproduce the main features of the circulation structure in the target area. Successively, once recognized that the hydrodynamic pattern of the basin is the first necessary step for further investigations, the MIKE 3D FM Oil Spilling model has been implemented to simulate a hypothetical case of an oil spill caused by a ship failure in the central area of the Mar Grande. The oil spill model has been driven by the outputs of the calibrated hydrodynamic model MIKE 3FM. Two simulation runs have been carried out, with the intent to assess the different way in which the contaminant is transported in this area during the winter and summer seasons of 2015. For both tests, the principal results show that the oil slick is governed primarily by the action of surface currents, in turn driven by the winds. In the winter case, with predominantly NW and NE winds, it travels the basin directed outside. In the summer case, under the action of winds with varying direction, the northern and western coasts of the Mar Grande are seriously exposed to an oil pollution load. As well, the eastern coast of the Cheradi Isles, characterized by the presence of important natural species, such as Posidonia Oceanica, is dramatically reached by oil pollution.

The proposed scenarios highlight the necessity to properly reproduce hydrodynamics in the basin, to provide useful information and predictions. This can be achieved through the joint use of high-quality field data and numerical models.

Author Contributions: Conceptualization, M.M. and F.D.S.; formal analysis, E.A. and D.D.P.; writing—Original draft preparation, F.D.S., D.D.P.; writing—Review and editing, F.D.S., D.D.P., E.A., M.B.M. and M.M.

Funding: The monitoring station examined in the present research was settled in the frame of the Italian Flagship Project RITMARE (chief scientist Michele Mossa of the research unit of the CoNISMa–Polytechnic University of Bari-Italy) and with funds from PON R&C 2007-13 Project.

Conflicts of Interest: The authors declare no conflict of interest.

References

1. Armenio, E.; Ben Meftah, M.; De Padova, D.; De Serio, F.; Mossa, M. Monitoring System in Mar Grande Basin (Ionian Sea). In Proceedings of the 2018 IEEE International Workshop on Metrology for the Sea; Learning to Measure Sea Health Parameters (MetroSea), Bari, Italy, 8–10 October 2018; pp. 104–109. [CrossRef]
2. Armenio, E.; Ben Meftah, M.; Bruno, M.F.; De Padova, D.; De Pascalis, F.; De Serio, F.; di Bernardino, A.; Mossa, M.; Leuzzi, G.; Monti, P. Semi enclosed basin monitoring and analysis of meteo, wave, tide and current data. In Proceedings of the IEEE Workshop on Environmental, Energy, and Structural Monitoring, Bari, Italy, 13–14 June 2016. [CrossRef]
3. Armenio, E.; De Serio, F.; Mossa, M.; De Padova, D. Monitoring system for the sea: Analysis of meteo, wave and current data. In Proceedings of the IMEKO TC19 Workshop on Metrology for the Sea, MetroSea, Naples, Italy, 11–13 October 2017; pp. 143–148.
4. Armenio, E.; De Serio, F.; Mossa, M. Analysis of data characterizing tide and current fluxes in coastal basins. *Hydrol. Earth Syst. Sci.* **2017**, *21*, 3441–3454. [CrossRef]
5. De Serio, F.; Mossa, M. Environmental monitoring in the Mar Grande basin (Ionian Sea, Southern Italy). *Environ. Sci. Pollut. Res.* **2016**, *23*, 12662–12674. [CrossRef] [PubMed]
6. Mossa, M. Field measurements and monitoring of wastewater discharge in sea water. *Estuar. Coast. Shelf Sci.* **2006**, *68*, 509–514. [CrossRef]
7. De Carolis, G.; Adamo, M.; Pasquariello, G.; De Padova, D.; Mossa, M. Quantitative characterization of marine oil slick by satellite near-infrared imagery and oil drift modelling: The Fun Shai Hai case study. *Int. J. Remote Sens.* **2013**, *34*, 1838–1854. [CrossRef]
8. Fay, J.A. *The Spread of Oil Slicks on a Calm Sea*; Springer: Boston, MA, USA, 1969; pp. 53–63.
9. Huang, J.C.; Monastery, F.C. Review of the State of the Art of Oil Spill Simulation Models. Available online: https://ci.nii.ac.jp/naid/10013314715/ (accessed on 30 March 2019).
10. Huang, J.C. A review of the state of the art of oil spill fate/behavior models. In Proceedings of the International Oil Spill Conference, San Antonio, TX, USA, 28 February–3 March 1983; pp. 313–322.
11. Beegle-Krause, J. General NOAA oil modeling environment (GNOME): A new spill trajectory model. *Int. Oil Spill Conf. Proc.* **2001**, *2001*, 865–871. [CrossRef]
12. De Dominicis, M.; Pinardi, N.; Zodiatis, G.; Lardner, R. MEDSLIK-II, a Lagrangian marine oil spill model for short-term forecasting—Part I: Theory. *Geosci. Model Dev.* **2013**, *6*, 1851–1869. [CrossRef]
13. Sven, A.; Uiboupin, R.; Verjovkina, S.; Raudsepp, U. SAR imagery and Seatrack Web as decision making tools for illegal oil spill combating—A case study. In Proceedings of the 2010 IEEE/OES Baltic International Symposium (BALTIC), Riga, Latvia, 24–27 August 2010; pp. 1–6.
14. Daniel, P. Operational forecasting of oil spill drift at Météo-France. *Spill Sci. Technol. Bull.* **1996**, *3*, 53–64. [CrossRef]
15. Korotenko, K.A.; Bowman, M.J.; Dietrich, D.E. Modeling of the circulation and transport of oil spills in the Black Sea. *Oceanology* **2003**, *43*, 367–378.
16. Reed, M.; Gundlach, E. A Coastal Zone Oil Spill Model: Development and Sensitivity Studies. *Oil Chem. Pollut.* **1989**, *5*, 411–449. [CrossRef]
17. Perivoliotis, L.; Nittis, K.; Charissi, A. An integrated servicefor oil spill detection and forecasting in the marine environment. In Proceedings of the Fourth International Conference on EuroGOOS, Brest, France, 6–9 June 2005; pp. 381–387.
18. MIKE 3 Flow Model: Hydrodynamic Module—Scientific Documentation. DHI Software. Available online: https://www.google.com.tw/url?sa=t&rct=j&q=&esrc=s&source=web&cd=2&ved=2ahUKEwjQqYKhlqnhAhXJKqYKHa4yAkMQFjABegQIBRAC&url=http%3A%2F%2Fmanuals.mikepoweredbydhi.help%2F2017%2FCoast_and_Sea%2FMIKE3HD_Scientific_Doc.pdf&usg=AOvVaw0pxx_Ab_XdVAv7oPyUsvVY (accessed on 30 March 2019).

19. Elhakeem, A.; Elshorbagy, W.; Chebbi, R. Oil Spill Simulation and Validation in the Arabian Gulf with Special Reference to the UAE Coast. *Water Air Soil Pollut.* **2007**, *184*, 243–254. [CrossRef]
20. Kankara, R.S.; Subramanian, B.R. Oil Spill Sensitivity Analysis and Risk Assessment for Gulf of Kachchh, India, using Integrated. *Modeling J. Coast. Res.* **2007**, *23*, 1251–1258. [CrossRef]
21. Vethamony, P.; Sudheesh, K.; Babu, M.T.; Jayakumar, S.; Manimurali, R.; Saran, A.K.; Sharma, L.H.; Rajanb, B.; Srivastavab, M. Trajectory of an oil spill off Goa, eastern Arabian Sea: Field observations and Simulations. *Environ. Pollut.* **2007**, *148*, 438–444. [CrossRef]
22. Elshorbagy, W.; Elhakeem, A. Risk assessment maps of oil spill for major desalination plants in the United Arab Emirates. *Desalination* **2008**, *228*, 200–216. [CrossRef]
23. Verma, P.; Wate, S.R.; Devotta, S. Simulation of impact of oil spill in the ocean-a case study of Arabian Gulf. Environ. *Monit. Assess.* **2008**, *146*, 191–201. [CrossRef] [PubMed]
24. Lončar, G.; Beg Paklar, G.; Janekovic, I. Numerical Modelling of Oil Spills in the Area of Kvarner and Rijeka Bay (The Northern Adriatic Sea). *J. Appl. Math.* **2012**, *2012*, 497936. [CrossRef]
25. Lončar, G.; Leder, N.; Paladin, M. Numerical modelling of an oil spill in the northern Adriatic. *Oceanologia* **2012**, *54*, 143–173. [CrossRef]
26. De Padova, D.; Mossa, M.; Adamo, M.; De Carolis, G.; Pasquariello, G. Synergistic use of an oil drift model and remote sensing observations for oil spill monitoring. *Environ. Sci. Pollut. Res.* **2017**, *24*, 5530–5543. [CrossRef]
27. Christiansen, B.M. 3D Oil Drift and Fate Forecasts at DMI. Available online: www.fargisinfo.com/referanser/LinkedDocuments/D050.pdf (accessed on 30 March 2019).
28. De Padova, D.; De Serio, F.; Mossa, M.; Armenio, E. Investigation of the current circulation offshore Taranto by using field measurements and numerical model. In Proceedings of the 2017 IEEE International Instrumentation and Measurement Technology Conference (I2MTC), Turin, Italy, 22–25 May 2017. [CrossRef]
29. De Serio, F.; Mossa, M. Analysis of mean velocity and turbulence measurements with ADCPs. *Adv. Water Resour.* **2015**, *81*, 172–185. [CrossRef]
30. De Serio, F.; Mossa, M. Meteo and Hydrodynamic Measurements to Detect Physical Processes in Confined Shallow Seas. *Sensors* **2018**, *18*, 280. [CrossRef] [PubMed]
31. Rodi, W. Examples of calculation methods for flow and mixing in stratified fluids. *J. Geophys. Res. Ocean.* **1987**, *92*, 5305–5328. [CrossRef]
32. Galperin, B.; Orszag, S.A. *Large Eddy Simulation of Complex Engineering and Geophysical Flows*; Cambridge University Press: New York, NY, USA, 1993; pp. 3–36.
33. Wilmott, C.J. On the validation of models. *Phys. Geogr.* **1981**, *2*, 184–194. [CrossRef]
34. Stolzenbach, K.D.; Madsen, E.E.; Adams, A.; Pollack, A.; Cooper, C. A Review and Evaluation of Basic Technique for Predicting the Behavior of Surface Oil Slicks. Available online: https://repository.library.noaa.gov/view/noaa/9623/noaa_9623_DS1.pdf (accessed on 30 March 2019).
35. Audunson, T. The fate and weathering of surface oil from the Bravo blow out. *Mar. Environ. Res.* **1980**, *3*, 35–61. [CrossRef]
36. Reed, M.; Turned, C.; Odulo, A. The Role of Wind and Emulsification in Modelling Oil Spill and Surface Drifter Trajectories. *Spill Sci. Technol. Bull.* **1994**, *1*, 143–157. [CrossRef]
37. Spaulding, M.; Kolluru, V.; Anderson, E.; Howlett, E. Application of three-dimensional oil spill model (WOSM/OILMAP) to hindcast the Braer spill. *Spill Sci. Technol. Bull.* **1994**, *1*, 23–35. [CrossRef]
38. Parenzan, P. Il Mar Piccolo e il Mar Grande di Taranto. *Thalass. Salentina* **1969**, *3*, 19–34.
39. Parenzan, P. *Puglia Marittima*; Congedo Editore: Galatina, Lecce, Italy, 1983.
40. Matarrese, A.; Mastrototaro, F.; Maiorano, P.; Tursi, A. Mapping of the benthic communities in the Taranto seas using side-scan sonar and an underwater video camera. *Chem. Ecol.* **2004**, *20*, 377–386. [CrossRef]

Article

Experimental Setup and Measuring System to Study Solitary Wave Interaction with Rigid Emergent Vegetation †

Davide Tognin [1,*], Paolo Peruzzo [1], Francesca De Serio [2,3], Mouldi Ben Meftah [2,3], Luca Carniello [1], Andrea Defina [1] and Michele Mossa [2,3]

1 Department of Civil, Environmental and Architectural Engineering, University of Padova, via Loredan, 20, 35131 Padova, Italy; paolo.peruzzo@unipd.it (P.P.); luca.carniello@unipd.it (L.C.); andrea.defina@unipd.it (A.D.)
2 Department of Civil, Environmental, Land, Building Engineering and Chemistry, Polytechnic University of Bari, via Orabona, 4, 70125 Bari, Italy; francesca.deserio@poliba.it (F.D.S.); mouldi.benmeftah@poliba.it (M.B.M.); michele.mossa@poliba.it (M.M.)
3 CoNISMa, National Interuniversity Consortium of Marine Sciences, Piazzale Flaminio 9, 00196 Roma, Italy
* Correspondence: davide.tognin@phd.unipd.it; Tel.: +39-049-827-5433
† This paper is an extended version of the conference paper: Tognin, D.; Peruzzo, P.; De Serio, F.; Ben Meftah, M.; Carniello, L.; Defina, A.; Mossa, M. Laboratory experiments on solitary wave interaction with rigid emergent vegetation: Some preliminary results, from the 2018 IEEE International Workshop on Metrology for the Sea, Bari, Italy, 8–10 October 2018.

Received: 19 February 2019; Accepted: 12 April 2019; Published: 14 April 2019

Abstract: The aim of this study is to present a peculiar experimental setup, designed to investigate the interaction between solitary waves and rigid emergent vegetation. Flow rate changes due to the opening and closing of a software-controlled electro-valve generate a solitary wave. The complexity of the problem required the combined use of different measurement systems of water level and velocity. Preliminary results of the experimental investigation, which allow us to point out the effect of the vegetation on the propagation of a solitary wave and the effectiveness of the measuring system, are also presented. In particular, water level and velocity field changes due to the interaction of the wave with rigid vegetation are investigated in detail.

Keywords: laboratory experiments; solitary wave; wave-vegetation interaction; advanced hydrometry

1. Introduction

It is widely recognized that vegetation plays a pivotal role in the preservation and restoration of coastal environments, since it controls the sedimentation and transport, as well as it contributes to dissipating wave energy [1–12]. Concerning the latter aspect, as a consequence of the catastrophic tsunami event on the coast of South-East Asia in 2004, many studies have focused on the protective action of the coastline provided by mangrove forests [13–18].

Mangroves, in fact, can effectively protect the coastline from the action of wind and tidal waves [19]; however, dedicated studies suggest that tsunamis and storm surges behave differently. For example, as water height of severe tsunami and surges increases, the attenuation provided by mangrove forests is likely to reduce. The long period of tsunami waves may also influence the mitigation provided by mangroves because plants could be already damaged or uprooted as the wave continues to propagate through the coastal forest [20,21].

To better understand the action exerted by mangrove forests, the quantification of the tsunami wave attenuation, as well as the drag arising within mangroves have been investigated both numerically and experimentally, but, regardless of the approach employed, the efficacy of the mangroves is still an open question [22–27]. For instance, Kathiresan and Rajendran [28] demonstrated the ability of mangrove vegetation to protect the coastline from tsunamis' fury; meanwhile, Kerr et al. [29] stated that this sense of protection from large tsunamis appears to be unrealistic.

These dissimilarities in the outcomes can be explained by the complexity of the phenomena that characterize the interaction between waves and mangroves, which is often simulated through a series of small amplitude waves propagating within a canopy of artificial vegetation, in most of the cases mimicked by rigid dowels [23–25,30–34]. Such modeling of the problem is quite far from reality since the tsunami approaching the coast can be seen as a solitary wave, rather than a small wave. The latter only propagates its form; meanwhile, the orbital paths of water particles in a solitary wave are open; thus, water mass transport occurs. The reproduction of different types of waves in shallow water in physical models is a quite demanding task and should be carefully addressed [35,36].

The use of laboratory-scale models, despite the considerable simplification necessary for the experiments being feasible, is a consistent approach to isolate the effects of different wave and plant characteristics, thanks to the controlled conditions, as well as to validate numerical models (i.e., [37,38]).

The present work aims to describe a properly-designed experimental setup to reproduce and study the propagation of a solitary wave in the presence of emerging rigid cylinders mimicking the propagation of tsunamis through mangrove forests. The complex process of drag and flow separation is analyzed by means of several types of equipment able to measure different quantities; namely, ultrasonic probes for the water level, Acoustic Doppler Velocimeters (ADVs) for local velocity, and a Particle Image Velocimetry (PIV) system for the velocity field in a frame. Many previous experiments have been carried out by using ADVs or PIV separately. The advantages of using ADVs are based on their high frequency rate ($100 \div 200$ Hz), providing also information about turbulent aspects of the phenomenon. However, it is an intrusive instrument presenting limitations especially close to obstacles and channel bottom/walls. PIV overcomes this limitation, but it is often insufficient for a detailed turbulence analysis because common PIV systems generally have relatively low acquisition frequency (in the range of $5 \div 50$ Hz). Therefore, the coupling of ADVs and PIV has an intrinsic value as it allows us accurate measurements near boundaries and obstructions with high-frequency sampling rates. To our knowledge, this is the first example of integrated systems that measure the dynamics of a solitary wave in this context.

2. Experimental Setup

Experiments are carried out in a laboratory rectangular flume at the Department of Civil, Environmental, Land, Building Engineering and Chemistry (DICATECh) of the Polytechnic University of Bari (Italy) (Figure 1). The channel is 25 m long, 0.40 m wide, and 0.50 m high, and it is made of Plexiglas to guarantee optical access. Water recirculates through the channel in two partially separated circuits, each one fed by a different tank. More precisely, the main circuit maintains steady flow conditions via a first constant head tank, whilst a secondary tank can discharge up to 80 L/s by regulating a software-controlled electro-valve, and thus generating a wave due to the flow rate change. A triangular sharp-crested weir at the downstream tank is used to estimate the steady flow rate, whereas a more accurate measure of the flow rate from the secondary tank is provided by an electromagnetic flow meter, placed upstream of the channel. The water level at the end of the flume is controlled by a sloping (1:50) gravel beach. For further details, see [39].

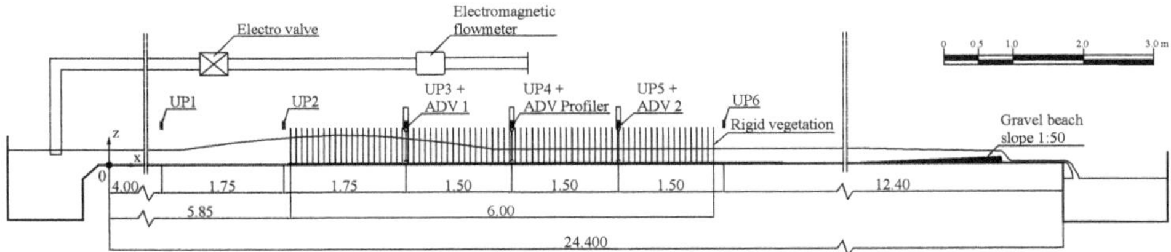

Figure 1. Side view of the experimental setup, with the positions of the Ultrasonic Probes (UPs) and the Acoustic Doppler Velocimeters (ADVs).

A six meter-long vegetation canopy is housed within the flume at 5.85 m from the channel inlet. Vegetation consists of a set of rigid steel cylinders with diameter $d = 3$ mm, inserted into six previously drilled Plexiglas panels. Different cylinder densities and patterns are obtained using the regular grid of holes longitudinally and transversally spaced with the same axis-to-axis distance (i.e., $s = 4$ cm).

The free surface variation in the flume is measured by six ultrasonic probes at a 100 Hz sampling rate. The probes are located at $x = 4.00$ m, 5.75 m, 7.50 m, 9.00 m, 10.50 m, and 12.00 m, so that the first two probes measure the approaching wave height, and the other four sensors measure the wave attenuation along the canopy (see Figure 1, which includes the x, z reference coordinate system).

Two different systems are used to measure the velocity within the canopy. The single point velocity is measured by means of two 3D Acoustic Doppler Velocimeters (ADVs), whose sampling rate, velocity range, and sampling volume are set equal to 100 Hz, ± 1.00 m/s, and 7 mm, respectively, in order to achieve a velocity accuracy of $\pm 1\%$. A further three-dimensional ADV Profiler, which measures velocity components in 33 cells with 1 mm of vertical extent at a 100 Hz sampling rate, is used to reconstruct nearly instantaneous vertical velocity profiles. During the ADVs' measurements, a correlation of the signal around 90% and a signal-to-noise ratio around 12 dB are achieved, thus proving the good quality of the signal itself. Figure 1 shows that the three ADV measurement systems are located in the canopy at the same position of the ultrasonic probes, i.e., $x = 7.50$ m, 9.00 m, 10.50 m, in order to obtain water level and velocity measures simultaneously.

The flow velocity field within the canopy is also measured by 2D Particle Image Velocimetry (PIV). The system consists of a timer box and a timer card cable box, triggering a FlowSense EO 4M-32 camera (frame rate of 32 Hz) with a dual power Laser Continuum Minilite (energy of 25 mJ at 15 Hz). The PIV, with the camera mounted to record from the bottom view, is setup to investigate turbulence structures in the horizontal plane at $x = 9.00$ m. The PIV system is handled in double-frame mode, and the total number of analyzed pair of frames is 381. The PIV sampling frequency is settled to its maximum (i.e., 16 Hz), while the time interval between two images of the same pair is 500 µs. Thus, the total acquisition time is 24 s for each measurement in the target area of the channel. The obtained images have dimensions of 2072×2072 pixels, corresponding to 132 mm $\times$ 132 mm, after the calibration. The interrogation area is 16×16 pixels, thus, the velocity vectors are computed on points spaced 1 mm, providing a very high spatial resolution in the measurement.

For simulating the tsunami propagation, we generate a solitary wave, reaching a flow peak of 85 L/s, as the sum of a steady flow of 10 L/s and an unsteady flow produced by linearly opening and closing the electro-valve for 10 s and 20 s, respectively.

Four scenarios are investigated. In the first scenario, the wave propagates in the absence of vegetation; the other three scenarios consider different vegetation densities n (i.e., $n = 156.25$, 312.5, and 625 cylinder/m^2) with different vegetation patterns (see Figure 2). For each scenario, we perform at least ten series of experiments, and each series of experiments counts 30 runs in order to analyze the data according to a phase-resolved approach.

An overview of the experimental set-up is shown in Figure 3.

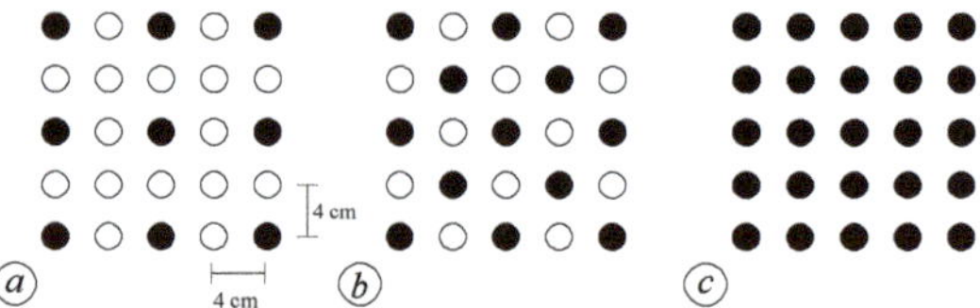

Figure 2. Vegetation patterns for the three tested configurations: (**a**) plant density $n = 156.25$ cylinder/m^2; (**b**) plant density $n = 312.5$ cylinder/m^2; (**c**) plant density $n = 625$ cylinder/m^2.

Figure 3. (**a**) View of the flume section with the Particle Image Velocimetry (PIV); (**b**) the particulars of an ultrasonic probe and ADV.

3. Results and Discussion

The first scenario is aimed at understanding the characteristics of the wave in the absence of vegetation. Examples of water level during the propagation of the solitary wave within the flume and without vegetation are shown in Figure 4, where the time-varying water level is measured by the four ultrasonic probes, located where the cylinder array will be housed in the scenarios reproducing the presence of vegetation. Each probe measures the typical symmetric, bell-shaped form of the water level at the passage of the solitary wave. As provided by the solitary wave theory, the propagation of the wave occurs entirely above the undisturbed water surface. The data collected clearly show the presence of a first reflected wave due to the sloping beach at the end of the flume (60 s $< t <$ 80 s) and a second one reflected by the upstream tank (80 s $< t <$ 95 s). However, these reflected waves do not affect the main incoming wave, since they both reach the area of interest after the main incoming wave has almost completely passed. Removing the sloping beach could reduce the reflection phenomena, but would make it impossible to control the downstream boundary condition.

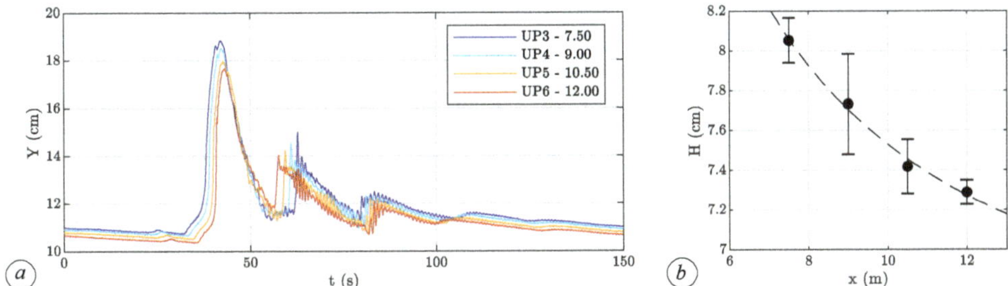

Figure 4. (**a**) Water level Y measured in 4 sections, without vegetation; (**b**) reduction of the wave height H without vegetation.

The spatial variation of the wave height in bare soil conditions (i.e., without vegetation) is shown in Figure 4b. The water level decreasing is more than linear along the considered 4.5 m long reach, achieving a maximum reduction of about 12%. In the presence of vegetation, the free surface varies similarly to the bare soil case, and for sake of brevity, this result is not reported here.

Experimental results referring to the vegetated scenarios show that the canopy density strongly affects the spatial reduction of the wave peak. Figure 5 shows the wave attenuation along the vegetated reach for the three investigated vegetation densities. In accordance with the literature [8,40–42], a stronger attenuation of the wave occurs with the increasing of the density n. In fact, the wave height at the end of the vegetated area compared to the incoming wave height reduces by about 40% for $n = 156.25$ cylinder/m^2 and up to more than 60% for $n = 625$ cylinder/m^2. The observed reduction of the solitary wave height is quite far from the hyperbolic decrease provided by the linear wave theory [41], confirming the idea that a reliable representation of the impact of mangroves on tsunamis cannot leave a careful wave generation out of consideration. Moreover, the wave height upstream of the canopy increases before impacting the vegetation. This supports the idea that vegetation could behave as a porous barrier for the incident solitary wave. These aspects deserve further theoretical and experimental analysis, in which the combined effect of water level and flow velocity needs to be taken into account to determine the wave attenuation.

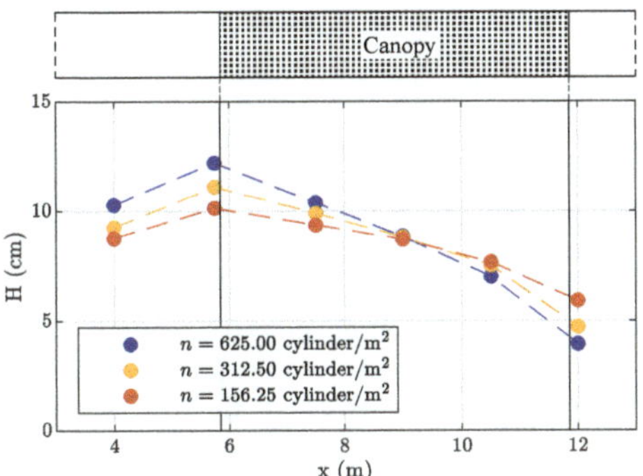

Figure 5. Reduction of the wave height H with different vegetation densities.

The ADV acquisitions allow us to couple water level and velocity data. In Figure 6, we can compare water level and the three velocity components at $x = 10.5$ m and $z = 2$ cm as a function of time for the bare soil (Figure 6a) and in the presence of vegetation with density $n = 156.25$ cylinder/m^2 (Figure 6b). In both cases, the peak of longitudinal velocity u slightly precedes the water level peak. The transverse velocity component, v, and the vertical one, w, are negligible compared to the longitudinal velocity. By comparing the two cases, we observe that the vegetation smoothing effect is more evident for water levels than for flow velocities.

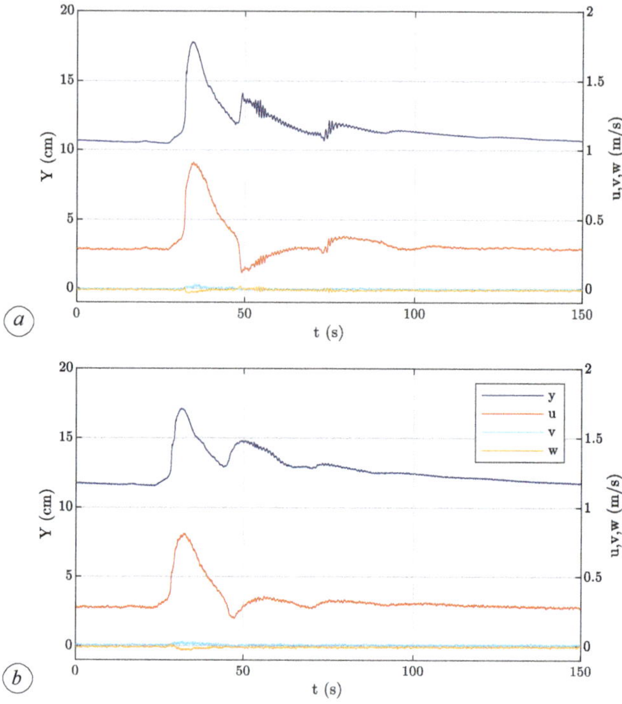

Figure 6. Water level (Y) and velocity components (u, v, w) measured at $x = 10.5$ m and $z = 2$ cm (**a**) without vegetation and (**b**) in the presence of vegetation with density $n = 156.25$ cylinder/m^2.

Measures at wave crest conditions point out the advantages of simultaneous water level and velocity measurements. Indeed, ADV outputs are meaningful only if the sensor's measurement volume is completely submerged, and this condition can be easily detected comparing ultrasonic probe measurements. In this way, we can recognize the true signal from the noise recorded when the ADV sensors are out of the water, and thus, we can investigate the velocity field even in points above the base flow water level. Figure 7 shows a comparison between the vertical profiles of the longitudinal velocity u at $x = 9.00$ m on the channel axis, in bare soil conditions and with vegetation density $n = 156.25$ cylinder/m^2, both during base flow conditions and at the wave peak. It is worthwhile noting that the velocity measurements in the second case are also reported in the wave crest, namely between 7 cm and 14 cm (see Figure 7b), discerned by means of the analysis of the signal provided by the water level probe. Measurements with the ADV Profiler overlap those of the single point ADV, but they have a finer spatial resolution. In bare soil conditions, the profile clearly follows a logarithmic law, with a maximum value of 0.3 m/s during base flow conditions and 0.9 m/s at the wave peak. Meanwhile, with vegetation, the velocity is nearly uniform over depth (equal to 0.27 m/s during the base flow condition and 0.75 m/s at the wave peak), except near the bottom, due to the presence of the boundary layer. However, it should be noted that the proximity of a boundary may affect the ADV probe output, so that the streamwise velocity component can be underestimated [43]. The ADV and ADV Profiler have almost the same precision and accuracy. In fact, the mean value and standard deviation of the streamwise velocity do not significantly differ, regardless of the instrument used. In both cases, the relative errors are lower than 5% (error bars are not reported in Figure 7 for the sake of clarity). Specifically, the relative error is around 1–2% in the lower and intermediate part of the water column, while it increases in the upper part where the interaction between the wave and vegetation disturbs the signals.

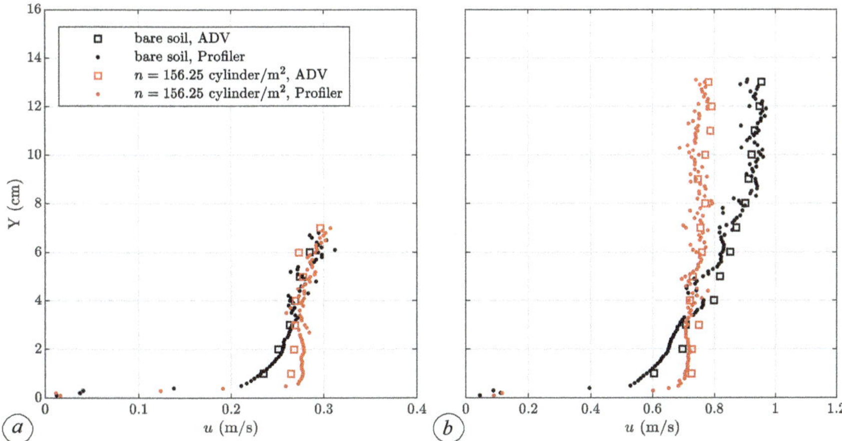

Figure 7. Vertical profile of the longitudinal velocity component u at $x = 9.00$ m on the channel axis, in bare soil conditions (black markers) and with vegetation (red markers), during base flow conditions (**a**) and at the wave peak (**b**). Small dots denote the ADV Profiler output, and squares denote ADV measurements.

The presence of vegetation affects the velocity distribution also in the transverse direction. In Figure 8, we report the transverse profiles of the longitudinal velocity u behind the cylinder located in the middle of the array with vegetation density of $n = 156.25$ cylinder/m^2, during base flow conditions and at the wave peak. The velocity profile reconstruction is carried out both using ADV and PIV measurements. The use of ADV is feasible only 4 cm apart from the rod, due to the minimum free volume required by the sensors to correctly measure the flow, whereas the PIV acquisition allows us to measure the flow velocity also very close to the cylinders and hence to overcome this drawback of the intrusive system. Figure 8b,d, where both ADV and PIV outputs are available in the same cross-sections, shows a very good agreement between the two velocity profiles, suggesting that the methodologies have likely the same accuracy. Such a result is also supported by the error analysis that gives a relative error of about 5% for ADV and of the same order of magnitude for PIV, thus making the two velocity measures equivalent. In both conditions, a relevant velocity reduction is detected downstream the cylinder, passing from 0.25 m/s to 0.10 m/s for the base flow (Figure 8a), and from 0.60 m/s to nearly 0 m/s at the wave peak (Figure 8c). The higher reduction of velocity observed between base flow and wave peak conditions can be ascribed to the larger wake area arising back at the cylinder.

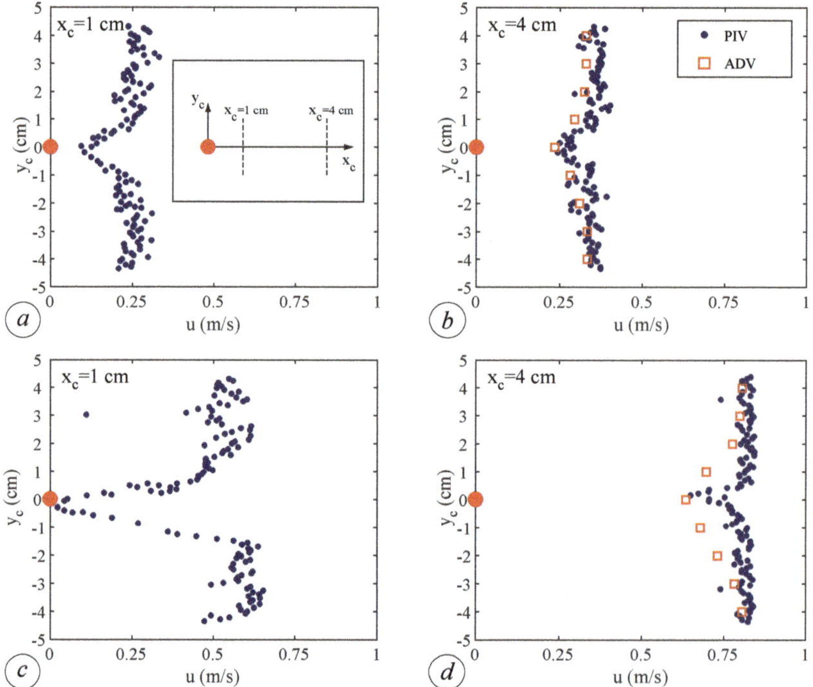

Figure 8. Transverse profiles of the longitudinal velocity component u, at $x = 9.00$ m and $z = 2$ cm, with vegetation density $n = 156.25$ cylinder/m^2. Base flow conditions at $x_c = 1$ cm (**a**) and at $x_c = 4$ cm (**b**). Wave peak conditions at $x_c = 1$ cm (**c**) and at $x_c = 4$ cm (**d**). The origin of the local reference system (x_c, y_c) coincides with the center of the cylinder, represented by a red dot (inset of Panel (**a**)).

The flow velocity field obtained by means of the PIV technique confirms the latter result. Figure 9 shows the instantaneous horizontal velocity and vorticity field in the same domain analyzed in Figure 8. The velocity field for the base flow, as well as at the wave peak has heterogeneous low values downstream the cylinder ($0 \div 0.1$ m/s for the base flow and $0 \div 0.4$ m/s at the passage of the wave), which transversally increases to an almost uniform higher value of 0.3 m/s and 0.75 m/s for the base flow and at the wave peak, respectively (see Figure 9a,c). Figure 9b,d shows the flow vorticity. The vertical component of the vorticity fluctuation is measured with the aim of relating it to the energy dissipation, hence with wave height attenuation. The vorticity map at the wave peak clearly shows two intense opposite vortexes downstream the cylinder, with vorticity ranging between -200 s^{-1} and 200 s^{-1} (Figure 9d). These values are almost four times greater than those estimated considering base flow conditions when the vorticity values range between -50 s^{-1} and 50 s^{-1} (Figure 9b). There are no substantial differences with the scenarios with higher vegetation density, save the mean value of the velocity.

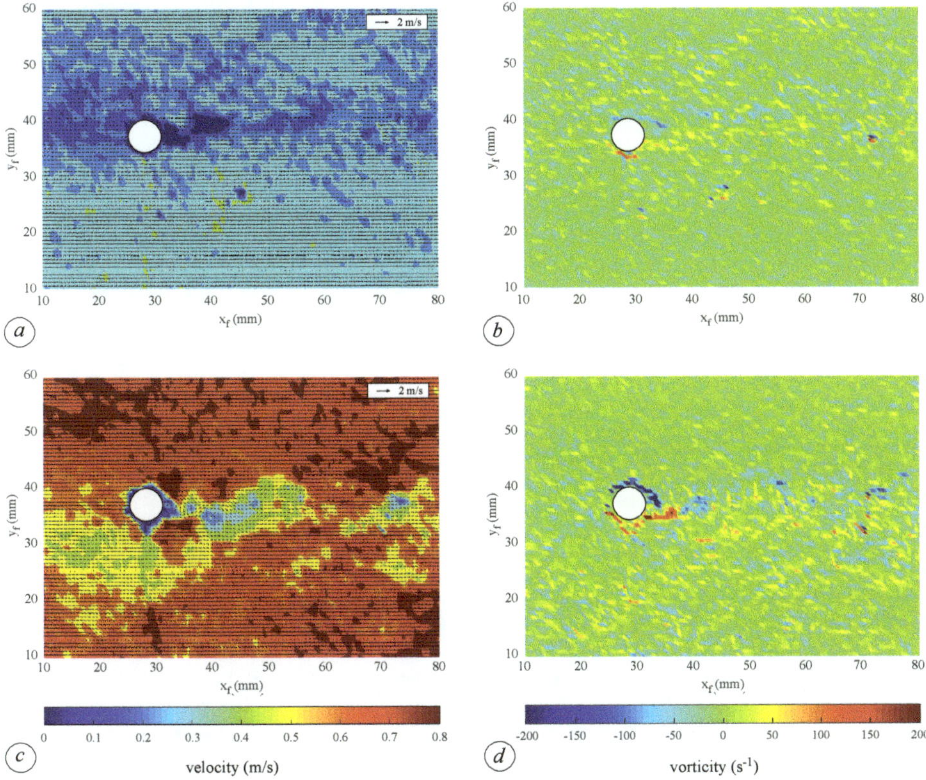

Figure 9. Example of PIV measurements at $x = 9.00$ m and $z = 2$ cm: velocity field during base flow condition (**a**) and at the wave peak (**c**); vorticity during base flow condition (**b**) and at the wave peak (**d**). The local coordinate system (x_f, y_f) refers to the frame. The flow is from left to right.

4. Conclusions

This study presents an experimental setup properly designed for reproducing a solitary wave by an impulsive flow rate increase that is regulated by a software-controlled electro-valve. This approach overcomes the limits of the use of small waves and a cylinder array to represent the dissipation of flood-waves or tsunamis due to different types of vegetation in laboratory flumes.

In our experiments, water levels, point, and field velocity measurements have been carried out by means of ultrasonic probes, ADVs, an ADV Profiler, and a PIV system. The coupling of ADV and ultrasonic probes' measurements allowed us to detect the velocity also during the significant rise of the water free surface, due to wave crest traveling. Moreover, the limits of ADV acquirements very close to the cylinders and to the walls were overcome by means of the non-intrusive nature of the PIV system. The results obtained using these different devices allowed us to reconstruct a detailed velocity field and, hence, the wake area arising back at the cylinder, quantifying the vorticity causing wave dissipation. Specifically, it was observed that the vegetation strongly affects the wave behavior, reducing the wave height proportionally with the vegetation density. During base flow conditions, the vertical velocity profile clearly showed a logarithmic profile in bare soil condition, whereas it turned into a rather uniform velocity distribution in the presence of vegetation, agreeing with many previous works (i.e., [40]). A wake area was observed downstream the cylinder rows, where a significant velocity reduction was noted both during base flow conditions and at the wave peak, even stronger in this second case. In fact, also the vorticity map field at the wave peak clearly showed the coexistence of two intense opposite eddies downstream the cylinder, with vorticity values almost four times greater than those estimated for the base flow condition.

The main aspects of the interaction between wave and vegetation can be fully obtained thanks to such a complete measurement system. However, the combined use of the different adopted sensors and techniques still has unexpressed potential in the cutting-edge investigation of this phenomenon and deserves further analysis.

Author Contributions: Conceptualization, P.P., F.D.S. and L.C.; investigation, D.T., F.D.S., M.B.M. and M.M.; formal analysis, D.T. and F.D.S.; writing, original draft, D.T. and P.P.; writing, review and editing, L.C., A.D. and M.M.; supervision, A.D. and M.M.

Funding: This research was partially funded by GII (Italian Hydraulic Group), under the GII Placement 2017 grant.

Acknowledgments: The technical staff of the Hydraulic Laboratory of the DICATECh of the Polytechnic University of Bari is gratefully acknowledged.

Conflicts of Interest: The authors declare no conflict of interest.

References

1. Keulegan, G.H. Gradual damping of solitary waves. *Nat. Bur. Sci. J. Res.* **1948**, *40*, 487–498. [CrossRef]
2. Ghisalberti, M.; Nepf, H. The structure of the shear layer in flows over rigid and flexible canopies. *Environ. Fluid Mech.* **2006**, *6*, 277–301. [CrossRef]
3. Defina, A.; Peruzzo, P. Floating particle trapping and diffusion in vegetated open channel flow. *Water Resour. Res.* **2010**, *46*, W11525. [CrossRef]
4. Defina, A.; Peruzzo, P. Diffusion of floating particles in flow through emergent vegetation: Further experimental investigation. *Water Resour. Res.* **2012**, *48*, W03501. [CrossRef]
5. Peruzzo, P.; Defina, A.; Nepf, H. Capillary trapping of buoyant particles within regions of emergent vegetation. *Water Resour. Res.* **2012**, *48*, W07512. [CrossRef]
6. Peruzzo, P.; Defina, A.; Nepf, H.M.; Stocker, R. Capillary interception of floating particles by surface-piercing vegetation. *Phys. Rev. Lett.* **2013**, *111*, 164501. [CrossRef]
7. Peruzzo, P.; Viero, D.P.; Defina, A. A semi-empirical model to predict the probability of capture of buoyant particles by a cylindrical collector through capillarity. *Adv. Water Resour.* **2016**, *97*, 168–174. [CrossRef]
8. Peruzzo, P.; de Serio, F.; Defina, A.; Mossa, M. Wave height attenuation and flow resistance due to emergent or near-emergent vegetation. *Water* **2018**, *10*, 402. [CrossRef]
9. Ben Meftah, M.; Mossa, M. Prediction of channel flow characteristics through square arrays of emergent cylinders. *Phys. Fluids* **2013**, *25*, 045102. [CrossRef]
10. Ben Meftah, M.; De Serio, F.; Mossa, M. Hydrodynamic behavior in the outer shear layer of partly obstructed open channels. *Phys. Fluids* **2014**, *26*, 065102. [CrossRef]
11. Carniello, L.; D'Alpaos, A.; Botter, G.; Rinaldo, A. Statistical characterization of spatiotemporal sediment dynamics in the Venice lagoon. *J. Geophys. Res. Earth Surf.* **2016**, *121*, 1049–1064. [CrossRef]
12. Mossa, M.; Ben Meftah, M.; De Serio, F.; Nepf, H.M. How vegetation in flows modifies the turbulent mixing and spreading of jets. *Sci. Rep.* **2017**, *7*, 6587. [CrossRef] [PubMed]
13. Danielsen, F.; Sørensen, M.K.; Olwig, M.F.; Selvam, V.; Parish, F.; Burgess, N.D.; Hiraishi, T.; Karunagaran, V.M.; Rasmussen, M.S.; Hansen, L.B.; et al. The Asian tsunami: A protective role for coastal vegetation. *Science* **2005**, *310*, 643. [CrossRef]
14. Dahdouh-Guebas, F.; Koedam, N. Coastal Vegetation and the Asian Tsunami. *Science* **2006**, *311*, 37–38. [CrossRef]

15. Danielsen, F.; Sørensen, M.; Olwig, M.; Selvam, V.; Parish, F.; Burgess, N.; Topp-Jørgensen, E.; Hiraishi, T.; Karunagaran, V.; Rasmussen, M.; et al. Response to "Coastal Vegetation and the Asian Tsunami". *Science* **2006**, *311*, 37–38.
16. Mazda, Y.; Magi, M.; Ikeda, Y.; Kurokawa, T.; Asano, T. Wave reduction in a mangrove forest dominated by Sonneratia sp. *Wetl. Ecol. Manag.* **2006**, *14*, 365–378. [CrossRef]
17. Das, S.; Vincent, J.R. Mangroves protected villages and reduced death toll during Indian super cyclone. *Proc. Natl. Acad. Sci. USA* **2009**, *106*, 7357–7360. [CrossRef]
18. Marois, D.E.; Mitsch, W.J. Coastal protection from tsunamis and cyclones provided by mangrove wetlands—A review. *Int. J. Biodivers. Sci. Ecosyst. Serv. Manag.* **2015**, *11*, 71–83. [CrossRef]
19. Mazda, Y.; Magi, M.; Kogo, M.; Hong, P.N. Mangroves as coastal protection from waves in the Tong King delta, Vietnam. *Mangroves Salt Marshes* **1997**, *1*, 127–135. [CrossRef]
20. Yanagisawa, H.; Koshimura, S.; Goto, K.; Miyagi, T.; Imamura, F.; Ruangrassamee, A.; Tanavud, C. The reduction effects of mangrove forest on a tsunami based on field surveys at Pakarang Cape, Thailand and numerical analysis. *Estuar. Coast. Shelf Sci.* **2009**, *81*, 27–37. [CrossRef]
21. Yanagisawa, H.; Koshimura, S.; Miyagi, T.; Imamura, F. Tsunami damage reduction performance of a mangrove forest in Banda Aceh, Indonesia inferred from field data and a numerical model. *J. Geophys. Res. Oceans* **2010**, *115*, C06032. [CrossRef]
22. Harada, K.; Imamura, F.; Hiraishi, T. Experimental Study on the Effect in Reducing Tsunami by the Coastal Permeable Structures. In Proceedings of the Twelfth International Offshore and Polar Engineering Conference, Kitakyushu, Japan, 26–31 May 2002; Volume 3, pp. 528–533.
23. Huang, Z.; Yao, Y.; Sim, S.Y.; Yao, Y. Interaction of solitary waves with emergent, rigid vegetation. *Ocean Eng.* **2011**, *38*, 1080–1088. [CrossRef]
24. Husrin, S.; Strusińska, A.; Oumeraci, H. Experimental study on tsunami attenuation by mangrove forest. *Earth Planets Space* **2012**, *64*, 973–989. [CrossRef]
25. Maza, M.; Lara, J.L.; Losada, I.J. Tsunami wave interaction with mangrove forests: A 3-D numerical approach. *Coast. Eng.* **2015**, *98*, 33–54. [CrossRef]
26. Horstman, E.M.; Dohmen-Janssen, C.M.; Narra, P.M.; van den Berg, N.J.; Siemerink, M.; Hulscher, S.J. Wave attenuation in mangroves: A quantitative approach to field observations. *Coast. Eng.* **2014**, *94*, 47–62. [CrossRef]
27. Maza, M.; Lara, J.L.; Losada, I.J. Solitary wave attenuation by vegetation patches. *Adv. Water Resour.* **2016**, *98*, 159–172. [CrossRef]
28. Kathiresan, K.; Rajendran, N. Coastal mangrove forests mitigated tsunami. *Estuar. Coast. Shelf Sci.* **2005**, *65*, 601–606. [CrossRef]
29. Kerr, A.M.; Baird, A.H.; Campbell, S.J. Comments on "Coastal mangrove forests mitigated tsunami" by K. Kathiresan and N. Rajendran [Estuar. Coast. Shelf Sci. 65 (2005) 601–606]. *Estuar. Coast. Shelf Sci.* **2006**, *67*, 539–541. [CrossRef]
30. Ozeren, Y.; Wren, D.G.; Wu, W. Experimental Investigation of Wave Attenuation through Model and Live Vegetation. *J. Waterw. Port Coast. Ocean Eng.* **2013**, *140*, 04014019. [CrossRef]
31. Anderson, M.E.; Smith, J.M. Wave attenuation by flexible, idealized salt marsh vegetation. *Coast. Eng.* **2014**, *83*, 82–92. [CrossRef]
32. Blackmar, P.J.; Cox, D.T.; Wu, W.C. Laboratory Observations and Numerical Simulations of Wave Height Attenuation in Heterogeneous Vegetation. *J. Waterw. Port Coast. Ocean Eng.* **2014**, *140*, 56–65. [CrossRef]
33. John, B.M.; Shirlal, K.G.; Rao, S. Effect of artificial vegetation on wave attenuation - An experimental investigation. *Procedia Eng.* **2015**, *116*, 600–606. [CrossRef]
34. Dupuis, V.; Proust, S.; Berni, C.; Paquier, A. Combined effects of bed friction and emergent cylinder drag in open channel flow. *Environ. Fluid Mech.* **2016**, *16*, 1173–1193. [CrossRef]
35. Stefanon, L.; Carniello, L.; D'Alpaos, A.; Lanzoni, S. Experimental analysis of tidal network growth and development. *Cont. Shelf Res.* **2010**, *30*, 950–962. [CrossRef]
36. Rampazzo, M.; Tognin, D.; Pagan, M.; Carniello, L.; Beghi, A. Modelling, simulation and real-time control of a laboratory tide generation system. *Control Eng. Pract.* **2019**, *83*, 165–175. [CrossRef]

37. Viero, D.P.; Pradella, I.; Defina, A. Free surface waves induced by vortex shedding in cylinder arrays. *J. Hydraul. Res.* **2017**, *55*, 16–26. [CrossRef]
38. Viero, D.P.; Peruzzo, P.; Defina, A. Positive surge propagation in sloping channels. *Water* **2017**, *9*, 518. [CrossRef]
39. Ben Meftah, M.; De Serio, F.; Malcangio, D.; Mossa, M.; Petrillo, A.F. Experimental study of a vertical jet in a vegetated crossflow. *J. Environ. Manag.* **2015**, *164*, 19–31. [CrossRef]
40. Nepf, H.M. Drag, turbulence, and diffusion in flow through emergent vegetation. *Water Resour. Res.* **1999**, *35*, 479–489. [CrossRef]
41. Mendez, F.J.; Losada, I.J. An empirical model to estimate the propagation of random breaking and nonbreaking waves over vegetation fields. *Coast. Eng.* **2004**, *51*, 103–118. [CrossRef]
42. Augustin, L.N.; Irish, J.L.; Lynett, P. Laboratory and numerical studies of wave damping by emergent and near-emergent wetland vegetation. *Coast. Eng.* **2009**, *56*, 332–340. [CrossRef]
43. Chanson, H.; Trevethan, M.; Koch, C. Discussion of to "Turbulence Measurements with Acoustic Doppler Velocimeters" by Carlos M. García, Mariano I. Cantero, Yarko Niño, and Marcelo H. García. *J. Hydraul. Eng.* **2007**, *133*, 1283–1286. [CrossRef]

Article

New Approach to Analysis of Selected Measurement and Monitoring Systems Solutions in Ship Technology

Boleslaw Dudojc * and Janusz Mindykowski

Gdynia Maritime University, Department of Marine Electrical Power Engineering, ul. Morska 81-87, 81-225 Gdynia, Poland; j.mindykowski@we.umg.edu.pl

* Correspondence: b.dudojc@we.umg.edu.pl; Tel.: +48-728-912-772

Received: 8 March 2019; Accepted: 9 April 2019; Published: 13 April 2019

Abstract: This paper is dedicated to certain types of measurement in ship systems, analyzed based on selected case studies. In the introductory part, a simplified structure of a modern cargo ship as an object of measurement and control is presented. Next, the role of measurement in the ship's operation process is described and commented on, with focus on specifics of local and remote control, both manual and automatic. The key part of the paper is dedicated to a short overview of selected examples of measuring and monitoring systems. The basic criteria for the aforementioned selection are the vital role of the considered systems for safe and effective ship operation as well as documented innovative contribution of Gdynia Maritime University (GMU) in development of the state-of-the-art in the analysed area of measurement. Based on these criteria, the monitoring of operational parameters of main engine and temperature measurement in the ships hazardous areas have been chosen. The aforementioned measurement and monitoring systems are analysed, taking into account both innovation of technical solutions together with their ship technology environment conditions and related legal requirements. Finally, some concluding remarks are formulated.

Keywords: ship systems; measurement; monitoring; main engine parameters; temperature measurement; ship environment; hazardous areas

1. Introduction

The core information of this paper is based on a recently published paper [1], describing the problem of the vital role of measurement and monitoring of the operational parameters of main engine (power system) and temperature measurement in the ship's hazardous areas (cargo system) for safe and effective ship operation as a whole.

Ships are very sophisticated maritime, industry objects. On the one hand, each ship or other floating object differs from the other, but on the other hand, all of them have a lot of common properties. A modern automated cargo ship as an object of measurement and control presents a complex structure of four main systems: navigation, power, cargo and crew living conditions. In each system there can be distinguished some subsystems, which are illustrated in Figure 1 (for each system there are only two corresponding subsystems, but there can be more). Among all those mentioned above, the electric power subsystem is very important and it has a significant influence on the other systems and subsystems. It is worth underlining that there is a close relation between electric power and propulsion systems. More and more often we can meet not only with traditional sources of electrical energy as a set of diesel-generator units for electric power, but other configurations are also used, e.g., additional shaft generators or generators supplied by system utilizing heat from exhaust gases of the main engine. The electrical generators can also be driven by classical engines powered by gas or gas turbines. In recent years, electric ships have been explored more intensively and high voltage electrical systems are often used to additionally raise efficiency of such solutions. Taking into consideration different aspects like efficiency, reliability, environmental protection and economical effectiveness, AC (Alternating Current) electrical systems are already partially supported by DC (Direct Current) creating hybrid electric power system, and in the near future they can even be replaced by these new solutions, especially on small ships using renewable energy sources.

It is worth mentioning that at the ships available, electrical power is comparable to the loads, which is automatically adjusted. Among all mentioned above, the electric power subsystem is very important and it has a significant influence on the other systems and subsystems. All of the very sophisticated systems of a ship require supervision by monitoring and control systems. Additional systems with safety functions are installed. Safety functions have higher priority than normal operation functions.

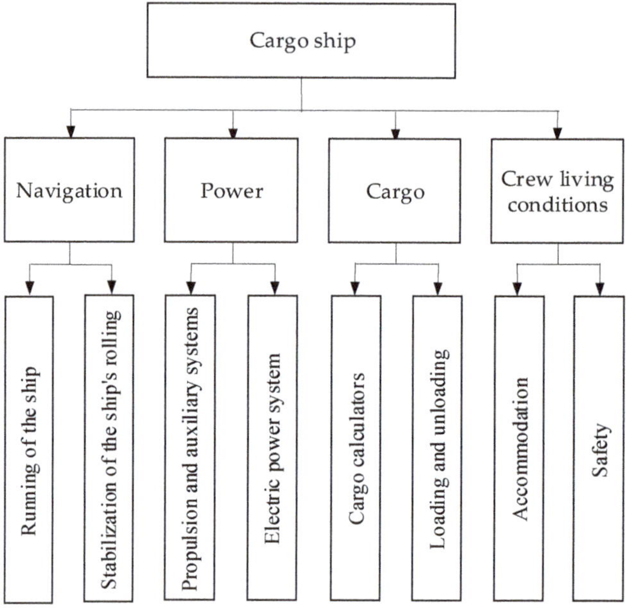

Figure 1. Systems and subsystems of typical cargo ship. (Updated version based on [2]).

At present, usually the distributed microprocessor systems are used for implementation of the measurement, monitoring, control and safety functions. Microprocessor systems are more susceptible to electromagnetic interference and disturbances of power supply than traditional analog systems.

In this paper two typical case studies related to ship practice are presented. These studies are addressed to critical operational parameters, like main engine parameters in ships' power plants or temperature in ship tanks. A novelty of this paper concerns the carrying out and analysis of the main engine signals as well as the experimental research of newly implemented SeaPerformer System—in the case of main engine research. In the second case, a new approach concerns in-depth theoretical analysis of the temperature measurement line of liquefied gas in tanks together with consequences resulting from it. This in-depth analysis is resulting in identifying and explaining the reasons of new additional error, which may appear during the temperature measurement in the ship's hazardous areas. To indicate unexpected alternating current component, disturbing a correct temperature measurement, there was patented a new impendence simulator of Pt-100 sensors, which enable to improve significantly the quality and reliability of commissioning of the intrinsically safe measurement line under investigation Following this chapter, the rest of this article is structured in this way. Section 2 describes the role of measurement in the ships' operation from different points of view. Section 3 is fundamental for this paper, being devoted to a short overview of selected examples of measuring and monitoring systems on sea ships, and presents, firstly, a ship performance monitoring system concept based on SeaPerformer System case study measurement of operational parameters of main engine, and finally, the closing subsection of Section 3 analyzes the temperature measurements in the ship's hazardous areas. Section 4 summarizes the conclusions of the study, stressing proposed novelties against the state-of-the-art.

2. The Role of Measurement in the Ship's Operation Process

To achieve complete overview of any condition on board considered sea ship, it is vital to measure not only affecting parameters such as flow, pressure, temperature and level [3–5], but also the operational parameters of main engine, or more widely—ship propulsion [6,7]. Special attention should be paid to the measurement systems supporting navigation area [8–10]. This information allows crew members to take action before changes affect equipment performance and cause downtime or failures.

Considering the ship as a measurement and control object, there are different tasks executed in the following areas:

1. Navigation;
2. Operation of power plant;
3. Operation of cargo system;
4. Ensuring safety of crew, equipment and environment.

In the navigation area, these tasks may include:

- Measurement, filtration and estimation of ship position coordinates (measurement of the position of the ship and foreign ships) for the purpose of controlling its movement along the desired trajectory, position stabilization and use in anti-collision systems;
- Measurement of parameters characterizing movement of the ship (determining the velocity vector, course, draft, trim, side swaying of the ship, wind direction and force).

Measurement tasks for safe and economical operation of the ship's power plant, i.e., the main engine and all the mechanisms and auxiliary devices consist of:

- Measurement of mechanical motion parameters of devices driving the generators and selected characteristic parameters, such as the quality of the fuel fed to them and the composition of exhaust gases;
- Measurement of parameters, in general temperatures, pressures, levels, viscosity and flows at selected points determining operational conditions of the power plant;
- Measurement of torque, power and revolution speed on the shaft of the Main Engine (ME) for the purpose of economical operation of the ship's propulsion;
- Measurement of quantities characterizing the power grid (voltages, currents, powers, frequency, cos φ, insulation resistance, THD—Total Harmonic Distortion);
- Fire protection system.

Cargo transportation requires the following measurement tasks:

- Measurement of cargo parameters that determine its qualitative features;
- Measurement of the amount/weight of the transported cargo;
- Load distribution measurement (loading calculators);
- Measurement of technological media parameters enabling transport of cargo;
- Measurement of water levels inside of ballast tanks.

Qualified crew members are required to operate the ships adequately. Ships have to provide proper accommodation and additional systems necessary to support good nutrition and health of the crew while ensuring adequate environmental protection. The following measurement tasks are done in order to fulfil that:

- Measurement of parameters of potable water system;
- Measurement of parameters of air conditioning system;
- Measurement of parameters of cooling system for food storage;
- Measurement of parameters of sewage system;
- Identification of smoke and temperature of fire alarm system.

Safety function of the ship as a whole requires control of limit values and trends in their changes, basic parameters determining safe operation of technical systems and safety of people, cargo and the marine environment.

The aforementioned measuring tasks include both operational and diagnostic measurement, but the authors are of the opinion that the latter ones are beyond the main scope of this paper. The purpose of operational measurement is to determine the current value of the parameters of a given system and to check the correctness of its operation and related processes.

Operational measurement is used for both manual and automatic remote control, and its main purpose is safe and effective operation of the ship. In case of emergency, vital systems can be controlled locally, where all the system operations are done by the crew directly, sometimes using only human power. Reading of parameters is made on the instrumentation placed directly on the controlled device or machine. Local control of the main engine or local control of the rudder are examples of such a solution.

In case of remote manual control, the results of operational measurement are read from the instrumentation installed in the Main Switchboard or on the relevant Engine Control Room (ECR), Cargo Control Room (CCR) desks or Bridge Control Console (BCC) navigation bridge console. They are used in multi-parameter regulation systems, where a human factor plays a basic role.

For the remote automatic control, the respective automatic control loops of controlled systems are used without human interference.

In both approaches presented above, the same set of measured parameters from sensors and transducers are used for systems control. Some measured parameters can be doubled or made according to the special solution to ensure adequate level of reliability.

All measured points of measured parameters are scattered on the ship. In spite of measured parameters, a connection from a sensor or a transmitter to a measurement unit has to be done by wires. A sensor or a transmitter (field devices), connecting wires and input circuits of measurement unit create independent electrical circuits which will be named measurement lines. The lengths of used wires can be from a few to hundreds of meters. The measurement lines of measurement systems are most sensitive to influence of interferences and disturbances. Additional requirements should be done in the case of parameters measured in hazardous areas. Modification to explosion protection can limit measurement properties or even introduce additional errors.

Hundreds or sometimes more than one thousand parameters can be measured on the modern ships. Most from them are measured by binary sensors (on-off type, even around 50%), analog transmitters (two wire current 4–20 mA standard, about 20–30%) and for measurement of temperature most often thermo-resistive Pt-100 sensors (also Pt-500 or Pt-1000) or rarely thermocouple are used. Measurement of temperature is realized by connection of sensors either to the measurement unit directly or to the transmitters with standard output electrical signals. In such cases, the two wire current 4–20 mA standard is still used very often in hybrid option. Hybrid option concerns a very popular solution where both the analog and digital communication are used simultaneously. Most common available transmitters are a solution with Highway Addressable Remote Transducer (HART) Protocol, invented and introduced by Rosemount Inc. The pure digital transmitters Manchester Coding, Bus Powered (MBP) for Foundation Fields or Profibus protocols are also available as the field devices but still less popular in shipbuilding applications.

3. A Short Overview of Selected Examples of Measuring and Monitoring Systems on Sea Ships

Measurement methods in marine technology have passed from the model of direct and indirect measurement into the system measurement phase. The most important characteristics of the system measurement are:

A. Measurement must be treated together with the measured object,
B. Measurement is accompanied by instrumentation, usually called a measuring and/or monitoring system, whose role is to implement a set of operations, among others:

 - Detection of signals from the measured object;
 - Analog-to-digital conversion of signals, which consists of sampling, quantization and digital coding;
 - Operational processing of signals, mainly digital, carried out using computer technology, consisting of forming signals carrying information about measured quantities; usually we deal with multi-path processing.

A short overview of selected examples presented below is based on the selection criteria concerning firstly a vital role of the considered system for safe and effective ship's operation and secondly well documented contribution of Gdynia Maritime University in the development of the state-of-the-art in the analyzed area of measurement.

3.1. Ship Performance Monitoring System

Ship performance monitoring systems [10–12] are implemented on modern vessels in order to support efficient and environmentally sound operation. Such systems integrate vast amounts of signals and process them with respect to ship energy efficiency. An example of such a system is SeaPerformer developed by Research and Development Company Enamor Ltd. Gdynia, Poland, which cooperates closely with Gdynia Maritime University. SeaPeformer has been developed within the framework of research project co-financed by Polish National Centre for Research and Development (research project POIR.01.01.01-00-0933/15). The system has been successfully implemented on dozens of deep sea ships.

SeaPerformer is a modular solution which consists of:

- Data processing and storage server based on marine industrial computer which serves the purpose of data acquisition, storage and automated processing. Server allows user interaction with the help of dedicated user interface and support ship-to-shore data exchange realized with the help of the existing satellite communication system;
- One or more data collecting units (typically there are two units, one located on the ship bridge for the purpose of navigation signals and one installed in the engine control room for collecting machinery signals);
- Power supply unit featuring system safe shut down in the case of blackout and automatic system start when power is restored;
- One or more high resolution and large size (24″ or more) marine industrial monitors with keyboard and trackball serving user interaction.

System interconnection is realized with the help of a dedicated or general ship Ethernet network. This solution simplifies system installation in the case of retrofit projects and allows data access with the use of other computers within the same LAN (Local Area Network).

Usually, such a system is capable of the simultaneous recording of a large amount of data. Depending on the type of ship, main engines and subsystems, SeaPerformer collects from a few hundred up to a couple thousand signals. Data collection is performed with the use of two general hardware interfaces: NMEA 0183 for the communication with navigation equipment such as ship GPS, log, echo-sounder, gyro-compass, weather station etc. and Modbus (RTU or TCP) for data collection from machinery e.g., main and auxiliary engines, boilers, fuel system, alarm and monitoring system, loading computer etc. Information collected with the use of the mentioned interfaces are stored in two databases. The high frequency database collects signals with frequency ranging from 1 Hz up to 100 Hz as appropriate, with respect to the nature of the signal. Due to the large amount of information, this database is kept onboard and can be transferred onshore only on request, usually when detailed signal examination is needed for the purpose of troubleshooting. Standard ship-to-shore data exchange is facilitated with used low frequency database created by 1-min averaging of high frequency database. This process, illustrated in Figure 2, allows for significant reduction of data amount and allows for data exchange with the use of a contemporary ship satellite communication system. Data measured directly onboard are supplemented additionally by weather data with the use of ship weather service. Weather data are interpolated along ship track and integrated with data measured onboard.

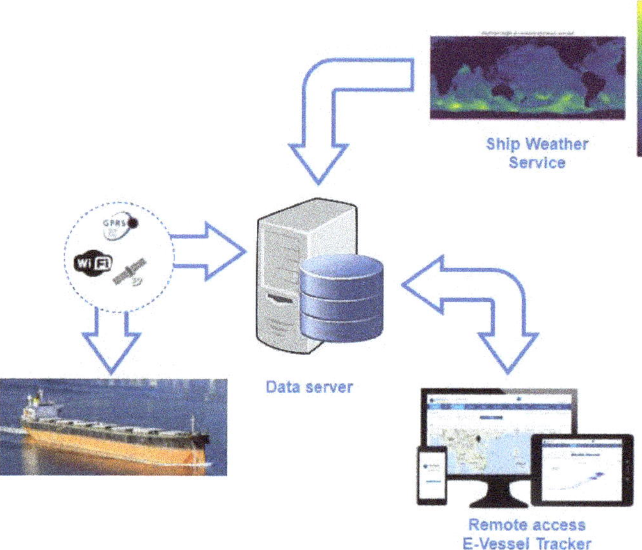

Figure 2. Scheme of ship to shore data transfer and online access to data with the use of web application.

In order to secure communication against unauthorized access, data package is encrypted with use of Rivesta-Shamira-Adlemana (RSA) and Advanced Encryption Standard (AES) algorithms. Data sent from the vessel are stored in cloud server and can be accessed by the owner of the technical office (usually superintendent or fleet manager) for analyses and comparison against the fleet.

Based on collected data, the ship performance monitoring system can be used for number of standard functions:

- Data reporting by aggregation, filtering and data grouping with respect to standard (noon, trip, Ship Energy Efficiency Management Plans (SEEMP), Monitoring, Reporting, Verification (MRV) [12], Data Collection System (DCS) for fuel oil consumption of ships) [13] and user defined reports;
- Faults detection in signal sources (sensors' faults or connections' faults) in order to maintain data access;
- Non-optimum operation warnings—system continuously evaluates selected signals according to specialized algorithms in order to recognize situations when vessel or her subsystems are operated in inefficient manner (e.g., concurrent operation of auxiliary engines at low output which can realized with lower number of units);

- Subsystems' operational performance trending—in order to monitor subsystem performance deterioration and plan maintenance actions (e.g., hull fouling trend or monitoring of main engine specific fuel oil consumption (SFOC) trend as presented in Figure 3 [14]. In the monitored time period, an increased value of SFOC index was observed. In order to evaluate main engine performance, the system computes actual SFOC based on momentary fuel consumption and main engine load. Based on additional data collected by the system, actual SFOC is corrected to reference conditions [12,15,16]. In parallel, theoretical SFOC in reference conditions is calculated based on data specific for the engine, usually based on results of factory acceptance tests. Finally, difference between actual and theoretical SFOC is plotted as the time trend allowing for detection of gradual engine wear or improper engine tuning);
- Operational optimization—based on actual data and reference model [11] system recommends optimum parameters of operation (e.g., optimum trim with respect to ship draught and speed through the water).

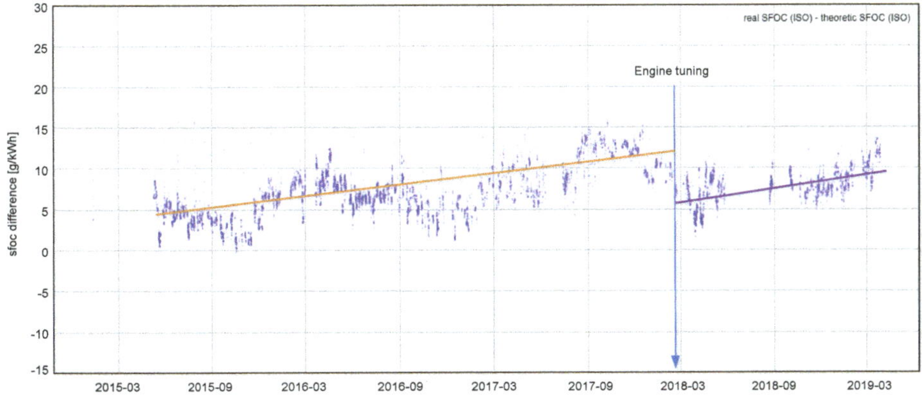

Figure 3. Main engine specific fuel oil consumption (SFOC) trend with visible gradual deterioration of engine performance [14]. (Courtesy of Enamor Ltd.).

Figure 3 presents typical SFOC analyses based on measurements taken onboard during vessel operation. It is visible that data collection is affected by considerable scatter. The reason for this is two-fold: an important factor influencing data scatter is measurement accuracy, secondly, the quality of reference SFOC model also influences results. First factor results in conditions of measurement onboard the ship during her operation. Harsh environmental conditions of ship engine room imply that measurement equipment usually compromise accuracy in order to get required robustness. Especially at low engine load where both fuel flow and shaft power meters operate at low rate, measurement errors are higher. As far as reference SFOC, it is usually obtained during engine shop tests where due to time constraints, only few (sometimes one) load points are exercised. In order to obtain a reference model for complete range of engine, loads interpolation and extrapolation techniques are used which introduce additional error. Despite obvious limitations of this approach, appropriate data processing in long term trending allows for taking practical conclusions. Linearization of time trend (indicated by orange and magenta lines in Figure 3) allows for making predictions of required engine maintenance and thus allowing for better planning. Correlation of rapid SFOC changes with engine tuning actions performed by crew in February 2018 (denoted on Figure 3 by a vertical arrow) in the case of electronically controlled engines allows for evaluation of tuning quality and its impact on engine performance. In this particular case, tuning resulted in drop of specific fuel consumption by ~6 g/kWh.

3.2. Measurement of the Operational Parameters of Main Engine

Main engine can be monitored based on general output or detailed signals describing engine operation performance such as cylinder combustion pressure. The latter requires special equipment since very high resolution of signal recording is required. In order to reproduce dynamics of combustion process, angular resolution of 0.1 degree of crankshaft position must be provided. For this purpose, FPGA processing unit in real time is employed. Data acquisition of following analog signals of engine operation are executed:

- Crankshaft angular position;
- Combustion pressure;
- Exhaust valve position (EVP);
- Injection quantity sensor position (for selected engines);
- Gas admission valves (GAV) positions (for selected engines).

The above data are post-processed and usually depicted in reference to cylinder top dead position as shown on Figure 4, in exemplary combustion pressure process monitoring.

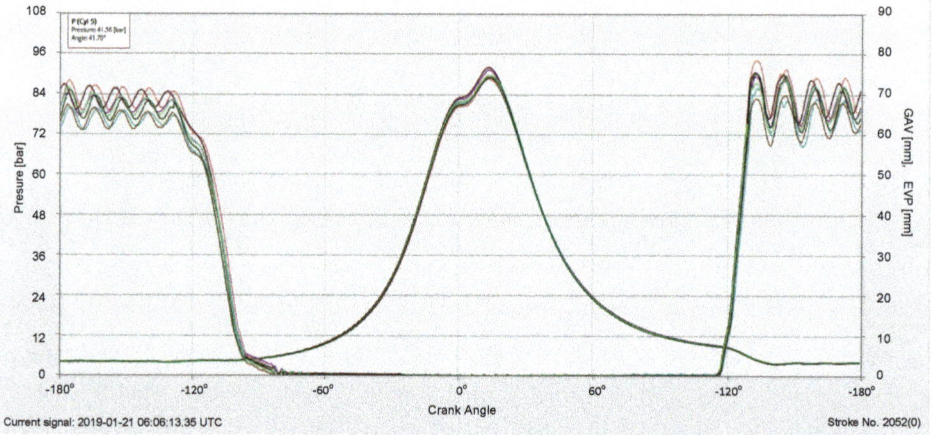

Figure 4. Combustion pressure monitoring of main engine [14]. (Courtesy of Enamor Ltd.).

Main engine general output may be monitored with use of shaft torque meter [14]. A good illustration of such a kind of measurement is propulsion control assistance system ETNP-10 for optimization of ship operation according to operation area of the main engine, being a part of information provided by SeaPerformer system, is depicted in Figure 5 [14]. Both cases actual engine load and rotational speed (dots) are presented in reference to so called engine layout or engine operational envelope. Engine layout presents standard operation area (green) over-speed area (yellow) and over-load area (red). Clear color indication allows ship crew to monitor engine operation and prevent conditions which may results in fault. As an extension to ETNP-10 user interface visualization which is devoted to actual engine conditions, SeaPerformer system may be used for engine monitoring in longer time range. The SeaPerformer system measures the torque on shaft and revolution (RPM-Revolution Per Minute) of the main engine. Other parameters like power, fuel consumption, energy and ship's speed are calculated.

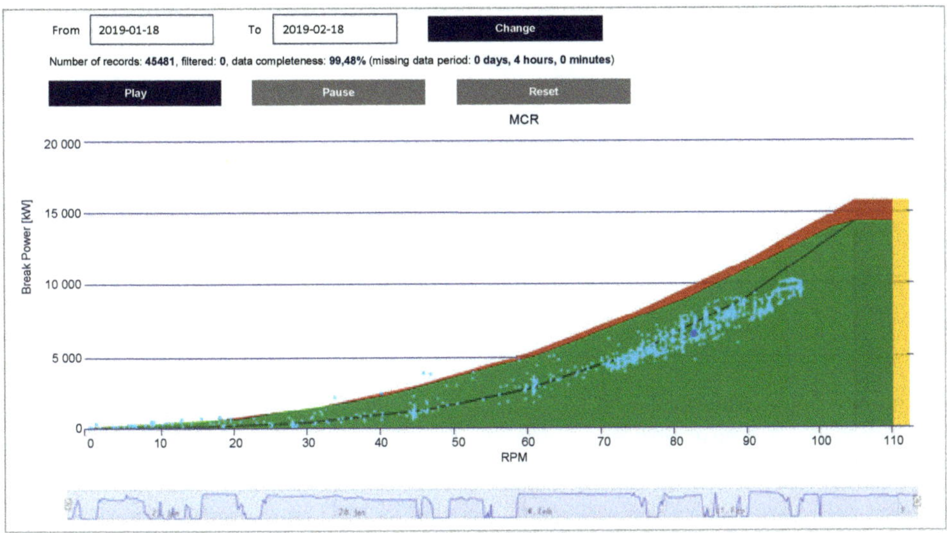

Figure 5. Main engine (ME) layout with actual operation points for one-month ship operation provided by SeaPerformer system; the records below the main engine layout present momentary changes of RPM on the ME shaft in the above-defined period (Courtesy of Enamor Ltd.).

Based on the state-of-the-art [5,17,18], measurement of torque can be done with use of measurement of the torsion angle φ of propeller shaft during running of engine. Different torque measurement methods can be applied. In many methods, strain gauges as the sensors are used [5]. This solution requires the strain gauges to be placed directly on the rotating shaft and additional measurement of RPM should be performed. The strain gauges are configured in bridge system equipped with slip rings and brush contact devices. The strain gauges bridge is fed through the two slip rings and output signal taken from the bridge diagonal is also provided with two slip rings. The all slip rings are fitted on the rotating shaft and they provide a continuous electrical connection through brushes on stationary contacts. At the same time, slip ring and brush contact solution are important disadvantages of this method.

The other solution, free of aforementioned weaknesses, is based on the photo-optical method [17–19] (Figure 6), which uses two specially designed teethed rings. Each ring is fixed separately in same distance "l" on the shaft, but the teeth are placed in the same plane which is presented in Figure 6a. Such a solution gives possibility for use of only one photo-optical detector. The relation between teethed rings is proportional to the torsion angle of the shaft. In this method measurement of RPM by additional sensor is not required. In the case when the shaft is without load, the torsion angle $\varphi = 0$. This situation is presented in Figure 6b. Electric signal received from photo-optical detector shows that pulse period T_1 and T_2 are equal. When the load of the shaft occurs, the torsion angle is higher than 0. In such a situation, presented in Figure 6c, the periods of electrical signal are different and they are proportional to the measured value of torque M according to the formula:

$$M = \frac{k_T(t_1 - t_2)}{n} \quad (1)$$

where k_T—coefficient depends on construction of shaft and teethed wheels, n—revolution, t_1, t_2—time of pulses received from photo-optical detector.

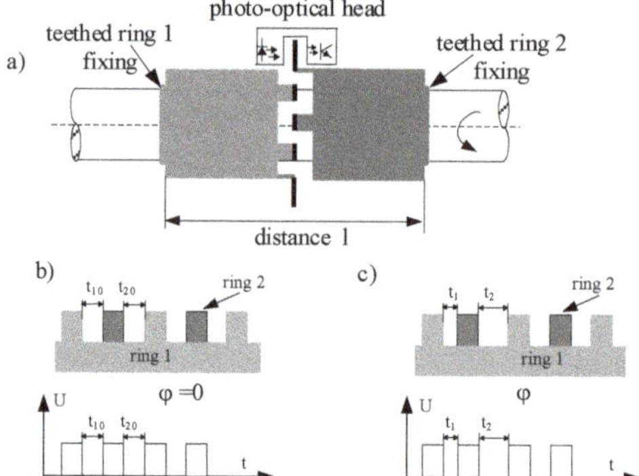

Figure 6. Photo-optical method for torque measurement, (**a**) concept of torque measurement, (**b**) relation between teeth of two rings and electrical signal for torsion angle $\varphi = 0$, (**c**) relation between teeth of two rings and electrical signal for torsion angle $\varphi \neq 0$.

The method was developed [17,18] and patented in GMU [19] as well as implemented on many ships in cooperation with ENAMOR Ltd. company. Contribution of GMU in state-of-the-art development consisted of elaboration of a new solution of torque meter related transmitter [19]. A novelty of the transmitter solution [19] is based on special construction including two mutually advantageous elements, placed on the rotating shaft and each of them consist of the spacer pipe (distance sleeve) and measuring ring with evenly and radially distributed at the circumference adequate teeth. An advantage of this transmitter construction is use of only one photo-optical detector, that limits the influence of changeable conditions and work parameters of transmitter and electronic sets of photo-optical detector circuit on measurement accuracy. The validation tests of the propulsion control assistance system ETNP-10 type have executed by Research and Development Company ENAMOR Ltd. [14]. The investigated system installed on propeller shaft is presented in Figure 7 [14]. Figure 7a shows two teeth rings with photo-optical detector that are installed on propeller. The local human-machine interface presents current measurement parameters (Figure 7b) or engine load based on engine layout (Figure 7c).

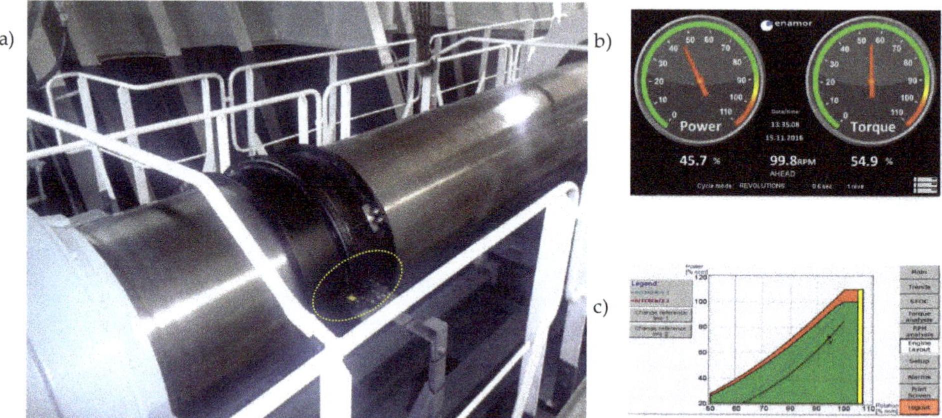

Figure 7. Propulsion control assistance system ETNP-10, where position of photo-optical detector is marked by ellipse, (**a**) two teeth rings installed on propeller shaft with photo-optical detector, (**b**) example of measurements, (**c**) actual engine load based on engine layout (Courtesy of Enamor Ltd.).

3.3. Temperature Measurements in the Ship's Hazardous Areas

Hazardous areas occur on almost all the ships. However, in certain ships, such as tankers (crude oil, chemicals, gas), mobile oil rigs or other offshore ships the hazardous areas are common. Due to technical conditions, the measurement of different parameters has to be made in the hazardous areas.

Hazardous areas are the places where the risk of appearing mixture of flammable materials with oxygen from air might occur [20]. According to International Electrotechnical Committee (IEC) approach hazardous areas are divided into three zones for gas hazard: zone 0, zone 1, zone 2 and three zones for dust hazard: zone 20, zone 21 and zone 22. Depending on the zone, adequate explosion-proof equipment should be used.

The IEC approach is commonly accepted by the majority of state maritime administrations and maritime classification societies [21–23].

Electrical equipment placed in the hazardous areas can be protected by different methods. For measurement and control, intrinsically safe solution "Ex i . . . " is the most adequate explosion-proof protection [24,25].

In case of long distance measurement or control lines, which are typical on ships especially for deck monitoring and control, a special approach has to be used in comparison to the standard solution. The basic assumption of intrinsically safe solution is that inside of the intrinsically safe circuit, the accumulated electrical energy is so low that in the case of its realization in a form of a spark or a thermal effect, it is incapable of igniting surrounding explosive mixture. To ensure this, specially certified equipment has to be used inside the hazardous area and additional, also certified, associated apparatus has to be used and placed in the safe area. When both are connected by cables, this creates an intrinsically safe system [19]. It is worth underlining that in comparison to a standard solution, an associated apparatus is in fact an additional equipment connected to the measurement line. The associated apparatus is connected between intrinsically safe equipment placed in the hazardous area and the standard monitoring system placed also in the safe area. It can be a source of some limitations or even additional measurement errors. As an illustration of the discussed problem, the measurement of temperature in tanks of gas tanker is presented. The general scheme of the tanks placement and related temperature measurement is shown in Figure 8.

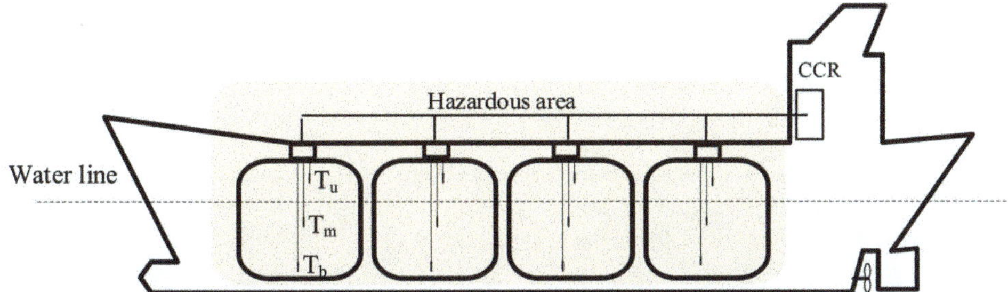

Figure 8. The temperature measurement of liquefied gas in tanks, CCR—Cargo Control Room, T_b, T_m—sensor in bottom and middle of tank to measure temperature of liquefied gas, T_u—upper sensor to measure temperature of vapors.

3.3.1. Accuracy Aspects of Temperature Measurement Line

Measurement of temperature can be performed in different ways [3,4], but platinum thermo-resistance Pt-100 is the most popular sensor for that. A predominant role of this kind of sensors results from the particular environmental and constructional conditions related to ship tanks, as well as with usually measured negative temperature range. Because of the long distance between measurement temperature point inside tanks and monitoring system placed in CCR, it is common to use a transmitter which converts resistance of a sensor to a standard current value. In our example, the two wires 4–20 mA standard is the most convenient to use [26]. The functional diagram of the discussed measurement line is presented in Figure 9.

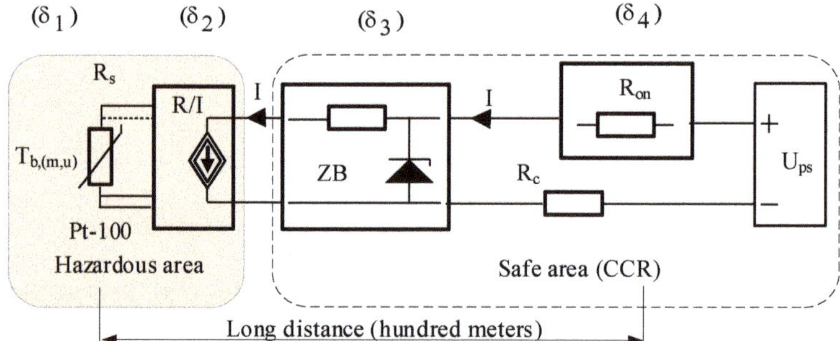

Figure 9. Functional diagram of temperature measurement line, where: R_s—resistance of Pt-100 sensor, R/I—transmitter-source of current 4–20 mA sink type, ZB—Zenner barrier as the associated apparatus, R_{on}—resistance of monitoring system, R_c—equivalent resistance of cables, U_{ps}—source of DC voltage, δ_1, δ_2, δ_3, δ_4 are relative errors.

In the analyzed case, combined standard uncertainty is based only on the uncertainty of B type and it is described by the formula:

$$u_c = \left(\sqrt{\frac{1}{3}\left[(\delta_1)^2 + (\delta_2)^2 + (\delta_3)^2 + (\delta_4)^2\right]} \right)(T_2 - T_1) \tag{2}$$

where δ_1, δ_2, δ_3, δ_4 are relative errors of Pt-100 sensor, transmitter, Zenner barrier and monitoring system itself, T_1, T_2 are min and max of the temperature span.

For the real data $\delta_1 = 1.6 \times 10^{-3}$, $\delta_2 = 3 \times 10^{-3}$, $\delta_3 = 1 \times 10^{-5}$, $\delta_4 = 2 \times 10^{-3}$, $T_1 = -120$ °C, $T_2 = 120$ °C and related value of uncertainty corresponds to $u_c = 0.57$ °C. A set of value for liquefied gas of tanks corresponds to −90 °C. Accuracy of the temperature measurement in tanks illustrated in Figure 8 plays a fundamental role in estimation of the transported cargo volume and finally the transport cost and price of the gas. Therefore, all disturbing circumstances, causing additional errors in the process of temperature measurement, are very important.

On top of that, an additional error was observed. This error was caused in the real measurement circuits by own capacities of long distances of sensors, screened cables, grounding and dynamic properties of a transmitter. In reality, all the elements of the measurement line connected together produce an active electrical circuit with positive feedback. The equivalent electrical diagram of such temperature measurement line is presented in Figure 10.

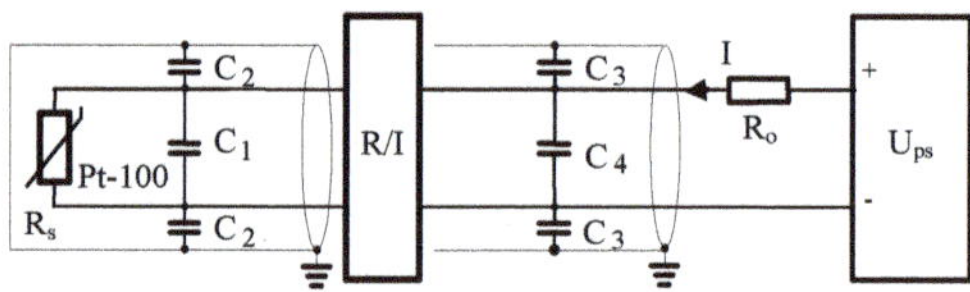

Figure 10. Equivalent electrical diagram of the temperature measurement line, where: R_s—resistance of Pt-100 sensor, R/I—transmitter 4–20 mA, R_o—equivalent resistance of all resistances in the 4–20 mA current loop I, $C_{1,2,3,4}$—equivalent capacities of Pt-100 sensor and connecting cables, U_{ps}—source of DC voltage.

In order to analyze the above circuit, a small signal model of the measurement line is presented in Figure 11. Dynamic properties of transmitter R/I should be defined in the model. It is worth mentioning that transmitter R/I is the same as converter U/I with internal source of current I_1 in Figure 11. Generally, the dynamic properties of the transducers are often omitted due to the fact that changes in the measured temperatures are relatively slow, especially when the industrial Pt-100 sensors are used.

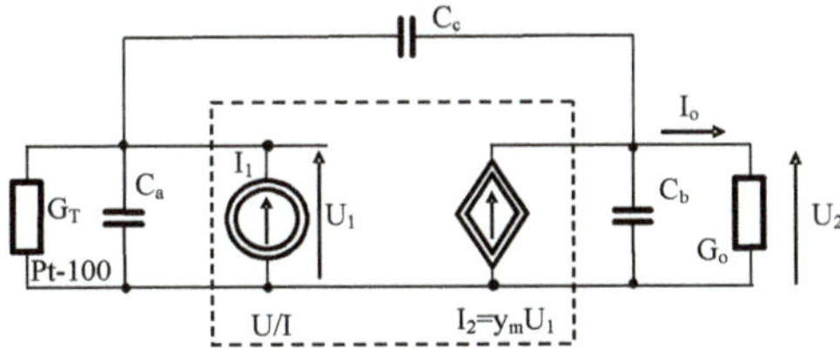

Figure 11. Small signal model of the temperature measurement line, where: G_T—conductivity of Pt-100 sensor, U/I—transmitter 4–20 mA, G_o—equivalent conductance of all resistances in the 4–20 mA current loop I_o, $C_{a,b,c}$—equivalent capacities of Pt-100 sensor and connecting cables, y_m—transadmittance of transmitter R/I.

Actually, electrical dynamic properties of transmitter should be taken under consideration. They are described by transadmittance y_m. Most often transmitters R/I are first-order inertial elements but sometimes they can be second-order oscillation elements.

3.3.2. Real Measurement Line with Transmitter as First-Order Inertial Element

For transmitters of first-order inertial type the transconductance y_m

$$y_m = g_m \frac{\omega_2}{s + \omega_2} \tag{3}$$

where g_m—transconductance depends on measured range of temperature, ω_2—3 dB frequency.

An analysis of stability of the circuit presented in Figure 7 was done for the set of real parameters given in Table 1 [1].

Table 1. Parameters of real temperature measurement line for first-order inertial transmitter.

Parameter [unit]	Value/Range	Parameter [unit]	Value/Range
C_a [nF]	4.5	G_o [mS]	2
C_{BS} [nF]	8	g_m [S]	0.043–0.83
C_c [nF]	1.7	ΔT^2 [°C]	50–1050
G_T [mS]	8.34	ω_2 [rd/s]	14,700
T_o [1] [°C]	50	-	-

[1] T_o measured temperature. [2] ΔT temperature measured span.

While performing computer simulations for the parameter values presented in Table 1, it was shown that the temperature measurement line with first-order inertial transmitter in special condition can work in an unstable manner and as a result an alternating current can occur.

The tendency to unstable working rises when the range is relatively small. That case is theoretical rather than practical because oscillating condition is fulfilled only for boundary conditions.

3.3.3. Real Measurement Line with Transmitter as Second-Order Inertial Oscillation Element

For transmitters of second-order oscillation inertial type, the transconductance y_m is given by the following formula

$$y_m = g_m \frac{\omega_o^2}{s^2 + 2\zeta\omega_o + \omega_o^2} \tag{4}$$

where g_m—transconductance depends on measured range of temperature, ω_o—natural frequency, ζ—damping ratio.

The analysis of stability of the circuit presented in Figure 11 was done for the set of real parameters given in Table 2 [1].

Table 2. Parameters of real temperature measurement line for oscillating second-order inertial transmitter.

Parameter [unit]	Value/Range	Parameter [unit]	Value/Range
C_a [nF]	4.5	G_o [mS]	2
C_b [nF]	8	g_m [S]	0.043–0.83
C_c [nF]	1.7	ΔT [°C]	50–1050
G_T [mS]	10	ω_o [rd/s]	20,400
T_o [°C]	0	ζ	0.212

Computer simulations performed for parameters given in Table 2 have shown that temperature measured line with the oscillating second-order inertial transmitter can be very unstable. As a result, the theoretically direct current I_o is disturbed by alternating part. Further detailed research has shown that the frequency and shape of the additional alternating current depends not only on measured range but also strongly on value of measured temperature (T_o) and equivalent load conductance (G_o).

The laboratory experiments of temperature measurement line with the oscillating second-order inertial transmitter R/I presented in Figure 12 were also performed and completely confirmed theoretical considerations.

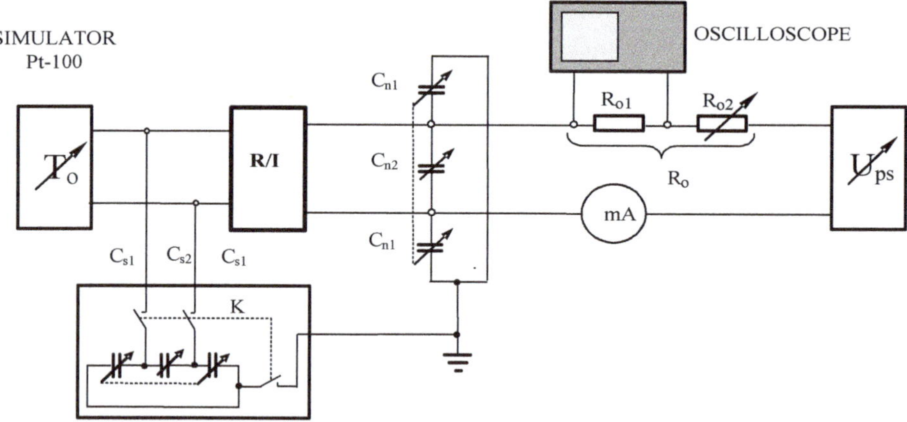

Figure 12. Laboratory experimental circuit of temperature measurement line with the possibility of taking account capacities of the Pt-100 sensor and connecting cables, where: R/I—second-order inertial transmitter, mA—DC ammeter, U_{ps}—source of voltage, Ro—equivalent of all resistances in the 4–20 mA current loop, $C_{s1,2}$—equivalents of own capacities of Pt-100 sensor, $C_{n1,2}$—equivalents of own capacities of cables, K- switch.

The experiments were performed for different simulated temperatures from measurement range of transmitter R/I (from −30 °C to 60 °C) and for different value of R_o for two cases. The first one where a capacitive feedback was broken (switch K-off position) and the second case where the own capacities of sensor by switch K in position on were taken into consideration. In the second case, the capacitive feedback was rebuilt. In both cases for the same simulated temperature, the current was measured by DC ammeter. While the capacitive feedback was done, sometimes the alternating current was observed by oscilloscope. An example of one such screenshot was presented in Figure 13 for a transmitter with a span $\Delta T = T_{max} - T_{min} = 60 - (-30) = 90$ °C at measured temperature $T_o = 50$ °C for equivalent load resistance $R_o = 600\ \Omega$.

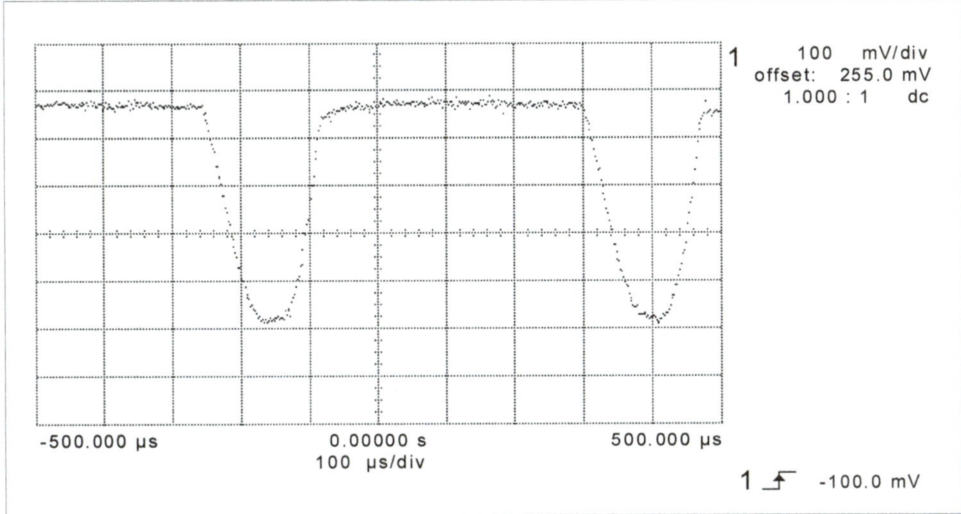

Figure 13. Exemplary of screenshot for a transmitter with a span $\Delta T = T_{max} - T_{min} = 60 - (-30) = 90$ °C at measured temperature $T_o = 50$ °C for equivalent load resistance $R_o = 600\ \Omega$.

In Figure 13 an unexpected alternating current i(t) disturbed a correct temperature measurement procedure was presented. Actually, the shape of the alternating current depends on the measured temperatures. From that, a mean value of measured current I_{m1} by DC ammeter differs from measured value I_{m2} when only DC current appears while the capacitive feedback is off.

To explain the nature of unexpected additional error while measuring temperature in the discussed cases, a collected set of alternating currents observed for different measured temperatures was presented in Figure 14. For legibility of the picture for a few selected temperatures, only a significant portion of the current waveforms has been included. For each graph, mean values for bath cases were pointed out. It is worth emphasizing that mean value of current I_{m2} depends on shape of current, which can be higher, lower or equal to the mean value of DC current in the case of lack of capacities feedback. Actually, that is observed by DC indicator as additional relative error δ_a of measured temperature and can be calculated according to the following formula:

$$\delta_a = \frac{(I_{m2} - I_{m1})[\text{mA}]}{16\ [\text{mA}]} \cdot 100[\%] \tag{5}$$

where value of 16 mA results from the 4–20 mA standard current transmitter.

To calculate the absolute value, simply multiply the relative error by the measuring range, which was mentioned above, in the discussed case is $\Delta T = 90$ °C. The adequate values of additional absolute errors were presented in Figure 14 [27].

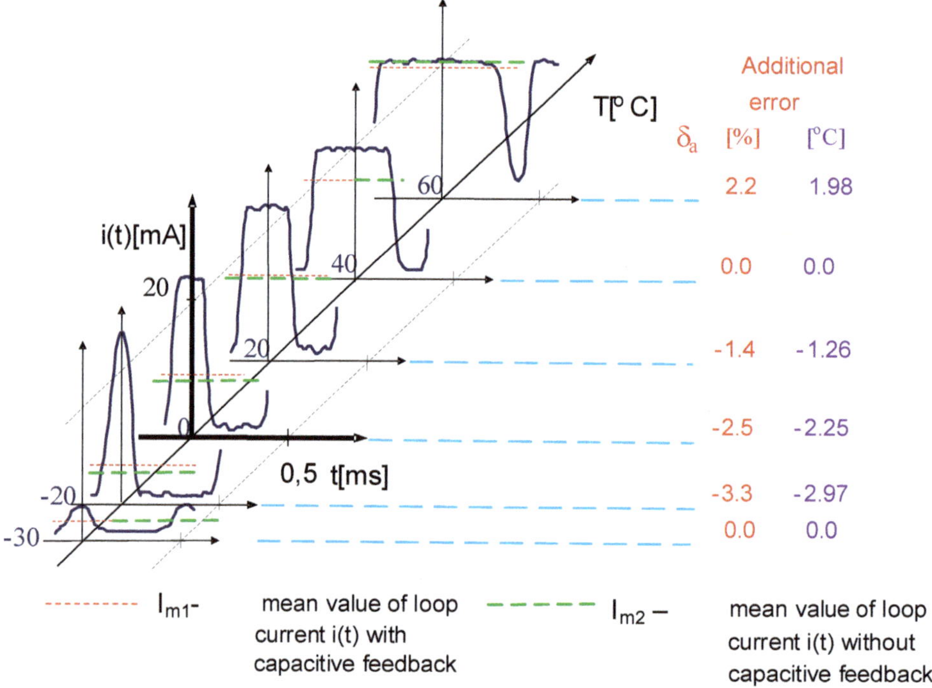

Figure 14. Graphical illustration of influence of unexpected alternating current on a measurement error.

As a result, a new error appeared, and what is worse, that error was not observed during commissioning procedures while using only resistance simulator sensor.

To avoid such situations, in GMU there was patented an impedance simulator of Pt-100 sensors [28] introducing a new concept. According to this, during commissioning it is possible to take into account also the sensor's own capacities. This solution improves significantly the quality and reliability of commissioning, and in consequence accuracy of evaluation of the quantity of transported liquefied gas.

4. Conclusions

In many cases, ship measurement methods require individual and specific approach, because of the marine environment influence. Such affecting factors like vibration, ship rolling and pitching, temperature, humidity, salinity or changeable weather conditions must be taken into account. In this context, well known currently existing and applied measurement methods and devices should be completed and developed. Two case studies, addressed to operational parameters of main engine and temperature measurement of liquefied gas in ship tanks, are shortly described from the perspective of a new approach to analysis of their operation in specific conditions, which are vibrating environment and hazardous area, respectively. This new approach is presented with stressing proposed novelties against the state-of-the-art and in relation to the operational parameters of main engine in ship power plant consists in, that:

- not only traditionally measured quantities on the main engine shaft like torque and RPM are measured but also some other main engine characteristic quantities, related to crankshaft angular position, combustion pressure, exhaust valve position, injection quantity sensor position and gas admission valve position are taken into account,
- in the presented photo-optical method based on the measurement of the torsion angle of the shaft, a measurement of RPM by additional sensor is not required, contrary to the method based on strain

gauges. Moreover, presented photo-optical method is free of slip rings and brush contact devices, which are a serious maintenance problem in previously applied torque measurement method,

- described photo-optical method, aided by new solution of torque meter related transmitter gives possibility for use of only one photo-optical detector and at the same time the influence of changeable conditions and work parameters of transmitter and electronic sets of photo-optical detector circuit on measurement accuracy is significantly reduced comparing with other previously used solutions equipped with two photo-optical detectors,
- extended application of the aforementioned photo-optical method in SeaPerformer system enables not only measurement the torque on shaft and revolution (RPM) of the main engine, but also a calculation of other parameters like power, fuel consumption, energy and propeller slip. In consequence, also monitoring of main engine specific oil consumption (SFOC) trend is executed. Aforementioned monitoring is supported by data acquisition of engine operation signals related to crankshaft angus or position, combustion pressure, exhaust valve position (EVP), injection quantity sensor position and gas admission valve (GAV) position respectively,
- Next, in wider perspective, engine performance evaluation based on momentary fuel consumption analysis and main engine load is carried out. Monitoring of the above listed operational parameters of main engine aided by a new class of systems like SeaPerformer gives possibility for operational optimization of main engine exploitation and constitutes a new perspective comparing with hitherto used solutions in considered field of research.

On the other hand, the new approach of analysis in relation to the temperature measurement in ship tanks, with placing emphasis on presented novelties consists in, that:

- Temperature measurement in ship's hazardous areas covers not only use of intrinsically safe solutions for design of temperature measurement line, but also takes into account the additional measurement errors caused in the real measurement circuits by own capacities of long distances of sensors (predominant solutions are based on the Pt-100 sensors), screened cables, grounding and dynamic properties of a transmitter.
- Predominant solutions of sensor due to specific conditions of temperature measurement in ship tanks are based on the Pt-100 elements applied in intrinsic safe version of the measurement line, often equipped with passive Zener Barrier; Passive Zener Barrier dependently of the circuit parameters may enhance additional measurement error and what is worse, that error was not observed during commissioning procedure while using only resistance simulator Pt-100 sensor,
- To avoid unexpected alternating current component disturbing a correct temperature measurement, there was patented a new impedance simulator of Pt-100 sensors,
- This new solution enables to take into account also the sensors own capacities and in consequence, to improve significantly the quality and reliability of commissioning.
- Finally, accuracy of evaluation of the quantity of transported liquefied gas (corresponding to cost of transport) is most reliable.

Future analyses, respectively in the two areas under consideration, will focus on:

- In the scope of operational parameters of the main engine in a ship power plant—on the further development and implementation of automatic detection of engine anomalies and failures and context-based information selection both in onboard and on-line applications,
- In the area of temperature measurement on ships—on carrying out an analysis of operating conditions, in which an additional temperature measurement error may occur, resulting from the dynamic properties of electronic measuring line elements, where the sensor is a thermocouple. The different types of T/C converters and the impact of the intrinsically safe solution will be taken into account.

Author Contributions: Conceptualization, B.D. and J.M.; Methodology, B.D. and J.M.; Validation, B.D. and J.M.; Formal analysis, B.D.; Investigation, B.D.; Resources, B.D. and J.M.; Writing—original draft preparation, B.D. and J.M.; Writing—review and editing, B.D. and J.M.; Supervision, J.M.; Funding acquisition J.M.

Funding: The studies were carried out within the framework of work N°WE/DS/429/2017, financed from the funds for science by the Polish Ministry of Science and Higher Education and within the research project POIR.01.01.01-00-0933/15 co-financed by Polish National Centre for Research and Development.

Acknowledgments: We appreciate our colleague Wojciech Górski from company ENAMOR Ltd. for his contribution in this study. The authors are also grateful to Research and Development Company ENAMOR Ltd. for providing us with related company databases and results of implementation tests of the main engine operational parameters measurement and monitoring systems.

Conflicts of Interest: The authors declare no conflict of interest.

References

1. Mindykowski, J.; Dudojć, B. Case-study based overview of selected measurement in ship system. In Proceedings of the 2018 IEEE International Workshop on Metrology for the Sea, MetroSea 2018, Bari, Italy, 8–10 October 2018; pp. 1–4.
2. Mindykowski, J. *Assesment of Electric Power Quality in Ships Fitted with Converter Subsystems*; Shipbuilding & Shipping: Gdańsk, Poland, 2004.
3. Sydenham, P.H.; Thorn, R. (Eds.) *Handbook of Measurement Science. Vol. 1 Theoretical Fundamentals, A Wiley-Interscience Publication*; John Wiley and Sons Ltd.: Hoboken, NJ, USA, 1991.
4. Sydenham, P.H.; Thorn, R. (Eds.) *Handbook of Measurement Science. Vol. 2 Practical Fundamentals, A Wiley-Interscience Publication*; John Wiley and Sons Ltd.: Hoboken, NJ, USA, 1991.
5. Miłek, M. *Electrical Metrology of Non-Electrical Quantities*; Oficyna Wydawnicza Universytetu Zielonogórskiego: Zielona Góra, Poland, 2006; ISBN 83-7481-023-8. (In Polish)
6. Ballegooijen, E.; Helsloot, T.; Timmer, M. Experience with Full-Scale Thrust Measurements in Dynamic Trim Optimisation. In Proceedings of the 3rd Hull Performance & Insight Conference, HullPIC'18, Redworth, UK, 12–14 March 2018; pp. 36–47.
7. Wienke, J. Thrust Measurements during Sea Trials—Opportunities, Challenges and Restrictions. In Proceedings of the 3rd Hull Performance & Insight Conference, HullPIC'18, Redworth, UK, 12–14 March 2018; pp. 141–148.
8. Borkowski, P. Fusion of data from GPS receivers based on a multi-sensor Kalman filter. *Transp. Probl.* **2008**, *3*, 5–11.
9. Kazimierski, W.; Stateczny, A. Radar and Automatic Identification System Track Fusion in an Electronic Chart Display and Information System. *J. Navig.* **2015**, *68*, 1141–1154. [CrossRef]
10. Kano, T. On the Effect of Navigation Support System—From Collecting Data to Operational Support. In Proceedings of the 3rd Hull Performance & Insight Conference, HullPIC'18, Redworth, UK, 12–14 March 2018; pp. 123–134.
11. Górski, W. Role of reference model in ship performance management. In Proceedings of the 1st HullPIC Conference, Castello di Pavone, Italy, 13–15 April 2016; pp. 202–214.
12. Górski, W. Automatic MRV Reporting with Use of Onboard Performance Management Systems. In Proceedings of the 2nd HullPIC Conference, Ulrichshusen, Germany, 27–29 March 2017; pp. 166–172.
13. IMO. RESOLUTION MEPC.278(70), Amendments to MARPOL Annex VI, Data Collection System for Fuel Oil Consumption of Ships, Adopted on 28 October 2016. Available online: http://www.imo.org/en/OurWork/Environment/PollutionPrevention/AirPollution/Documents/278(70).pdf (accessed on 11 April 2019).
14. *Results of the Validation Tests of the Propulsion Control Assistance System ETNP-10 Installed on Propeller Shaft*; Research and Development Company Enamor Ltd.: Gdynia, Poland, unpublished works.
15. ISO 15550:2009. *Internal Combustion Engines—Determination and Method for the Measurement of Engine Power—General Requirements*; ISO: Geneva, Switzerland, 2009.
16. ISO 3046-1:2009. *Reciprocating Internal Combustion Engines—Performance—Part 1: Declarations of Power, Fuel and Lubricating Oil Consumptions, and Test Methods—Additional Requirements for Engines for General Use*; ISO: Geneva, Switzerland, 2009.

17. Makowski, L.; Morawski, L.; Tański, J.; Sikora, M. The torqumeter TNP-7 for measuring of power and rotating speed on the propeller shaft of ship. In Proceedings of the Word Congress ENGINE'96, Łódź, Poland, 22–24 October 1996; pp. 169–172.
18. Morawski, L.; Szuca, Z. The monitoring of ship propulsion by torque and rotational speed measurements on the propeller shaft. *J. Pol. CIMAC* **2009**, *4*, 105–110.
19. Makowski, L.; Morawski, L.; Szuca, Z. Transmitter of Torque Meter. Patent Number PL 210,430, 31 January 2012. (In Polish)
20. IEC 60079-0:2017. *Explosive Atmospheres—Part 0: Equipment—General Requirements*; Head of Sales & Business Development, IEC Central Office: Geneva, Switzerland, 2011.
21. IACS Interpretations: SC70 (1985) (Rev.1May 2001) (Rev.2 Nov 2005) IACS Int. 1985/Rev. 2 2005 Area Classification and Selection of Electrical Equipment (Reg. II-2/11.6.2). Available online: https://pl.scribd.com/doc/33454082/Interpretations-Solas (accessed on 11 April 2019).
22. *IECEx System for Certification to Standards Relating to Equipment for Use in Explosive Atmospheres*, 7th ed.; IECEx Publication: Geneva, Switzerland, 2015.
23. IACS Guidelines and Recommendations No.35 (1992) (Rev.1 Mar 2006, Corr.1 June 2015) Inspection and Maintenance of Electrical Equipment Installed in Hazardous Areas. Available online: http://www.iacs.org.uk/publications/recommendations/21-40/rec-35-rev1-corr1-cln/ (accessed on 11 April 2019).
24. IEC 60079-11:2011. *Explosive Atmospheres—Part 11: Equipment Protection by Intrinsic Safety "i"*; Head of Sales & Business Development, IEC Central Office: Geneva, Switzerland, 2011.
25. IEC 60079-25:2010. *Explosive Atmospheres—Part 25: Intrinsically Safe Electrical Systems*; Head of Sales & Business Development, IEC Central Office: Geneva, Switzerland, 2010.
26. Dudojć, B.; Mindykowski, J. Additional errors and their detection in temperature measurements in ship's hazardous areas. *Electron. Telecommun. Q.* **2001**, *47*, 323–354. (In Polish)
27. Dudojć, B.; Mindykowski, J. Appliance for Checking the Temperature Measurement Line. Patent Number PL 179,503, 29 September 2000. (In Polish)
28. Dudojć, B.; Mindykowski, J. Checking current analogue measurement channels. In Proceedings of the XVI IMEKO World Congress, IMEKO 2000, Vienna, Austria, 25–28 September 2000; pp. 175–180.

Article

Optimization of the Maritime Signaling System in the Lagoon of Venice †

Fabiana Di Ciaccio [1,*], Paolo Menegazzo [2] and Salvatore Troisi [1]

1 Department of Sciences and Technologies, Parthenope University of Naples, 80143 Naples, Italy; salvatore.troisi@uniparthenope.it
2 Strategic Planning, Port Authority of Venice, 30123 Venice, Italy; paolo.menegazzo@port.venice.it
* Correspondence: fabiana.diciaccio@uniparthenope.it; Tel.: +39-328-093-5198

† This paper is an extended version of paper published in Di Ciaccio, F.; Menegazzo, P.; Troisi, S. Optimization of the Maritime Signaling in the Venetian Lagoon. In Proceedings of the 2018 IEEE International Workshop on Metrology for the Sea, Bari, Italy, 8–10 October 2018.

Received: 28 December 2018; Accepted: 7 March 2019; Published: 10 March 2019

Abstract: Aids to Navigation (AtoN) are auxiliary devices intended to support maritime navigation. They include both traditional signals (e.g., buoys and lights) and electronic aids, as for example those transmitted to ships through automatic tracking systems. In both cases, international organizations together with local authorities define technical specifications and standards on their use. Work still being finalized in the Venetian Lagoon made it necessary an assessment of the existing signaling system to guarantee the maximum level of safety in the waterways. Considering the severe atmospheric conditions to which the Lagoon is frequently subjected and the bathymetry restrictions affecting the navigation, an alternative aid system has been formalized for the first time in Italy. It is based on electronic and identification devices employed to virtualize the AtoN that will not be located at sea but only remotely identified by their coded messages, thus guaranteeing the continuity of port operations in any visibility conditions. This paper presents the procedures followed to reach a solution in line with the safety and efficiency standards given for the AtoN systems, considering position and luminous characteristics of physical signals in the first case, theoretical and statistical studies on Virtual AIS AtoN placement in the second case.

Keywords: virtual AtoN; AIS; AtoN; ECDIS; e-navigation; safety; route planning; visibility

1. Introduction

The International Maritime Organization (IMO) issues standards on security, safety and environmental performance of shipping, encouraging innovation and efficiency. IMO works, among the others, with the International Association of Marine Aids to Navigation and Lighthouse Authorities (IALA), a non-profit technical organization which aims to harmonize Aids to Navigation (AtoN) worldwide considering the needs of mariners and authorities as well as the technological development. To assure the maximum level of safety and efficiency of maritime navigation, IALA publishes recommendations and guidelines on the use of auxiliary devices that improve maritime operations, such as Vessel Traffic Services (VTS), Automatic Identification System (AIS) and marine AtoN signal lights [1].

The development of ports and fairways, together with the recent progresses of technologies and the increasing demands on navigation services make a dynamic optimization of the AtoN systems necessary. The consistency of the AtoN placement will directly influence the safety of marine traffic, so consideration should be given either to the Collision Regulations (COLREG) and local traffic rules.

The knowledge of the weaknesses and problems of the existing AtoN system is of central importance to enhance the provided service.

In this regard, Chen et al. [2] proposed the Success Degree-Fuzzy comprehensive evaluation method to post evaluate the existing AtoN system to clarify the needs and to point out its optimization. This model works with a series of index, derived in accordance with the AtoN system characteristics and elaborated on the basis of the consideration of users to guarantee its accuracy. The final algorithm allows the valuation of the overall score of the system in a rank from 0 to 100.

The analysis carried out by our team arises from the immediate need of the Venetian institutions and local seafarers to redesign the AtoN system to assure the highest level of safety in the Venice Lagoon. In fact, the installation of the Experimental Electromechanical Module (MOSE or MOdulo Sperimentale Elettromeccanico) at the inlets of the lagoon led to a series of changes in the navigating area (such as the new artificial island at Lido), with remarkable effects on local traffic.

For this reason, we did not make an a priori study on the maritime traffic safety to evaluate the risks of collision or accidents in general as it was not required to proceed with the optimization of the system. We went directly to the problems expressed by the users to find a solution within the shortest possible time: firstly, we analyzed the physical disposition of the AtoN in the lagoon, then we searched for a more flexible solution to the restricted visibility problem which could guarantee the safe conduction of the navigation in any situation.

We followed a heuristic method which combined IALA, IMO and local normative with the practical needs in order to reach the optimal AtoN disposition: in fact, the iterative and constant update of the system has been one of the key points of the process.

Afterwards, the effective visibility of the signals has been verified using the theoretical considerations given by the IALA Recommendations E-200-2 [3]. The resulting data showed that, although complying with the safety requirements, the light of each AtoN cannot always assure their visibility.

Basing on E-navigation solutions, an alternative AtoN system has been developed. According to the IMO, E-Navigation can be defined as the harmonized analysis and exchange of marine information by electronic means, with the aim to enhance safety and security of the navigation and protect the marine environment.

In 2014 the MSC 94 approved the e-Navigation Strategy Implementation Plan (SIP) to complete a list of tasks by 2019. These tasks mainly regard the improvement of the communication systems (in particular the Vessel Traffic Service Portfolio, not limited to the shore stations) and the bridge design and its equipment, to enhance its reliability and resilience and make the interface more user-friendly [4]. This plan, moreover, bases on estimating the effect of e-Navigation applications on reducing navigational accidents, including collisions and groundings of ships falling under the International Convention for the Safety of Life at Sea (SOLAS) [5].

In this context it is worth to emphasize the role played by e-Navigation in relation to the development of Maritime Autonomous Surface Ships (MASS), driverless systems that are increasingly used in many applications (surveillance, research, monitoring, anti-mining, etc.). In this case, the collection of navigational information plays an important role in the decision-making process and can be achieved by collaboration with systems as the AIS [6]. In particular, hazardous scenarios in restricted water, mainly related to collision-avoidance, are identified as one of the main challenging tasks for autonomous navigation systems, which are expected to operate together with conventional vessel thus respecting the COLREG as well [7].

Following the work carried out by the IMO we identified the potential e-Navigation solution: the resulting approach uses the AIS to virtualize the AtoN by the transmission of a coded message containing the required information: the AtoN does not have to be physically located at sea as it will be shown on appropriate electronic system, usually Electronic Chart Display and Information System (ECDIS).

The employment of Virtual AtoN has already been tested to mark the entrance to a Traffic Separation Scheme (TSS), provide emergency wreck and obstruction markings and to mark offshore structures. In 2012 the General Lighthouse Authorities (GLA) established a port hand virtual buoy at the Rigg Bank, South of Mew Island at the entrance to Belfast Lough, to warn about a nearby sandbank with a depth of 8.4 m. With the Irish Lights Notice to Mariners No. 03 of 2012 the Authority underlined that the information available to mariners will be dependent on their display system and not all transmitted information may be displayed, so users are always encouraged to keep the system up-to-date. Moreover, mariners are requested to give feedback on their experience of AIS to help the Authority to improve the service checking if the AtoN data is available [8].

Moreover, Wright and Baldauf proposed to supplement physical AtoN with V-AtoN in the many inaccessible, remote and ecologically sensitive regions of the world, like for example the Arctic where ice can collide and displace buoys from locations, the tropics where the AtoN placement could damage coral reefs, and other areas devoid of navigational infrastructure [9].

In our case, with reference to IALA Guideline 1081 and IALA Guideline 1062, we decided to use Virtual AtoN (V-AtoN) of safe water to give safe route indications. These AtoN were strategically positioned on the points in which the ship changes course or Waypoints (WP). This information on the route is digitized, so it can be modified and sent according to the needs, thus making this type of AtoN a considerable support to the navigation. This method was first tested in Italy in the Venetian Lagoon to provide indications to help the crew during planning operations [10]. Here, situations of restricted visibility usually led to the closure of ports by the Authorities in order to prevent hazards and risks in general. This E-Navigation solution combines the theoretical consideration with the actual characteristics of the navigation in the Lagoon to improve the safety levels and to allow its proper conduction even in restricted visibility conditions. In fact, the V-AtoN thus obtained can be set as Waypoints on the ECDIS onboard to identify the safe route to the berthing point, giving indications to be followed in any situations which would otherwise prevent the normal conduction of navigation.

This paper outlines the establishment of the optimal AtoN system in the Lagoon of Venice and the consequent study on its effective visibility based on IALA guidelines. It follows an overview of the AIS, then the procedure adopted to establish Virtual AtoN in the Venetian Lagoon is presented. The first theoretical approach is based on the maneuvering characteristics of the ships, opportunely contextualized within the Lagoon, and the related safety requirements. The definitive setting of the system is then given through the statistical analysis of the AIS tracks of ships entering the lagoon recorded by the Coast Guard AIS system.

This system is still being tested: feedback by the users will help to improve and enhance the service, and future studies based on models or simulations will allow to extend this work to any coastal area having the same necessities, generalizing this work for every situation.

2. Classic AtoN System

To study the optimal AtoN disposition we made a first overview on their characteristics: as already stated, they are standardized among all the IALA members. The only distinction provided by the Maritime Buoyage System applies to lateral signals in relation to the membership of each Country of the IALA A or IALA B system: the countries of Region A use red to indicate the left and green for the starboard while those of Region B do the opposite. Other signals are standardized in terms of destination, shape, color and luminous characteristics. For example, the Safe Water signal indicates an area where navigation is safe. This AtoN has white and red vertical stripes, a red spherical mirage (in case of minor buoys) and a 12-s-period white light with a single 10-s flash.

One of the most important features of the luminous signal is its range, indicating the maximum distance at which the light is perceived. In navigation, the luminous range D in nautical miles is used; it is defined as the maximum distance to which the detection of the light beam is guaranteed (not the light source, still hidden due to the terrestrial curvature), considering the meteorological visibility v and the illuminance required at the observer's eye or threshold, E_T. The introduction of some photometric quantities is necessary to identify the illuminance.

From a quantum point of view, the light is characterized by a power W in Watt that is function of its own wavelength λ: the product of the power for the visibility factor function V_f, expressing the actual perception of light to the human eye, gives the luminous flux Φ as in Equation (1):

$$\Phi = V_f\ (\lambda)\ W(\lambda). \tag{1}$$

The luminous intensity I is the quantity of flux measured in a certain direction and is calculated as the ratio between the luminous flux Φ and the solid angle ω of emission as in Equation (2). It is measured in candles:

$$I = \frac{\mathrm{d}\Phi}{\mathrm{d}\omega}. \tag{2}$$

The illuminance E is the ratio between the luminous flux Φ and the area of the surface A on which it impacts, see Equation (3). It is measured in lumens on squared meters or lux:

$$E = \frac{\mathrm{d}\Phi}{\mathrm{d}A}. \tag{3}$$

This illuminance E can be expressed as a function of meteorological visibility v and the distance to the observer d through Allard's law as in Equation (4):

$$E(d) = \frac{I}{3.43 \times 10^6} \frac{0.05^{\frac{d}{v}}}{d^2}. \tag{4}$$

From Equation (5) it is possible to estimate the luminous range of a signal D, which can also be obtained with the diagrams given by the IALA (Figure 1) [3,11,12]:

$$I = \left(3.43 \times 10^6\right) E_{\mathrm{T}} D^2\, 0.05^{\frac{d}{v}}. \tag{5}$$

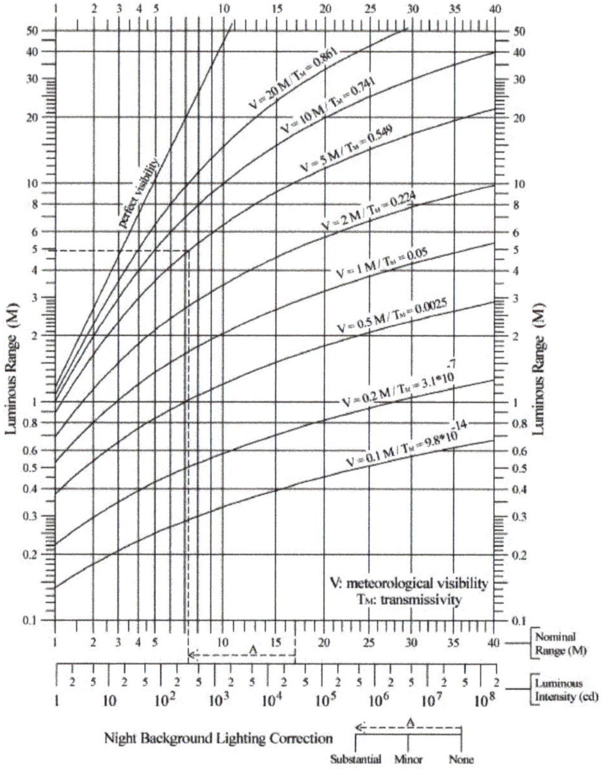

Figure 1. Luminous and nominal range at night.

Considering both IALA and local regulations together with seafarers needs, we proposed a maritime signaling system that combines structural efficiency with the highest level of navigational safety for the Venetian Lagoon.

To reach this solution, our team made an accurate survey of the actual AtoN disposition to check the correspondence between the functional AtoN at sea and those reported by the official cartography. This allowed the compilation of an updated database containing all the existing signals with their position and characteristics, according to the IALA standards.

We did not a formal safety diagnosis examination to identify potential safety hazards which may affect the navigation from the initial design phase as done in the Marine Traffic Safety Diagnostic Scheme (MTSDS). In that case, the procedural scheme starts from the analysis of the project and continues with the investigation by a predisposed audit team to review the maritime traffic. After an assessment phase, the audit team will describe all the potential safety problems and suggest all the possible measures to eliminate or mitigate these problems. The most important part of the MTSDS is the process of risk assessment, in which several models can be adopted. The main are the Environmental Stress (ES) Model, the IALA Waterway Risk Assessment Program (IWRAP), the Potential Risk Assessment Model (PARK). The first is based on the acceptance criteria of the stress value based on mariners' perception of safety, evaluating the difficulty arising from restrictions in maneuvering water area due to the traffic congestion. Three indexes are used: the ES_L (Environmental value for Land) calculated on the basis of the time to collision (TTC) with any obstacles, the Environmental stress value for Ship (ES_S) based on the TTC with ships, and the Environmental stress value for Aggregation (ES_A), combination between the previous two. However, this model reflects the users' perception of risks so it would not correct some problems.

The IALA IWRAP quantifies the risks involved with vessel traffic in specific geographical areas by calculating the annual number of collision and grounding in the specified location; the results are similar to the ES model [13]. The PARK model was developed based on the Korean mariner risk perception and calculates the risk through internal elements (characteristics of the vessel) and external elements such as approaching position of each ship, speed and distance between ships [14].

Due to the urgency of a solution, we decided to directly assess the current situation of the signaling system in the Lagoon through an elementary simulation.

Ji-Min Yeo, et al. studied the implementation of an AtoN database including the integrated system design to develop an AtoN simulator system [15]. In the same way, we set an elementary simulation process using Google Earth Pro (v. 7.3.2.5491) as supporting tool. Among the other features, this software allowed the overlap of the Lagoon official bathymetric and cartographic data to the satellite images. We could then import all the AtoN from our database to get an overall view on the area, thus allowing a more intuitive detection of the current issues and the consequent optimization of the entire AtoN system.

We began with the theoretical analysis of the area, searching for the best solution in line with IALA standards considering the morphology of the canals as well. Having presented this plan to the Authorities, we integrated the actual needs of the seamen together with those of the involved institutions, trying to reach a trade-off between benefits and costs.

Some areas needed particular attention. The main examples are the inlets of the Lagoon, where the presence of the MOSE had to be signaled. As previously mentioned, in the specific case of the Port of Lido the artificial island supports the gates, then requires a functional AtoN placement to prevent hazardous conditions. We proposed the installation of yellow flashing lights on the Port entrance lighthouses, to be activated simultaneously to the gates of the MOSE. On the island, four yellow structures could be positioned on its corners; their lights will be red and green for the starboard and the port side of the island respectively (correctly showing the starboard side of the main fairway and port side of the secondary one) and will turn in a yellow flashing light when the barriers are raised, alerting both the units entering and leaving the Lagoon. Since lateral AtoN should be placed in pairs to better indicate the fairway, we assured to the AtoN on the island the correspondent one alongside the canal to identify the main and the secondary canals (Figure 2).

Figure 2. View of the artificial island (Port of Lido). SketchUp models simulate the expected situation.

We also proposed some optional improvements, as for example a Preferred Channel signal located before the artificial island to distinguish the main channel on the port side (St. Nicolò Canal) from the secondary one on the starboard side (Treporti Canal). The same has been proposed in the area of Malamocco, where the main fairway turns on the right of the St. Leonardo canal. These buoys are modified Lateral signals: in the first case it will be a horizontally striped green-red-green buoy with a green flashing light while in the second case the stripes will be red-green-red with a red flashing light.

On the basis of the resulting proposals, we used SketchUp 2018 to create a 3D model of each AtoN; then, we imported them on Google Earth to simulate the expected situation in the Lagoon (Figure 2). To verify the efficacy of the proposed AtoN (whose range was given in accordance to those already present) even in poor visibility conditions, we made a purely theoretical analysis based on IALA criteria. The mean distance between each AtoN d and their nominal range D_N were known, so we could calculate the intensity produced by each signal as in Equation (5), using the illumination values required by IALA [3]:

(1) $E_T = 2 \times 10^{-7}$ lx at night;
(2) $E_T = 1 \times 10^{-3}$ lx in the daytime hours (to overpower sunlight).

We used the standard visibility of 10 nautical miles, as it is related to the nominal range. From the resulting data of Intensity, we derived the actual illuminance produced by each AtoN as in Equation (4) in three different visibility conditions (obtained from the data recorded by the visibilimeters placed in the Malamocco Canal): perfect, medium and poor visibility. We could then check the effective AtoN visibility using Equation (6):

$$E_d > E_T. \quad (6)$$

As we expected, the results showed the efficacy of the proposal in all the situations except for those of extremely poor visibility, thus making the system not adequate to guarantee the maximum levels of safety in these occasions.

We searched for an innovative tool complementary to the existing AtoN system to manage the problem of the ban on navigation in the Lagoon in case of thick fog: this condition, which frequently occurs in the Lagoon in winter and autumn (Figure 3), do not assure the required safety level according to the Authorities, forcing them to interdict the area and close the ports.

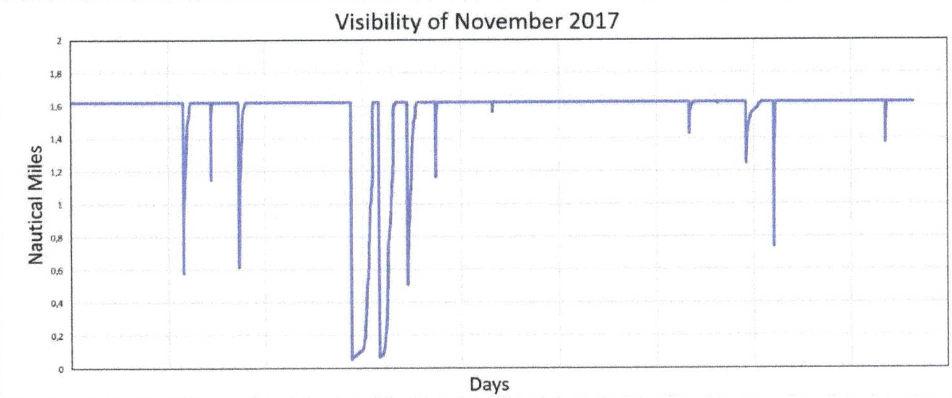

Figure 3. Visibility recorded in the canal of Malamocco in November 2017.

The increasing development of the E-navigation concept and its related technologies led our team to analyze all the different tools to find a solution which could solve this issue. We identified the AIS as the most reliable and versatile mean to send information to all the units in the working area, thus providing an active service in any situation.

3. An Overview on the AIS

Automatic Identification Systems (AISs) and Electronic Chart Display and Information Systems (ECDISs) have become mandatory for ships weighing more than 300 gross tons and for all the passenger ships, irrespective of their size, by the Maritime Safety Committee (MSC) as stated in the International Convention for the Safety of Life at Sea (SOLAS), Regulation V/19.2.4 [16].

The ECDIS is the system authorized by the International Maritime Organization (IMO) to display Electronic Nautical Charts (ENC), that are continuously updated by competent Authorities. This system is supported by the integration with other tools as for example the Global Navigation Satellite Systems (GNSS), Radar and AIS, and shows the details of the ship and the real time area within it is navigating. Unlike traditional charts, the ECDIS displays also bathymetry information and emits an alarm to warn the ship officer about dangerous situations.

AIS communication systems are Very High Frequency (VHF) radio transceivers that send ship's data to other units in the working area. Data are collected in real time by the sensors mounted on board and transmitted in broadcast without the need to be operated by ship staff.

The IMO distinguishes two AIS categories in line with ITU-R M.1371 standards:

1. Class A AIS meets IMO requirements [16,17]. During on-board assembly, the ship's static characteristics are set (name, dimensions, Maritime Mobile Service Identity-MMSI code, etc.); the AIS is then connected with the onboard GNSS and other related interfaces;
2. Class B AIS units are used by vessels that are not regulated by the IMO, such as leisure craft and fishing vessels. Compared with Class A AIS units, they have reduced functionality and are less powerful. Class B AIS units are designed to co-operate with Class A systems, which is why they do not need to respect IMO performance requirements [18].

3.1. AIS Messages

AIS messages consist of a set of digital data packets, for which the ITU defines technical characteristics and global frequencies. AIS Messages, in line with ITU-R M.1371, have a unique identification number from 1 to 27. Standard AIS Messages are divided into four macro information sectors:

1. Static or fixed information, which is set during the on-board installation phase, includes name, destination, dimensions, MMSI code and IMO ID of the ship and the antenna's position. It is transmitted once every six minutes or when requested by on-shore authorities.
2. Voyage-related information, which is manually entered, updated during the voyage and transmitted every six minutes. It includes the type of hazardous cargo, draught, destination and estimated time of arrival (ETA).
3. Dynamic information is automatically updated from the sensors onboard connected to the AIS. It includes Course Over Ground (COG), Speed Over Ground (SOG), position, time and navigation status.
4. Short safety related messages, broadcast or addressed to a specific unit as required.

AIS units can also transmit a series of messages whose data content is defined by the application: they do not affect the basic operation of the AIS as they do not have the same priority. As a result, secondary information does not cross primary information on the main channels. These messages are known as Application Specific Messages (AIS-ASM) and are generally used to send meteorological communications and warnings in general, but they are not intended to substitute messages sent by standard navigation support services (GNSS, Global Maritime Distress and Safety System-GMDSS, etc.) [19,20].

An example of AIS-ASM is the Route Information Message (ID: Message 8): it can be broadcast or addressed to a specific unit and allows communication of pertinent vessel routing information using waypoints (e.g., mandatory or recommended routes that are not given by official publications). The information has a starting date and time and a duration to respect. As defined by IMO SN.1/Circ. 289, up to five slot messages can be created, but no more than three are recommended, resulting in a maximum of 16 waypoints for each AIS station [19,21]. This limitation was one of the reasons why in this analysis we chose standard messages over AIS-ASM, as explained below. Other specifications can be found in the following documents: IMO SN.1/Circ.289, IMO SN.1/Circ.290, IALA Guideline N° 1028, IALA Recommendation A-124 and annex, Rec. ITU-R M.1371-5.

3.2. *AIS AtoN*

AIS transmitter can also be installed on AtoN to send their position, status and other relevant information, including that provided by ASM, such as meteorological and marine conditions.

The ITU recognizes their potential, as they can work with the VTS and enhance communications in areas where they are installed; the only drawback is the excessive application cost.

New concepts of AIS AtoN have been then introduced alongside the real one:

1. Synthetic AIS AtoN are physically located at sea but not equipped with a transmitter like the real one. The AIS message, carrying its data and position, is sent by an AIS shore station instead (depending on its range);
2. Virtual AIS AtoN, on the contrary, is not physically at sea: it is only shown in terms of position and other specific information by the AIS Message.

This means that each AIS AtoN message must specify the category to which the AIS AtoN belongs. AIS AtoN can use Message 6, 8 or 14, but the AIS AtoN specific Message has the ID 21: the "Position Report", in fact, carries all the information needed to locate and recognize the AtoN and its functionalities.

The main identifier for AIS units is the Maritime Mobile Service Identity (MMSI) code; with regards to the Rec. ITU-R M.585-7, this code consists of nine digits, assigned according to the unit (aircrafts, shore stations, etc.). Ships have the following MMSI:

$$M_1 I_2 D_3 X_4 X_5 X_6 X_7 X_8 X_9$$

The first three digits are the Maritime Identification Digits (MID), which denote the country the ship is allocated to (e.g., Italy's MID is 247). The other digits range from 0 to 9.

The first two AIS AtoN MMSI digits are always 9 9 and are followed by the MID.

$$9_1 9_2 M_3 I_4 D_5 X_6 X_7 X_8 X_9$$

The sixth digit indicates the category of the AtoN as described below:

1. 99MID**1**XXX, totalling 999 between real and synthetic AIS AtoN;
2. 99MID**6**XXX, for 999 virtual AIS AtoN.

The last three digits also range from 0 to 9 [22]. It is possible to visualize the unit identified by each MMSI code by consulting the ITU Maritime Mobile Access and Retrieval System (MARS) database, daily updated [23].

3.3. *Virtual AtoN*

V-AtoN are an excellent navigation support due to their versatility: being virtual objects, they do not have to be physically placed at sea. They operate through the transmission of a coded digital message sent by authorized systems onshore (the AIS stations, for example) containing the information which is then received by the AIS onboard and displayed on the screen, usually on the ECDIS.

IALA defines the guidelines on their use and, together with the ITU and the IHO, codifies the technical specifications of the digital messages. V-AtoN approved by the competent authorities play an important role in informing seafarers about visibility conditions, hazards, safe waters, etc.

The interesting feature of this device is the dynamism of the transmitted information, which can be modified depending on the purpose: hence the definition of temporary and permanent Virtual AtoN. While permanent V-AtoN have the same validity as real AtoN, temporary V-AtoN are used to transmit information about temporarily-varying conditions and are sent only when needed. If the temporary use of Virtual AtoN exceeds six months, according to the IHO, it will be considered permanent and shown on the relevant nautical traditional chart [18].

Advantages and Disadvantages

V-AtoN messages carry the same information with the same validity as real AtoN, so they are strongly recommended in areas subject to frequent variations, heavily congested routes and critical issues, or where positioning real AtoN could be complex or dangerous.

In fact, V-AtoN can be placed anywhere, regardless of the morphology of the area and with relatively low installation and maintenance costs. Moreover, the on-screen display ensures accurate positioning, while the instantaneous and easy-to-read notifications specific to the area of interest avoids the overload of the system with unnecessary information [24].

However, one of the main issues of this technology is that V-AtoN cannot be received and displayed by all the users. The installation of the AIS in not mandatory for Non-SOLAS unit and vessels under 20 m in length, but they are encouraged to voluntarily fit and use Class B AIS.

Since it has been designed primarily to provide basic data in a cost effective, reliable and user-friendly product, Class B AIS will not be able to receive and decode the AtoN report contained in the Application Specific Messages or in the AIS Message. So Authorities must be aware that neither the units which do not voluntarily install the Class B AIS nor the units following the recommendation to mount it onboard will have the capability to receive virtual AIS AtoN.

The different symbology used to depict V-AtoN may cause confusion, although it is expected to be definitively harmonized by 2020. Furthermore, ship officers may be unaware of a V-AtoN thus ignoring their messages or perceiving it as a real AtoN. Therefore, careful consideration is required before replacing a real AtoN with a Virtual one [18].

Accordingly to the Irish Authorities, this issue is almost frequent: for this reason, they provide all the necessary details about V-AtoN and their display modality to make all the users able to recognize them on the screen. Summarizing, communications by the users is of essential importance to assess the critical points and enhance the service [8].

The dependence of AIS on systems as GNSS should not be underestimated, since it makes the AIS vulnerable to the problems connected to data transmission. In fact, it is not advisable to rely on a single source of information which may be subject to spoofing, intentional interference that misleads the receiver to track counterfeit signals [25].

4. Navigate with Virtual AtoN

Assuming that ships usually navigate in the deepest part of the canal, we plotted the safe routes from the inlets of the Lagoon to the inner areas: the coordinates of the resulting Waypoints could have been transmitted as AIS-ASM "Route Information Message", but associated restrictions made a "change of course" necessary. According to Annex A of IALA Guideline 1081, Safe Water AtoN are recommended as temporary V-AtoN when poor visibility conditions (fog, heavy rain, etc.) prevent the safe navigation of the ship. Considering the safe routes previously plotted, V-AtoN were defined in two different ways described below [24].

4.1. The Analytical Method

The first method considers the relationship between the WP coordinates and the ship manoeuvring parameters, together with the geometric route relations [26–28]. A 324-m cruise ship has been used in the simulation: according to the Safety Management System (SMS) of its company, the Rate of Turn (ROT) of the ship should not exceed 12 deg/min in confined water (7 deg/min in open water) to ensure passenger comfort. The ROT is the average angular velocity of a ship, measured in deg/min. It depends on its Turning Radius T_R and speed V, see Equation (7):

$$T_R = \frac{0.955\ V}{ROT} \tag{7}$$

Figure 4 shows a simplified representation of the manoeuvring elements of the ship. We imported the official bathymetric data (white thin lines in Figure 5) provided by the Hydrographic Institute of the Italian Navy into Q-GIS 2.14.15, an open source software, to plot the safe routes in the canals and measure the turning angles Θ thus defined (straight red lines in Figure 5). The same has been done for the turning radius T_R of each suitable turning circle corresponding to the manoeuvres, passing through deep and safe water (green circles in Figure 5).

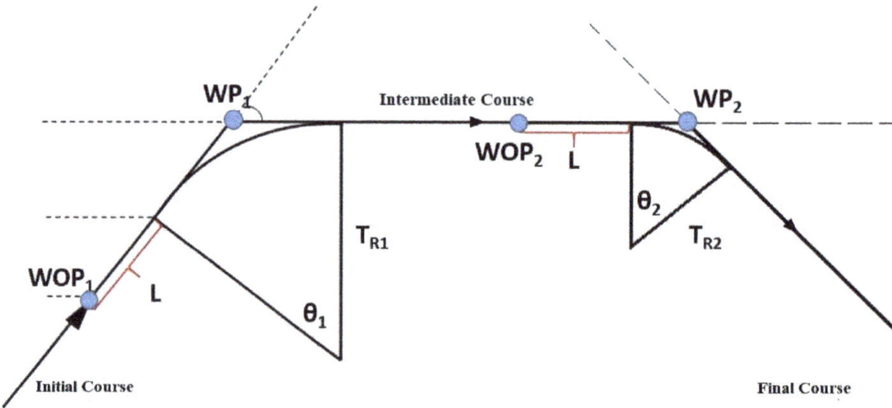

Figure 4. Simplified manoeuvring scheme.

Figure 5. Manoeuvring scheme in the Venetian Canal.

Moreover, since the manoeuvring process of a ship is not instantaneous, the distance on the route between the Wheel Over Point (WOP) and the previous WP should be considered. In correspondence of the WOP the wheel is put over, but only after a distance in general equal to the ship's length (L) the ship begins to turn. Having identified with A this second point and with C the WP (Figure 6), the segment AC can be calculated using geometric considerations as shown in Equations (8) and (9).

$$T_R\ (1 - \cos\ \Theta) = \overline{\mathrm{AC}} \sin\ \Theta; \tag{8}$$

$$\overline{\mathrm{AC}} = T_R \tan\left(\frac{\Theta}{2}\right). \tag{9}$$

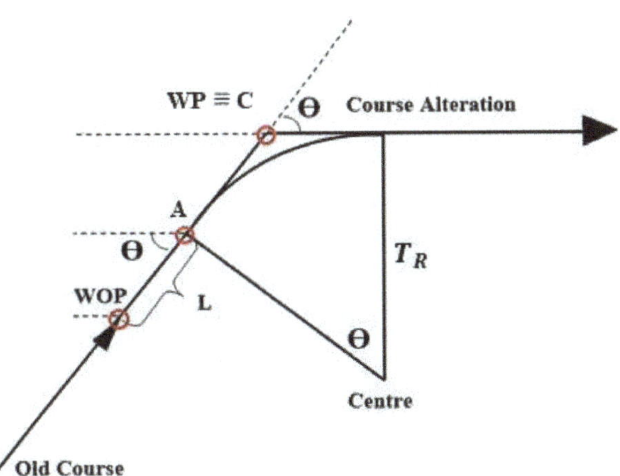

Figure 6. Simplified manoeuvring scheme.

The required distance between WOP and WP is then obtained using Equation (10):

$$\overline{\mathrm{WOP\_WP}} = \overline{\mathrm{AC}} + L. \tag{10}$$

We tested three requirements to position the V-AtoN.

The first derives from the relationship between ROT, T_R and V, see Equation (7):

1. The maximum speed limit, as stated in the Port of Venice's Ordinance N° 175/09, is 6 knots in the Lagoon, while the maximum *ROT* for the ship in question is 12 deg/min. That allows to determine the minimum T_R ensuring a safe manoeuvre.

   The second and the third requirements are related to geometrical consideration on the planned route [24]:

2. The sum of the turning radii of two successive manoeuvres (T_{R1} and T_{R2}) should not exceed the length of the leg, that is the arc of the great circle (or rhumb line) between two consecutive WP, $WP_1\_WP_2$ as in Equation (11):

$$\overline{WP_1\_WP_2} > T_{R_1} + T_{R_2}. \quad (11)$$

3. At the same time, the length of the leg $WP_1\_WP_2$ should allow the ship to safely conclude the first manoeuvre before starting the second one (Figure 7); that means the distance between two following WP ($WP_1\_WP_2$) should be greater than the distances between WOP and WP related to both the manoeuvres, see Equation (12):

$$\overline{WP_1\_WP_2} > \overline{WOP_1\_WP_1} + \overline{WOP_2\_WP_2}. \quad (12)$$

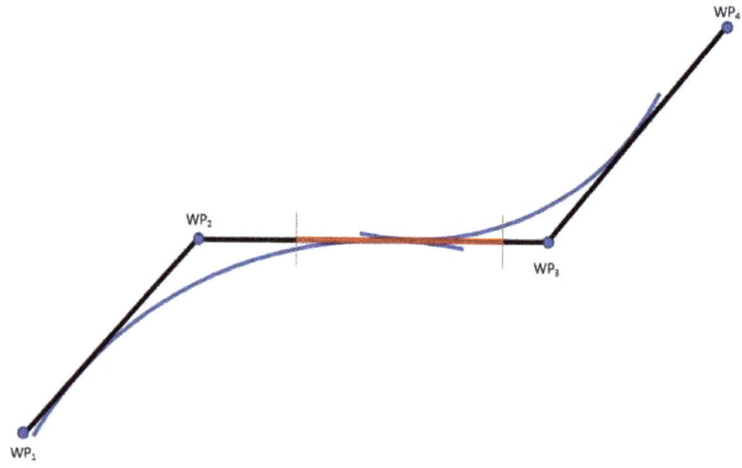

Figure 7. Simplified example of geometrical uncorrected manoeuvre.

Going step-by-step through these requirements, we iteratively established the ideal manoeuvring parameters for the simulative ship.

4.2. Numerical Approach

Before continuing the analysis, it is important to remember that the ship's manoeuvrability is significantly affected, amongst other things, by its interaction with the bottom of the waterway; the resulting turning circles will have different characteristics that should then be considered [29].

Deep and shallow water are classified according to the ratio between water depth and ship's mean draught: deep and unrestricted waters have a ratio of at least 3, while shallow waters do not exceed the same value. We set our study and perform the consequent calculus in shallow waters, since the canals in the Venetian Lagoon are not more than 15 m deep and the given ship has a mean draught of 9 m.

In the first step, we determined the minimum value of T_R in the waterways using the relationship defined in Equation (7), valid for each turning circle considering the maximum value of speed and ROT.

$$T_{R_{min}} = \frac{0.955 * 6}{12} \cong 0.48 \text{ nm}$$

However, the complex structure of the canals makes it difficult to manoeuvre with these radii, especially if considering the safety requirements previously illustrated.

An example of the problem is shown in Figure 5 where the three manoeuvres in the Giudecca area (green circles) solely evaluated on the basis of the bathymetry are very closed to each other. In this situation the largest T_R, equal to 0.30 nm, has a value that is shorter than the minimum of 0.48 nm already calculated (Table 1).

Table 1. First and last set of measures in the area of Lido.

	R	T_R	$\overline{WP_1\_WP_2}$	$\frac{R_1+R_2>}{\overline{WP_1\_WP_2}}$	θ	$\overline{AC}$	$\overline{WOP\_WP}$	$\frac{\overline{WP_1\_WP_2}>}{\overline{WOP_1\_WP_2}}$
	(deg)	(nm)	(nm)		(deg)	(nm)	(nm)	
1	301	0.27	1.61	TRUE	85	0.25	0.42	TRUE
2	216	0.39	1.28	TRUE	106	0.53	0.70	TRUE
3	322	0.33/0.27	0.75	TRUE	89	0.33/0.26	0.50/0.43	FALSE/TRUE
4	233	0.30/0.27	0.66	TRUE	56	0.16/0.14	0.33/0.31	TRUE
5	289	0.18/0.06	0.37	TRUE	17	0.03/0.01	0.20/0.18	FALSE/TRUE
6	272	0.15/0.06			34	0.05/0.02	0.22/0.19	
	306							

ROT and T_R have established values: for this reason, the only way to fulfil the criteria provided in Equation (7) is to significantly reduce the speed of the ship; see Equations (13) and (14) respectively for the first and second curve in the Figure 5.

$$V_1 = \frac{T_{R_1} * \text{ROT}}{0.955} = \frac{0.30 * 12}{0.955} \cong 3.77 \text{ kts} \tag{13}$$

$$V_2 = \frac{T_{R_2} * \text{ROT}}{0.955} = \frac{0.18 * 12}{0.955} \cong 2.26 \text{ kts} \tag{14}$$

This implies other issues: in fact, this decrease in speed is unacceptable for a cruise ship since it does not allow to manoeuvre and, in addition, it does not meet the geometrical requirements as the radii are too large.

We used Matlab R2017b (The MathWorks, Inc., Natick, MA, USA) to calculate the ideal manoeuvring parameters for the simulation in question. Starting from the known values of the legs between the WP, the first set of assumed radii and the array of Θ, our code iteratively calculated the radii which best fit the conditions: comparing two following radii, the code decreased the largest one of the 10% of its length, until both finally met the criteria. In our case, 21 iterations were required to achieve an optimal solution which combines all three of the requirements (Table 1, Figure 8).

Afterward, we studied the ideal position of the Safe Water Virtual AtoN (Figure 9). We chose three points for each circle: the first and the third on the point of tangency between the circumference and the route (yellow and red respectively), and the second at the intersection of the circle with the segment passing through its centre and the WP (blue dots in Figure 9).

What is important to notify is that all the considerations made to date are based on inaccurate theoretical conditions that do not consider the manoeuvring requirements of the ship. The resulting V-AtoN AIS network, based on fixed geometries and specific evolutionary circumferences, despite complying with the IALA and IMO safety requirements, is then not versatile. In fact, it uses data and information related to the characteristics of a single ship, so the analysed situation and the consequent results are not valid, for example, for a ship with an overall length of 100 m.

In conclusion, the Virtual AtoN thus obtained cannot be considered a reliable AtoN system for every category of ship. We moved then to a more adaptive solution.

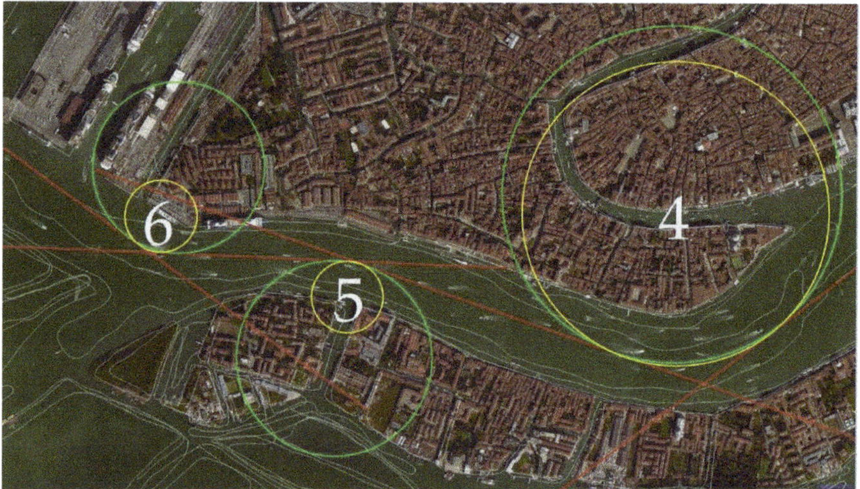

Figure 8. First (green, larger) and last (yellow, smaller) evolution circles in the Venetian Canal.

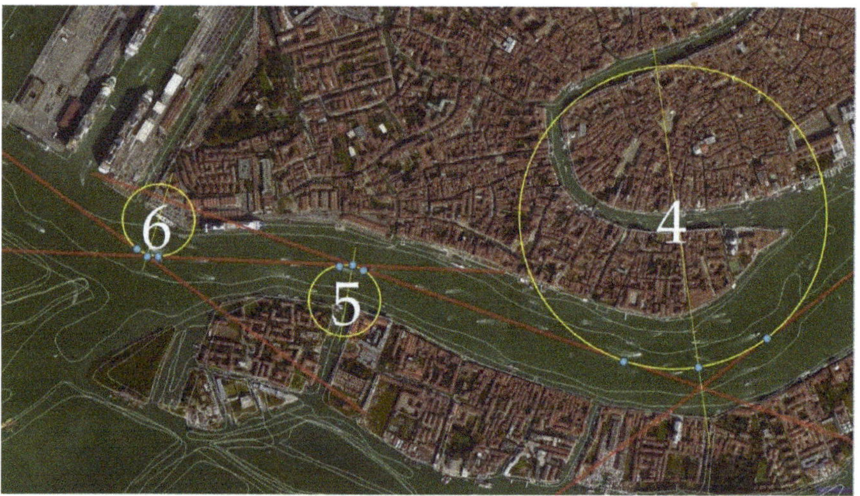

Figure 9. First proposal of Virtual AtoN in the Venetian Canal (light blue dots).

4.3. The Statistical Method

We studied the AIS tracks recorded for all the ships navigating in the areas of Lido, Malamocco and Chioggia and we got a solution that combines the simplicity of an easy-to-use service with the enhanced level of safety guaranteed also in adverse weather conditions.

As previously said, each ship sent its position through the on-board AIS. The coastal station received and stored significant amounts of data, but we selected only those of ships exceeding 50,000 tons, being then able to track their navigation in the Lagoon. We imported this information in Q-GIS (red dots in Figure 10) and estimated the most followed route from the cloud of points using the least-square method.

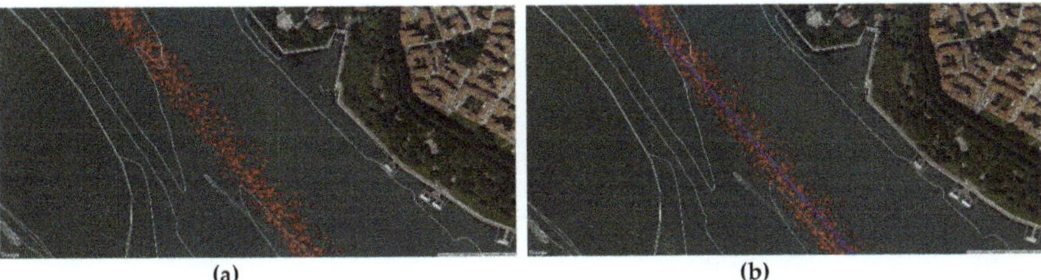

Figure 10. (**a**) Close-up of the AIS position data of ships over 50,000 tons in the Lagoon; (**b**) Resulting statistical followed route.

We positioned a V-AtoN on each resulting Waypoint (Figure 11) to define the extremes of safe navigable routes, valid even in poor visibility conditions: displayed on the on-board ECDIS and set as WP, these V-AtoN allow the system to calculate the manoeuvring parameters, taking into consideration the dynamic characteristics of the ship.

Figure 11. (**a**) Statistical followed route in the area of Venice; (**b**) Resulting Virtual AtoN.

Analysing the WP obtained by both the approaches, we found a maximum difference of 0.1 nm: this is not an excellent result, but it does not compromise the safety of the solution since all the V-AtoN are positioned in safe waters (Figure 12). Hence, they will transmit temporarily AIS Messages when needed, avoiding saturation of the transmission band, and their position will be easily modified according to the variations in width and depth of the waterways, facilitating a safe voyage planning.

Figure 12. Differences between the WP of the analytical method (light blue dots) and statistical method (white diamonds).

5. Final Application

As already stated, the conditions of poor visibility that frequently affect the Lagoon of Venice make in this case the navigation not safe enough to be conducted. For this reason, the Authorities are obliged to close the interested area and stop the port operations, resulting in unfavourable situations mainly under the economic aspect.

The establishment of a system whose operativity is guaranteed in any meteorological conditions is then the solution to this problem. The transmission of the Virtual AtoN WP as previously illustrated helps all the ships to get to the mooring point without the need for the user to worry about the visibility conditions: they will not be able to see the surrounding area, but the information provided by the AIS and displayed on the ECDIS, integrated with the classical aid (as for example the radar) will give all the necessary indications to safely reach the berth, avoiding risks of collisions or grounding. This system can be improved also in relation to autonomous ships, as they can utilise the information provided by V-AtoN (adequately integrated with meteorological and traffic data) to optimize the predictions and the consequent route decisions during navigation.

As a result, we defined fourteen V-AtoN in the Giudecca Canal, to allow vessels coming from the Lido inlet to get to the mooring (Figure 13). Similarly, nine V-AtoN have been defined to lead the ships to the Evolution Basin N° 1 in Marghera from the Malamocco inlet (Figure 13).

For the Port of Chioggia, eight V-AtoN were required to plan a safe route to the inner Basin coming from the inlet, where a Synthetic AtoN was placed (Figure 14).

To verify the integrity of the system with respect to the manoeuvring aspect, different enhanced method can be introduced. An example is the Fast-Time Manoeuvring Simulation Technology (FTS) developed at the Institute for Innovative Ship Simulation and Maritime Systems (ISSIMS): it uses a ship-dynamic simulation model to calculate and predict all the manoeuvres the ship can carry out and its motion status. In this way the officer can monitor the manoeuvring actions with the possibility to check for corrections. This new type of support is called Simulation-Augmented Manoeuvring Design and Monitoring (SAMMON) and allows the monitoring of even series of manoeuvring segments [30].

Next step will be the validation of the result. To verify its effective reliability, the Authority of Venice is transmitting the coordinates of the WP as Virtual AtoN every day. The transmission is currently limited to the daily hours to avoid harmful or dangerous situations: in this way, all the users will have the possibility to have a safe approach to this new system and give their considerations with respect to its versatility, user-friendliness and ease of interpretation and, in general, its actual contribution to the navigation during poor visibility conditions. The permanent application of the service will be gradually ultimate once feedback will be received: in fact, the most efficient way to evaluate this kind of solution is to rely on comments by the users to start an optimization process based on their needs.

The implementation of this E-Navigation system is expected to reduce the number of casualties in the Lagoon improving the safety conditions of the navigation. We did not have precise statistics relative to the traffic conditions in the area, so we could not use a model to assess the effective contribution of the system to the general improvement of the level of safety. From this perspective, further work will be done to overcome this lack of data based on the Marine Traffic Safety Diagnostic Scheme (MTSDS).

Moreover, it is important to notify that this solution has been essentially structured for SOLAS ships. However, non-SOLAS ships are more vulnerable to accidents, mainly due to the lack of navigational equipment on board leading to less safety information. For this reason, it is necessary to extend this study to all the units. The SMART-navigation approach proposed by Baldaufab and Hong could be used as starting point: it implements the IMO E-Navigation concept including services for non-SOLAS ships. The aim is to identify the units vulnerable to accidents by real time relevant statistics registered by the onshore stations and provide them with additional specific services to prevent the potential accident causes in advance by proactively supporting them. The introduction of new technologies, rules and operational procedures shall be accompanied by adequate training for the staff, so any new application needs to be carefully illustrated to the users [5]. The Virtual AtoN system proposed in this article can be improved and extended to non-SOLAS ships following this line.

Figure 13. Virtual AtoN for the areas of Lido and Malamocco.

Figure 14. S-AtoN (the first on the right) and V-AtoN in the area of Chioggia.

6. Conclusions

The analysis carried out in this paper aims to maximize the safety levels of the maritime traffic in any circumstance, in step with the E-Navigation progresses. This derives from the need to guarantee the proper conduction of the navigation even when adverse meteorological conditions affect the normal visibility, mainly during winter and autumn: in this case, the main trouble is the inability to use the classic navigational aids. This scenario is almost frequent in the Lagoon of Venice: here, the only extreme solution to prevent accidents is the closure of the ports by the predisposed Authorities.

Moreover, the installation of the MOSE at the inlets of the Lagoon and the consequent changes in the morphology of the area and in the traffic in general made it necessary an urgent assessment of the actual signaling system. We based our evaluation solely on the needs of the users since we did not have time to simulate different situations with other means. Combining this with the IALA normative and IMO requirements we defined an optimal disposition of the maritime signaling system in the Lagoon; then we checked its effective visibility. Results shown that in case of restricted visibility even this improved AtoN system results inadequate to guarantee a safe conduction of port operations. A different solution was then necessary.

The recent studies on the E-Navigation concept developed by the IMO and the consequent improvements made on these technologies suggested to deepen their potential to find a reliable and efficient solution to the problem. Studying the various IALA publications on the AIS, we focused on the concept of Virtual AIS AtoN: unlike the Real AIS AtoN, Virtual ones can be placed anywhere, without considering the morphology of the area. In addition, their characteristics can be modified to reflect both the channel and the maritime traffic changes, without high installation and maintenance costs. They can be quickly detected, thanks to the integration between the ECDIS screen and the GNSS instantaneous positioning service; lastly, they can be set as temporary AtoN by transmitting the AIS message only when required, preventing data overload with unnecessary information. Based on the statistical study of the safe routes followed by ships in the Lagoon, we could define a set of maneuvering points, the waypoints, to be sent to the AIS onboard by the predisposed stations onshore. The high number of course alterations needed to reach the mooring from the open sea makes the limit of 16 WP given by the AIS Application Specific Message not enough to use the AIS-ASM method. Therefore, we decided to virtualize and send the WP as Safe Water AtoN. We provided their coordinates to the authorities, which are responsible for authorizing AIS stations to send Message 21 (specific to the AtoN), appropriately set and coded.

In this way, the on-board AIS will receive a single message containing the AtoN information (including position, type, description), directly displayed on the ECDIS screen. The MMSI code is particularly useful as it provides a unique identifier for each AtoN.

These Virtual AIS AtoN allow the ship officer to detect the recommended route, making a safe navigation possible in any situation. Ship officers should be trained to recognize V-AtoN systems: IALA currently recommends that electronic systems must be used alongside traditional systems, rather than replacing them.

To maximize the effectiveness of this V-AtoN system, its transmission should be protected: due to its dependence on GNSS, the possibility of spoofing should be considered as it is the AIS major vulnerability. Eventually, the use of application specific messages could be focused to find a solution to the limitations on the number of AtoN to be sent.

For further improvements, three points should be analysed: the first one is the research of a scientific method to accurately investigate and assess the validity of this kind of solution and its results over the years, following methods of risk assessment. The second one regards the possibility to improve the system with respect to autonomous ships, which could be enhanced thanks to the information provided by the V-AtoN. Finally, the third is related to the availability of the service for both SOLAS and non SOLAS ships; this is of fundamental importance, since units without an AIS system onboard are currently not allowed to take advantage of this V-AtoN system, while it aims to ensure adequate safety levels also to the smallest unit.

In the meantime, the use of a Virtual AIS AtoN system thus defined represent a valid support to the navigation, being the only safe alternative to closing ports in case of restricted visibility conditions. It is important, however, that they are used in an informed way, adopting all the necessary precautionary actions.

Author Contributions: Conceptualization, F.D.C., P.M. and S.T.; methodology, F.D.C. and S.T.; software, F.D.C.; validation, F.D.C. and P.M.; formal analysis, F.D.C. and S.T.; investigation, F.D.C., P.M. and S.T.; resources, P.M. and S.T.; data curation, F.D.C.; writing—original draft preparation, F.D.C.; writing—review and editing, F.D.C., P.M. and S.T.; visualization, F.D.C.; supervision, S.T.; project administration, S.T.; funding acquisition, S.T.

Funding: This research was funded by the Port Authority System of the Northern Adriatic Sea (grant number CIGZ851E17165).

Conflicts of Interest: The authors declare no conflict of interest.

References

1. IALA. *The IALA Maritime Buoyage System*; Recommendation R1001, Ed. 1.0; IALA: St Germain en Laye, France, 2017.
2. Chen, J.; Chen, T.; Liu, D.; Shi, C.; Ying, S. Study on Navigation-aid System Optimization Based on Post-evaluation Theory. In Proceedings of the Fourth International Conference on Intelligent Computation Technology and Automation, Shenzhen, China, 28–29 March 2011.
3. IALA. *Marine Signal Lights Part 2—Calculation, Definition and Notation of Luminous Range*; Recommendation E-200-2, Ed. 1; IALA: St Germain en Laye, France, 2008.
4. IMO. E-Navigation. Available online: http://www.imo.org/en/OurWork/safety/navigation/pages/enavigation.aspx (accessed on 15 November 2017).
5. Baldauf, M.; Hong, S.B. Improving and Assessing the Impact of e-Navigation applications. *Int. J. e-Navig. Marit. Econ.* **2016**, *4*, 101016. [CrossRef]
6. Perera, L.P.; Carvalho, J.P.; Guedes Soares, C. Autonomous guidance and navigation based on the COLREGs rules and regulations of collision avoidance. In Proceedings of the International Workshop on Advanced Ship Design for Pollution Prevention, Split, Croatia, 23–24 November 2009; pp. 205–216.
7. Burmeister, H.C.; Bruhn, W.; Walther, L. Interaction of Harsh Weather Operation and Collision Avoidance in Autonomous Navigation. *Int. J. Mar. Navig. Saf. Sea Transp.* **2015**, *9*, 31–40. [CrossRef]
8. The Commissioners of Irish Lights, Rigg Bank Virtual AtoN. Available online: https://www.irishlights.ie/misc/rigg-bank-virtual-aid-to-navigation.aspx (accessed on 15 February 2019).
9. Wright, R.G.; Baldauf, M. Virtual Electronic Aids to Navigation for Remote and Ecologically Sensitive Regions. *J. Navig.* **2017**, *70*, 225–241. [CrossRef]
10. Di Ciaccio, F.; Menegazzo, P.; Troisi, S. Optimization of the Maritime Signaling in the Venetian Lagoon. In Proceedings of the 2018 IEEE International Workshop on Metrology for the Sea, Bari, Italy, 8–10 October 2018.
11. IALA. *Marine Signal Lights Part 3—Measurement*; Recommendation E-200-3, Ed. 1; IALA: St Germain en Laye, France, 2008.
12. IALA. *Marine Signal Lights Part 4—Determination and Calculation of Effective Intensity*; Recommendation E-200-4, Ed. 1; IALA: St Germain en Laye, France, 2008.
13. Park, J.S.; Park, Y.S.; Cho, I.S.; Kim, D.W. Marine Traffic Safety Diagnostic Scheme in Korea. In Proceedings of the 12th Annual General Assembly of IAMU: Green Ships, Eco Shipping, Clean Seas, Gdynia, Poland, 12–14 June 2011; Available online: http://iamu-edu.org/wp-content/uploads/2014/07/Marine-Traffic-Safety-Diagnostic-Scheme-in-Korea1.pdf (accessed on 15 February 2019).
14. Nguyen, T.X.; Park, Y.S.; Smith, M.V.; Aydogdu, V.; Jung, C.H. A Comparison of ES and PARK Maritime Traffic Risk Assessment Models in a Korean Waterway. *J. Korean Soc. Mar. Environ. Saf.* **2015**, *21*, 246–252.
15. Yeo, J.M.; Yu, Y.S.; Han, J.S.; Park, S.; Kim, J.U. Design and Establishment of Database on the AtoN Properties for AtoN Simulator. *Int. J. e-Navig. Marit. Econ.* **2015**, *3*, 123–131. [CrossRef]
16. IMO. Safety of Navigation. In *SOLAS*; IMO Publishing: London, UK, 2012; Chapter V.
17. IMO. *Adoption of New and Amended Performance Standards*; Resolution MSC.74(69); IMO Publishing: London, UK, 1998.
18. IALA. *Navguide, Aids to Navigation Manual*; Ed. 7; IALA: St Germain en Laye, France, 2014.
19. ITU-R. *Technical Characteristics for an Automatic Identification System Using Time Division Multiple Access in the VHF Maritime Mobile Frequency Band*; Recommendation ITU-R M.1371-5; ITU-R: Geneva, Switzerland, 2014.
20. IMO. *Guidance for the Presentation and Display of AIS Application-Specific Messages Information*; SN.1/Circ.290; IMO Publishing: London, UK, 2010.
21. IMO. *Guidance on the Use of AIS Application-Specific Messages*; SN.1/Circ.289; IMO Publishing: London, UK, 2010.
22. ITU-R. *Assignment and Use of Identities in the Maritime Mobile Service*; Recommendation ITU-R M.585-7; ITU-R: Geneva, Switzerland, 2015.

23. ITU. Maritime Mobile Access and Retrieval System (MARS). Available online: https://www.itu.int/en/ITU-R/terrestrial/mars/Pages/default.aspx (accessed on 11 January 2018).
24. IALA. *Provision of Virtual Aids to Navigation*; Guideline 1081, Ed. 1.1; IALA: St Germain en Laye, France, 2013.
25. IALA. *Provision of Virtual Aids to Navigation*; Recommendation O-143, Ed. 1.1; IALA: St Germain en Laye, France, 2013.
26. Norris, A. Integrated Bridge Systems. In *ECDIS and Positioning*; The Nautical Institute: London, UK, 2010; Volume 2.
27. Aarsæther, K.G.; Moan, T. Combined Maneuvering Analysis, AIS and Full-Mission Simulation. *Int. J. Mar. Navig. Saf. Sea Transp.* **2007**, *1*, 31–36.
28. Becker-Heins, R. *Voyage Planning with ECDIS, Practical Guide for Navigators*; Geomares Publishing: Lemmer, The Netherlands, 2016.
29. Maimun, A.; Priyanto, A.; Muhammad, A.H.; Scully, C.C.; Awal, Z.I. Manoeuvring prediction of pusher barge in deep and shallow water. *Ocean Eng.* **2011**, *38*, 1291–1299. [CrossRef]
30. Benedict, K.; Kirchhoff, M.; Gluch, M.; Fischer, S.; Schaub, M.; Baldauf, M. Simulation-Augmented Methods for Safe and Efficient Manoeuvres in Harbour Areas. *Int. J. Mar. Navig. Saf. Sea Transp.* **2016**, *10*, 193–201. [CrossRef]

Article

Signal in Space Error and Ephemeris Validity Time Evaluation of Milena and Doresa Galileo Satellites †

Umberto Robustelli *, Guido Benassai and Giovanni Pugliano

Department of Engineering, Parthenope University of Naples, 80143 Napoli, Italy; guido.benassai@uniparthenope.it (G.B.); giovanni.pugliano@uniparthenope.it (G.P.)

* Correspondence: umberto.robustelli@uniparthenope.it

† This paper is an extended version of our paper published in the 2018 IEEE International Workshop on Metrology for the Sea proceedings entitled "Accuracy evaluation of Doresa and Milena Galileo satellites broadcast ephemeris".

Received: 14 March 2019; Accepted: 11 April 2019; Published: 14 April 2019

Abstract: In August 2016, Milena (E14) and Doresa (E18) satellites started to broadcast ephemeris in navigation message for testing purposes. If these satellites could be used, an improvement in the position accuracy would be achieved. A small error in the ephemeris would impact the accuracy of positioning up to ±2.5 m, thus orbit error must be assessed. The ephemeris quality was evaluated by calculating the $SISE_{orbit}$ (in orbit Signal In Space Error) using six different ephemeris validity time thresholds (14,400 s, 10,800 s, 7200 s, 3600 s, 1800 s, and 900 s). Two different periods of 2018 were analyzed by using IGS products: DOYs 52–71 and DOYs 172–191. For the first period, two different types of ephemeris were used: those received in IGS YEL2 station and the BRDM ones. Milena (E14) and Doresa (E18) satellites show a higher $SISE_{orbit}$ than the others. If validity time is reduced, the $SISE_{orbit}$ RMS of Milena (E14) and Doresa (E18) greatly decreases differently from the other satellites, for which the improvement, although present, is small. Milena (E14) and Doresa (E18) reach a $SISE_{orbit}$ RMS of about 1 m (comparable to that of the other Galileo satellites reach with the nominal validity time) when validity time of 1800 s is used. Therefore, using this threshold, the two satellites could be used to improve single point positioning accuracy.

Keywords: Galileo; Doresa; Milena; broadcast ephemeris; precise ephemeris; sp3; SISE Signal In Space Error; broadcast ephemeris validity time

1. Introduction

On 15 December 2016, with 18 satellites in orbit, Europe's satellite navigation system Galileo was declared operational and started offering its initial services to public authorities, businesses and citizens. When Galileo is fully operational, the constellation will consist of 24 satellites plus spares in Medium Earth Orbit (MEO) at an altitude of 23,222 km. Eight active satellites will occupy each of three orbital planes inclined at an angle of 56° to the equator. The satellites will be spread evenly around each plane and will take about 14 h to orbit the Earth. Two further satellites in each plane will be a spare on standby should any operational satellite fail.

Currently, (February 2019), Galileo is in its full operational capability (FOC) phase; 22 FOC satellites were launched up to the start of 2019 in addition to the four IOV launched between 2011 and 2012. The navigation signals of these satellites are transmitted on five frequencies: E1, E5a, E5b, E5 and E6. Only three IOV satellites are operational because E20 has been declared unavailable since 27 May 2014 [1] when a power anomaly led E5 and E6 signals to a permanent loss of power. After this failure, all IOV satellites are backed-off and thus their signals have less transmitted power than the FOC satellites. Nineteen FOC satellites are declared operational, one (E22) was removed from active service in December 2017 for constellation management purposes, while the first two FOC satellites, Milena

(E14) and Doresa (E18), result under test. These two satellites were launched on 22 August 2014 at 09:27 local time in French Guiana by a Soyuz ST rocket. The two satellites were left in a non-nominal highly elliptical orbit characterized by an apogee of 25,900 km and a perigee of 13,713 km, with an inclination with respect to the equator of 49.69° instead of planned circular medium-Earth orbits at an altitude of 23,222 km with an inclination of 55.04°. The wrong injection orbits made the two satellites not usable for navigation mission, thus ESA planned a salvaged mission to make the two satellites usable [2,3]. From November 2014 to February 2015, the satellites made a series of maneuvers to raise the low point of their orbits by 3500 km and make their orbits more circular. Satellites in new orbits overfly the same location on the ground every 20 days. This is different from the nominal Galileo repeat pattern of 10 days, but makes possible a synchronization of theirs ground track with the rest of the Galileo constellation. Thus, the revised orbit (reported in Table 1) allowed ESA to switched on satellite's navigation payload. On 5 August 2016, beginning 00:00 UTC, GSAT0201 named Doresa (E18) and GSAT0202 named Milena (E14) started broadcasting navigation message for testing purposes. (For details, see notice advisory to Galileo users https://www.gsc-europa.eu/notice-advisory-to-galileo-users-nagu-2016030).

Table 1. Galileo satellite nominal orbit parameter.

Satellite	Semi-Major Axis (Km)	Eccentricity (Km)	Inclination (deg)
IOV and FOC	29,599.8	0.000	56.0°
E14 and E18	27,977.6	0.162	49.850°

At the end of the mission, the European Space Agency on 8 October 2016 requested to provide feedback on usage of these satellites (http://galileognss.eu/tag/sat-6/). After the ESA request, Sosnica et al. [4] in 2016 analyzed the accuracy of the Galileo system orbits including Milena (E14) and Doresa (E18) satellites, focusing on the precise orbits. Giorgi et al. [5] tested the general relativity studying the relativistic offset (redshift and Doppler shift) by using data provided by E18 Galileo satellite signals. In 2018, Paziewski et al. [6] investigated the potential use of Milena (E14) and Doresa (E18) satellites for positioning purposes. They evaluated the applicability of the Milena (E14) and Doresa (E18) satellites to precise GNSS positioning focusing on relative kinematic positioning as an instantaneous solution in multi-baseline mode. Robustelli and Pugliano [7] analyzed the code multipath error of GNSS satellites by using the short time Fourier transform and Wavelet analysis [8], showing that multipath performance of Milena (E14) and Doresa (E18) satellites are the same as the other Galileo FOC satellites when a mobile phone is used as receiver [9]. In 2018, Nicolini and Caporali [10] analyzed Milena (E14) and Doresa (E18) satellites' orbits during Week 1950 from 21 May 2017 to 27 May 2017. The Galileo system is constantly evolving and the dataset analyzed by Nicolini and Caporali does not fully cover the ground track of the satellites, thus a re-evaluation is justified. Moreover, they considered a period of validity of the Galileo ephemeris equal to 1 h, less than that fixed by the ESA of 4 h.

GNSS errors can be divided into errors that can be corrected by using differential techniques or not. The multipath error belongs to first group [11] while the ephemeris error to the second. GNSS satellites travel in precise and well known orbits. Unfortunately, the orbits have slight variations that lead to significant errors in the calculated positions. For these reasons, the GNSS ground control system continually monitors the satellite orbit. In the event that the orbit of a satellite changes, the ground control system sends a message to the satellite through which the content of the broadcast ephemeris is updated. However, even with the corrections from the GNSS ground control system, there are still small errors in the orbit that can result in up to ± 2.5 m of position error [12]. The ephemeris errors are the differences between the true satellite position and the position computed using the GNSS navigation message. The radial residual satellite position error (dr) is a vector that is depicted in Figure 1 where ρ represents the pseudorange measurement and P the point where receiver is located. The typical magnitudes of ephemeris error is in the range of 1–6 m [13]. To make the Milena (E14) and Doresa (E18) satellites usable, it is necessary that the broadcast ephemeris must have an accuracy comparable to that of the other Galileo satellites. Actually, it is very important to have two additional satellites available since currently Galileo system is not yet fully operational and the accuracy in the position increases as the number of visible satellites increases. We are focusing our attention on the broadcast ephemeris because they are the only ones that can be used in real time for single point positioning, a lack of accuracy on them will directly impact the accuracy of Galileo positioning. For this reason, the evaluation of the accuracy of the ephemeris broadcasted by the Milena (E14) and Doresa (E18) satellites is of fundamental importance and is the main purpose of this paper. In a first analysis conducted by the authors [14], the position error between the broadcast orbit and the IGS precise orbit was determined for each second for the period starting from 21 February (DOY 52) to 12 March (DOY 71) of 2018 by using ephemeris decoded by receiver located YEL2 station. The analysis showed that the orbits of the satellites Milena (E14) and Doresa (E18) are affected by an error comparable to that of the other Galileo satellites with the exception of DOY 62 during which two anomalies were found. The two anomalies were located in the time range between 528,000 and 531,000 s and between 556,000 and 559,000 s for Milena (E14) and Doresa (E18) satellites, respectively.

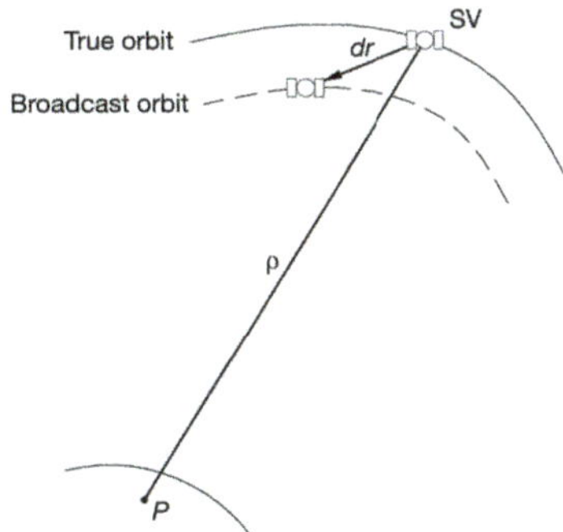

Figure 1. Ephemeris error: dr is the radial residual satellite position error, ρ is the pseudorange measurement and P is the point where receiver is located.

This study investigated these anomalies by comparing the position of each Galileo satellite obtained by MGEX multi-GNSS broadcast ephemeris (BRDM) with precise orbit every second for the above mentioned dataset and for an additional twenty-day dataset from 21 June 2018 (DOY 172) to 10 July 2018 (DOY 191). The BRDM ephemeris are different from received one: they are generated by Technische Universitat Munchen (TUM) and DLR by merging real-time streams of a number of selected MGEX stations. For the first period (DOY 52–71), two different types of transmitted ephemerides were used: those received by the receiver placed in IGS YEL2 station located in Canada and the BRDM ones. During the first period, 16 satellites were working while during the second there were 19 satellites.

2. Materials and Methods

2.1. Methodology

Galileo broadcast ephemeris contains Keplerian elements to describe the motion of satellite. Galileo system broadcast four different navigation messages: free accessible navigation (F/NAV) provided by E5a-I signal for open service, integrity navigation (I/NAV) provided by E5b and E1-B signal corresponding to both open and commercial service, Commercial Navigation Message C/NAV provided by E6-B signal supporting commercial service and the Governmental Navigation Message (G/NAV) provide by E1A and E6A signals. Currently, only F/NAV and I/NAV are available since the G/NAV navigation message does not belong to the public domain and the C/NAV is not yet defined.

Since the Keplerian parameters transmitted in the two messages are the same, in this work, we indifferently used F/NAV or I/NAV messages paying attention to use the one that has a minor age.

The Signal In Space Error (SISE) is considered a key performance indicator of all navigation systems. It provides the instantaneous difference between the Galileo satellite position/clock offset as obtained from the broadcast Navigation message and the "true" satellite position/clock offset. It can be used to assess the quality of clock and ephemeris contained in the navigation message transmitted by Galileo satellites. It is a function of time and user location within the satellite coverage area. SISE is defined as the difference of the satellite position and time as broadcast by the navigation message and the true satellite position and time, projected on the user-satellite direction [15]. It can be computed by comparing the predicted satellite position and time, based on the broadcast navigation message, with a posteriori precise clock and orbit estimations. Its analytic expression can be derived from the Galileo satellite's orbit error components expressed in along-track (A), cross-track (C) and radial (R) frame and its total Signal In Space (SIS) clock prediction error (CLK) by means of the following formula.

$$SISE = \sqrt{0.9673 \cdot R^2 + CLK^2 + 0.01632 \cdot (A^2 + C^2) + 1.967 \cdot CLK \cdot R} \quad (1)$$

The ACR coordinate system is depicted in Figure 2. Its origin is located at the centroid of the space object. The radial axis always points from the Earth's center along the radius vector toward the satellite as it moves through the orbit; the along-track axis (or transverse) points in the direction of (but not necessarily parallel to) the velocity vector and is perpendicular to the radius vector; the cross-track axis is normal to the plane identified by velocity and radius vectors [16].

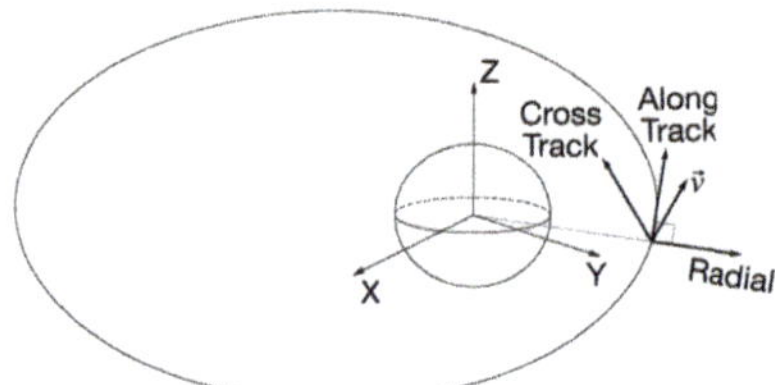

Figure 2. Along-track (A), cross-track (C) and radial (R).

We evaluated the accuracy of Milena (E14) and Doresa (E18) Galileo satellites broadcasted ephemeris by using the so-called in orbit SISE ($SISE_{orbit}$). It is derived by SISE without considering the clock error thus considering CLK= 0 in Equation (1). $SISE_{orbit}$ can be expressed as follows:

$$SISE_{orbit} = \sqrt{0.9673 \cdot R^2 + 0.01632 \cdot (A^2 + C^2)} \quad (2)$$

To calculate $SISE_{orbit}$, we used as reference the IGS precise orbits. They are generally more accurate than the broadcast orbits by almost two orders of magnitude [17].

Thus, we assumed that the precise orbits could be considered the truth and any difference between the two is attributed to broadcast orbit error.

The first step was to determine the position and velocity of the Galileo satellites at a certain time instant by the propagation of the Keplerian parameters contained in broadcast ephemeris using the algorithm described in Galileo Navigation Signal in Space Interface Control Document (ICD) [18]. At the end of this stage, we had satellites position and velocity expressed in the Earth Centered Earth Fixed (ECEF) frame. As recommended in European GNSS (Galileo) quarterly performance report [19], we performed a quality check on transmitted messages discarding parameters relative to satellites whose Signal Health Status (SHS) bits are set to 01 or to 10 (see Table 2) or whose age is beyond ephemeris validity time (VT) that is the maximum usability period of broadcasted navigation parameters. It should be emphasized that we did not discard the satellites with the SHS parameter set to "currently under test; because for the Milena (E14) and Doresa (E18) satellites SHS were set to "Test" and the Data Validity Status (DVS) flags to WWG (working without guarantee) since 8 October 2016 (https://galileognss.eu/galileo-satellites-in-eliptical-orbit-broadcasting-navigation-messages/).

Table 2. Signal status bits definitions [18].

Signal Health Status Dec	Signal Health Status Bits	Definition
0	00	Signal OK
1	01	Signal out of service
2	10	Signal will be out of service
3	11	Signal Component currently in Test

Currently, VT is 4 h as reported in [19,20]. The age of ephemeris were estimated according to methodology described in Annex B "estimating the age of ephemeris" [20] as follow:

$$t_{aoe} = t_{tom} - t_{oe} \tag{3}$$

where t_{aoe} is the age of ephemeris, t_{tom} is the transmission time of the navigation message and t_{oe} is the time when ephemeris parameters had been calculated. To determine if the navigation parameters can be used, the value of t_{aoe} must satisfy the following inequality:

$$0 \leq t_{aoe} \leq VT \tag{4}$$

Thus, the ephemerides that do not respect the inequality in Equation (4) were discarded. The next step consisted in the determination of the position and the velocity of satellites for each epoch starting from precise orbit parameters. The sp3 format stores the ECEF coordinates and velocities of all satellites every 900 s in the IGS14 frame, a frame adopted by IGS in order to align IGS products to International Terrestrial Reference Frame 2014 (ITRF2014) frame [21].

The satellites coordinates, at epoch t were obtained by interpolating parameters contained in sp3 format with a Lagrange polynomial function of 16th order in the time window [t − 2h, t + 2h] contained in the precise ephemeris sp3 files.

The satellite position in the broadcast ephemeris is referenced to the GTRF (Galileo Terrestrial Reference Frame) while the precise ephemeris is provided in the IGS14 frame aligned with the ITRF2014. According to Cai et al. [22], the difference in the two coordinate systems is only about 1–3 cm, thus the Galileo broadcast reference frame can be considered aligned with ITRF2014 as already done Nicolini and Caporali [10].

It is important to underline that the broadcast orbits are determined with respect to the satellite's antenna phase center while the precise orbits are relative to satellite's center of mass. Thus, the coordinates obtained by broadcasted ephemeris must be referred to the center of mass [23] using the correction values calculated in [24] and reported in Table 3.

Table 3. Antenna offsets for CoM correction of broadcast ephemeris [24].

Satellite Block	X(m)	Y(m)	Z(m)
IOV	+0.20	0.00	+0.60
FOC	−0.15	0.00	+1.00

Now, the difference between the satellites positions obtained starting from broadcast and precise orbits can be calculated and expressed in ECEF frame as:

$$\Delta P = \begin{bmatrix} X^{brdc} - X^{pre} \\ Y^{brdc} - Y^{pre} \\ Z^{brdc} - Z^{pre} \end{bmatrix} \tag{5}$$

$$\Delta V = \begin{bmatrix} V_X^{brdc} - V_X^{pre} \\ V_Y^{brdc} - V_Y^{pre} \\ V_Z^{brdc} - V_Z^{pre} \end{bmatrix} \tag{6}$$

where the brdc and pre superscripts indicate that the coordinates and velocities were obtained starting from the transmitted and precise ephemeris, respectively.

The errors reported in Equations (5) and (6) (referred to a rotating system) are the same as in an inertial system since the inertial and rotating coordinates differ for an additive term, thus, after calculating the unit vectors of the three axes of ACR coordinate system in ECI frame according to these relations:

$$\hat{R}_{ECI} = \frac{\vec{r}_{sat}^{ECI}}{\|\vec{r}_{sat}^{ECI}\|}; \quad \hat{C}_{ECI} = \frac{\vec{r}_{sat}^{ECI} \times \vec{v}_{sat}^{ECI}}{\|\vec{r}_{sat}^{ECI} \times \vec{v}_{sat}^{ECI}\|}; \quad \hat{A}_{ECI} = \hat{C}_{ECI} \times \hat{R}_{ECI}; \tag{7}$$

we can express ΔP in ACR frame by using:

$$\begin{bmatrix} \Delta P_{radial} \\ \Delta P_{along} \\ \Delta P_{cross} \end{bmatrix} = \begin{bmatrix} \hat{R}_X^{ECI} & \hat{R}_Y^{ECI} & \hat{R}_Z^{ECI} \\ \hat{A}_X^{ECI} & \hat{A}_Y^{ECI} & \hat{A}_Z^{ECI} \\ \hat{C}_X^{ECI} & \hat{C}_Y^{ECI} & \hat{C}_Z^{ECI} \end{bmatrix} \begin{bmatrix} \Delta P_X^{ECI} \\ \Delta P_Y^{ECI} \\ \Delta P_Z^{ECI} \end{bmatrix} \tag{8}$$

where ΔP_X^{ECI}, ΔP_Y^{ECI} and ΔP_Z^{ECI} are the difference between the satellites coordinates obtained from the transmitted ephemeris and those obtained from the precise ephemeris along the X, Y and Z axes of the ECI frame, respectively. We developed a suitable software tool in MATLAB® environment able to compute all Galileo satellites orbit error in ACR frame and relative $SISE_{orbit}$. The errors and $SISE_{orbit}$ were determined every seconds for the entire study using six different ephemeris validity time thresholds: 14,400 s, 10,800 s, 7200 s, 3600 s, 1800 s and 900 s.

2.2. Experimental Setup

The experiment was carried out using data produced by International GNSS Service (IGS). It is a voluntary association made up of universities, space agencies and geodetic agencies whose mission is "to provide the highest-quality GNSS data and products in support of the terrestrial reference frame, Earth rotation, Earth observation and research, positioning, navigation and timing and other applications that benefit society" [25]. IGS products are generated at several analysis centers (ACs) with strict quality control and proper weighting, thus having the highest quality and internal consistency. Due to the growth of GNSS constellations with the advent of the European Galileo, the Chinese Beidou, the Japanese QZSS and the Indian IRNSS, in 2012 IGS started the Multi-GNSS Experiment (MGEX) whose objective is to promote the generation of dedicated multi-GNSS precise orbit and clock products and the development of advanced processing algorithms [26,27]. The ground track of the Galileo FOC satellites Milena (E14) and Doresa (E18) repeats every 20 sidereal days, and therefore Milena (E14) and Doresa (E18) satellites do not reach the whole range of elevations during one single day. Three different datasets were analyzed.

The first dataset consists of the ephemeris decoded by receiver placed in YEL2 IGS station in the days from 21 February 2018 (DOY 52) to 12 March 2018 (DOY 71). It is related only to satellites that are visible in that location, however the elaborations related to the this dataset (already discussed in [14]) have been redone in order to calculate the orbit error in the ACR reference system. The second dataset is composed by the BRDM ephemeris relative to the same time range as the first one. The third dataset consists of BRDM ephemeris ranging from 21 June 2018 (DOY 172) to 10 July (DOY 191) 2018.

Precise orbits were obtained from the IGS website (ftp://igs.ensg.ign.fr/pub/igs/products/mgex/) that maintains precise orbit records from 1992 through to the present. Data are stored in

SP3 (Standard Product 3) version c format as compressed (zip) files. It is an ASCII file that contains information about the precise orbital data and the associated satellite clock corrections.

Broadcast ephemeris were obtained from the Federal Agency for Cartography and Geodesy (BKG) GNSS data center website (https://igs.bkg.bund.de/dataandproducts/rinexsearch) referred to YEL2 station located in Canada. It is a regional IGS data center that collects observational and navigational data from several operational centers and stations, maintains a local archive of the data received and provides online access to these data to the user community.

The merged, multi-GNSS broadcast ephemeris containing all the unique broadcast navigation messages for the day were downloaded from the MGEX campaign archive at CDDIS web-site ftp://ftp.cddis.eosdis.nasa.gov/gnss/data/campaign/mgex/daily/rinex3.

3. Results

The results relating to the first dataset have already been discussed in [14], and constitute a subset of the results obtained by processing the second dataset. Therefore, in this section, we report only the results obtained by analyzing the second and third datasets.

Figure 3 reports Root-Mean-Square value of $SISE_{orbit}$ calculated using an ephemeris validity time of 14,400 s (4 h) for each satellite. RMS has been calculated over time range of 20 days (DOYs 52–71). Milena (E14) and Doresa (E18) satellites (represented in figure in blue and green respectively) show a RMS of 3.56 and 1.88 m, respectively, which is four and two times the RMS shown by the other satellites (about 0.90 m).

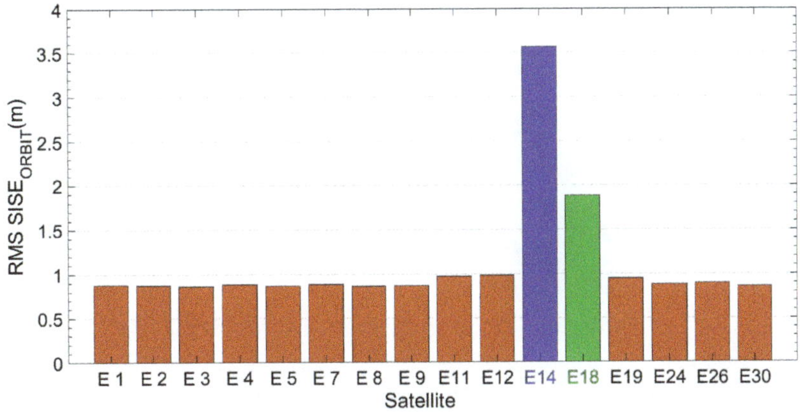

Figure 3. RMS $SISE_{orbit}$ calculated on the dataset ranging from DOY 52 to 71 for all Galileo satellites. Blue and green are Milena (E14) and Doresa (E18) satellites, respectively. Ephemeris validity time threshold is set to 14,400 s.

The time evolution of the $SISE_{orbit}$, radial, cross and along components of the two satellites are analyzed and compared with that of a satellite (E1) used as representative of all the others.

Figure 4 shows the evolution of the $SISE_{orbit}$ calculated with a period of validity of the ephemeris of 14,400 s (4 h) with respect to time for the satellites E1, Milena (E14) and Doresa (E18) reported in Figure 4a–c, respectively. Since 20 days of data are represented, it was not possible to use the Galileo epoch as univocal epoch because it is reset every new week. Thus, we used an epoch obtained by adding to its representation in Galileo time a number of seconds that takes into account the change of week. The E1 satellite was chosen as representative of all other Galileo satellites because they show very similar behaviour towards the orbit error. Figure 5 reports the radial, cross and along error components.

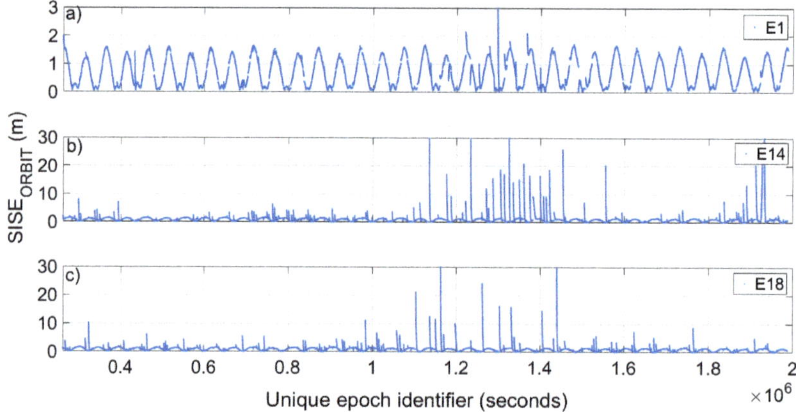

Figure 4. $SISE_{orbit}$ versus time obtained with an ephemeris validity time of 14,400 s (4 h). Satellite 1 (E1), Milena (E14) and Doresa (E18) are plotted in (**a–c**), respectively. To have a better representation, in (**b**,**c**), the maximum value of the $SISE_{orbit}$ shown is 30 m.

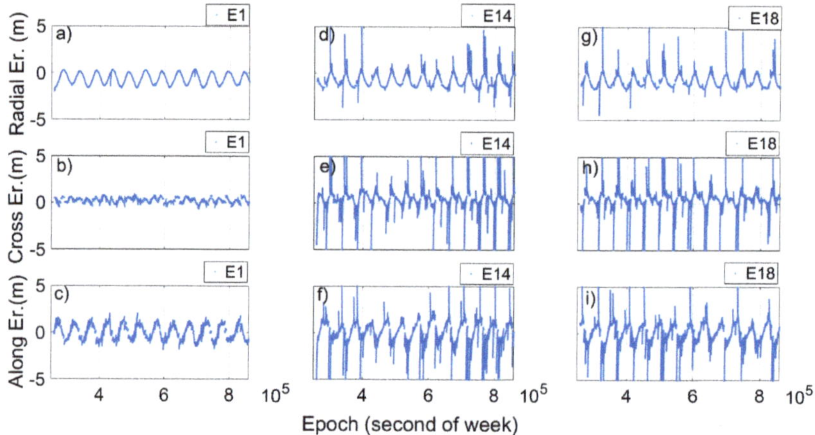

Figure 5. Radial, cross and along error for satellite 1 (E1), Milena (E14) and Doresa (E18) satellites plotted versus time. The ephemeris validity time is 14,400 s (4 h). To have a better readability for (**d–i**), the maximum value of errors is limited to 5 m.

Looking at Figures 4 and 5, two things are evident. The first is the extreme variation shown by $SISE_{orbit}$ in the case of the Milena (E14) and Doresa (E18) satellites; the second is related to the clear periodicity that the $SISE_{orbit}$ and error components show for all three satellites. Actually, Figure 4b,c shows a $SISE_{orbit}$ up to 30 m, well below the real value, i.e. 236.1 and 49.5 m, respectively. Figure 5d–i shows zoomed in images to achieve a better readability. The maximum errors obtained are 7.8, 32.76, 11.34, 9.53, 29.08, and 12.24 m in Figure 5d–i, respectively.

Figure 4 shows how the $SISE_{orbit}$ varies with time. In particular, it can be noted that in some days it is very high while in others it is lower; therefore, instead of calculating an RMS of the $SISE_{orbit}$ on the twenty days, a daily RMS was calculated, as reported in Figure 6. The figure shows clearly how the Milena (E14) and Doresa (E18) satellites have a higher $SISE_{orbit}$ compared to the other satellites, with very high discrepancies on DOYs 61–65 and 71.

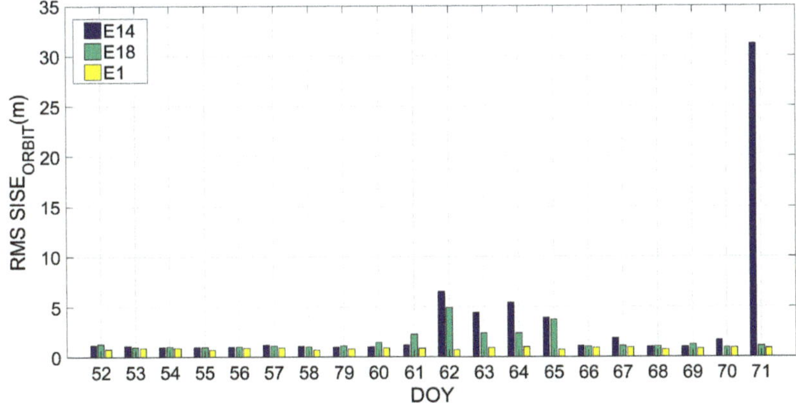

Figure 6. Daily RMS $SISE_{orbit}$ calculated on the dataset ranging from DOY 52 to 71 for satellite 1 (E1), Milena (E14) and Doresa (E18) satellites reported in yellow, blue and green, respectively.

The same analysis was conducted on the third dataset. Results are reported in the Figures 7–9. Figure 7 depicts the RMS value of $SISE_{orbit}$ calculated using an ephemeris validity time of 14,400 s (4 h) for each satellite. Figure 8 shows the evolution of the $SISE_{orbit}$ with respect to time for the satellites E1, Milena (E14) and Doresa (E18) reported in Figure 8a–c, respectively, while Figure 9 depicts the Daily RMS $SISE_{orbit}$ for satellite Milena (E14) and Doresa (E18) compared with satellite E1.

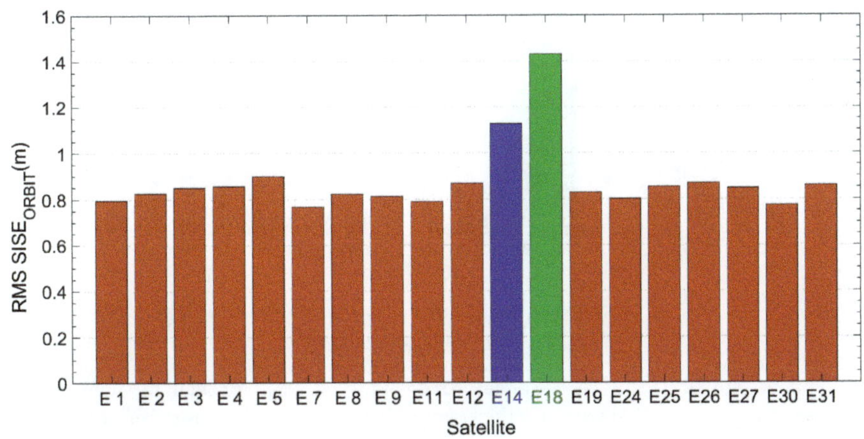

Figure 7. RMS $SISE_{orbit}$ calculated on the dataset of DOY 172–191 for all Galileo satellites. Blue and green are Milena (E14) and Doresa (E18) satellites, respectively. Ephemeris validity time threshold is set to 14,400 s.

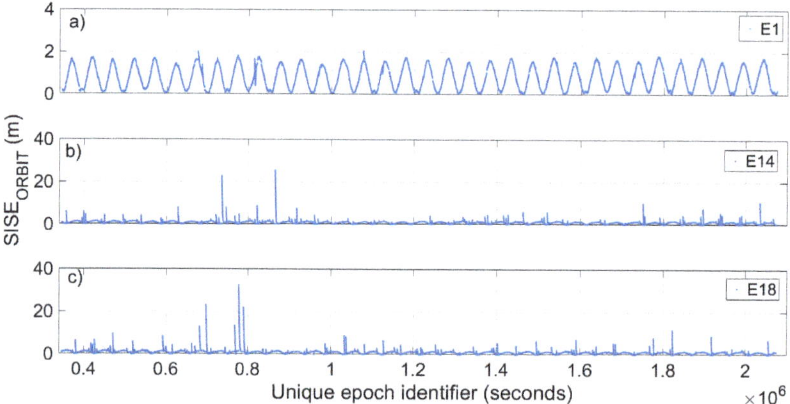

Figure 8. $SISE_{orbit}$ versus time obtained with an ephemeris validity time of 14,400 s (4 h). Satellite 1 (E1), Milena (E14) and Doresa (E18) are plotted in (**a**–**c**), respectively. To have a better representation, in (**b**,**c**), the maximum value of the $SISE_{orbit}$ shown is 3 m.

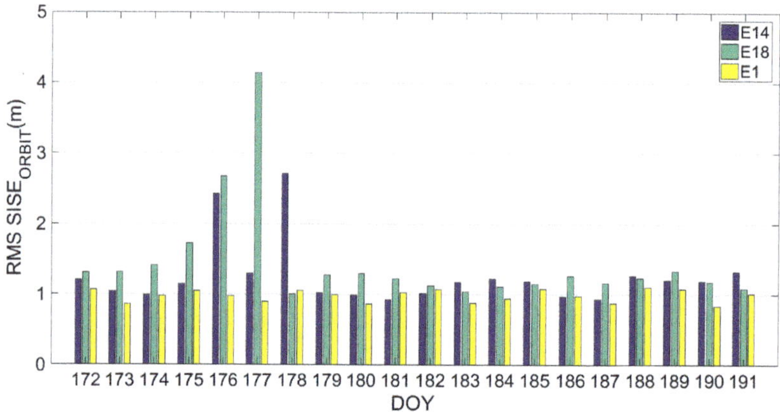

Figure 9. Daily RMS $SISE_{orbit}$ calculated on the dataset of DOY 172–191 for satellite 1 (E1), Milena (E14) and Doresa (E18) satellites reported in yellow, blue and green respectively.

If we compare Figures 3 and 7, we see an improvement in $SISE_{orbit}$: for satellite Milena (E14) the RMS is reduced from 3.56 to 1.12 m while for the Doresa (E18) satellite it decreases from 1.88 to 1.43 m. This improvement is also confirmed by what is shown in Figures 8 and 9, nevertheless the $SISE_{orbit}$ for the two satellites under analysis continues to be higher than that of the other Galileo satellites. Thus, this behaviour cannot have been caused by an anomaly because both analyzed datasets highlight it.

The validity time is a parameter that can be changed. Its reduction means using Keplerian parameter closer to the reference epoch and therefore the orbit error described by $SISE_{orbit}$ decreases. However, the reduction of the validity time has a negative effect, as it will increase the number of epochs for which the parameters of the ephemeris will be too old, thus the satellite to which those parameters refer will be discarded in the navigation solution.

Figures 10 and 11 show the daily RMS $SISE_{orbit}$ calculated by using a validity time of 7200 s, 3600 s, 1800 s and 900 s in Figures 10a–d and 11a–d, respectively, for the two different datasets. Milena (E14) satellite is represented by the blue bar, Doresa (E18) by the green one and satellite 1 (E1) represented by the yellow one. In the figures, it is evident how setting the ephemeris validity time equal to 1800 allows obtaining a $SISE_{orbit}$ comparable with that of the other satellites.

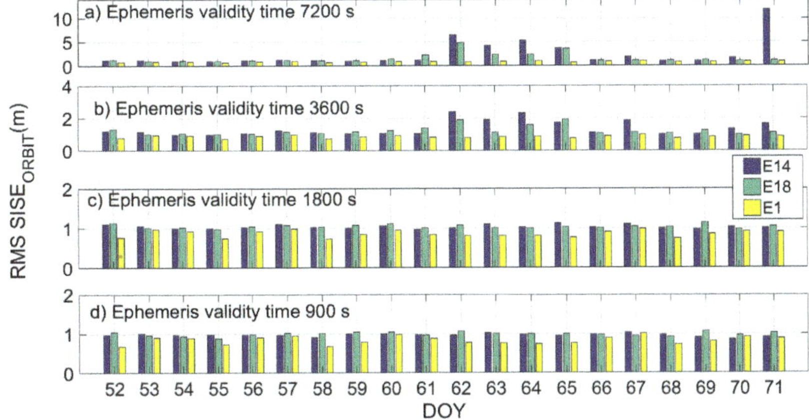

Figure 10. Daily RMS $SISE_{orbit}$ for satellite Milena (E14) and Doresa (E18) and satellite 1 (E1) represented by blue, green and yellow bars, respectively. $SISE_{orbit}$ was calculated using a validity time of 7200 s, 3600 s, 1800 s and 900 s in (**a**–**d**), respectively.

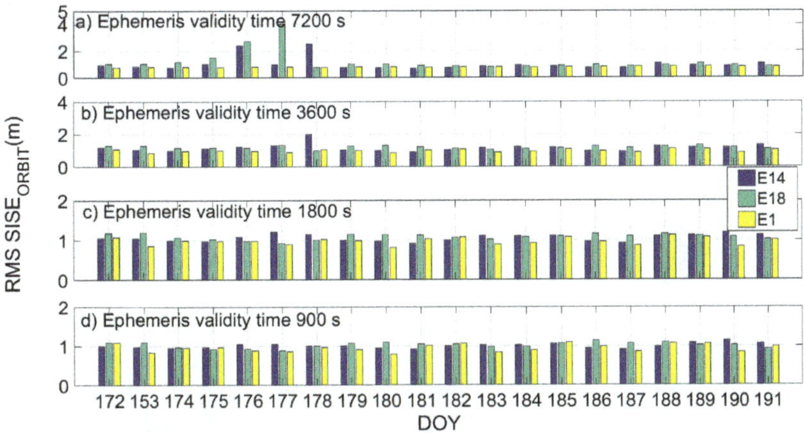

Figure 11. Daily RMS $SISE_{orbit}$ for satellite Milena (E14) and Doresa (E18) and satellite 1 (E1) represented by blue, green and yellow bars, respectively. $SISE_{orbit}$ was calculated by using a validity time of 7200 s, 3600 s, 1800 s and 900 s in (**a**–**d**), respectively. Figure refers to DOYs 172–191.

The reduction in validity time has a positive effect on the Milena (E14) and Doresa (E18) satellites; observing the days with the higher RMS (DOYs 61–65, 71, and 175–178), we can see how this value decreases when the validity time decreases. This effect does not occur on the other satellites for which the improvement of the $SISE_{orbit}$ with the diminution of the validity time is negligible. If a validity time of 1800 or 900 s is used, Milena (E14) and Doresa (E18) reach a $SISE_{orbit}$ RMS of about 1 m comparable to that of the other Galileo satellites.

Table 4 reports $SISE_{orbit}$ RMS value calculated on days with the higher RMS (DOYs 61–65, 71, and 175–178) using the six validity time thresholds.

Table 4. $SISE_{orbit}$ RMS (m) in DOYs with higher errors (DOYs 61–65, 71, and 175–178) calculated at different validity time thresholds.

Sat	VT: 14,400 s	VT: 10,800 s	VT: 7200 s	VT: 3600 s	VT: 1800 s	VT: 900 s
E1	0.92	0.92	0.72	0.90	0.89	0.86
E14	10.48	10.48	5.06	1.70	1.06	0.98
E18	2.90	2.90	2.82	1.39	1.00	0.97

Figure 12 shows the RMS $SISE_{orbit}$ calculated on DOYs showing higher errors (DOYs 61–65, 71, and 175–178) plotted versus the six different validity time thresholds (values are reported in Table 4). By observing Figure 12, it can be seen that: the SISE of the satellite E1 does not change when the threshold used changes; for all three satellites, the errors with the threshold set at 14,400 s are identical to those obtained with the threshold set at 10,800 s, the errors of the Milena (E14) and Doresa (E18) satellites become comparable with those of the E1 satellite starting from the threshold of 3600 s; these errors become very similar when the thresholds are set at 1800 and 900 s. It should be highlighted that the use of such a low validity time has the consequence of discarding a certain number of epochs for which the validity of the message is below the threshold considered. This loss is calculated, as shown in Tables 5 and 6 for the second and the third datasets, respectively.

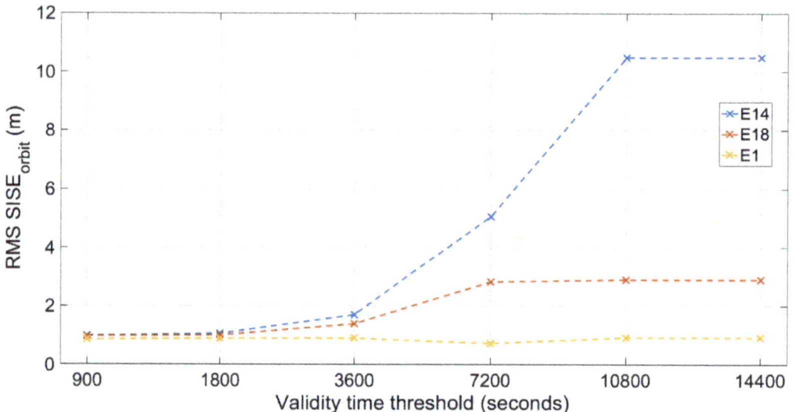

Figure 12. RMS $SISE_{orbit}$ on DOYs with higher errors (DOYs 61–65, 71, and 175–178) plotted versus the validity time threshold used. Milena (E14) and Doresa (E18) and satellite 1 (E1) satellites are represented by blue, red and yellow lines, respectively.

Table 5. Percentage of discarded epochs, DOYs 52–71.

Sat	VT: 14,400 s	VT: 10,800 s	VT: 7200 s	VT: 3600 s	VT: 1800 s	VT: 900 s
E1	0.04%	0.04%	0.32%	5.00%	15.83%	38.71%
E14	0.04%	0.04%	0.53%	5.35%	16.60%	37.82%
E18	0.04%	0.04%	0.04%	2.16%	11.36%	33.02%

Table 6. Percentage of discarded epochs, DOYs 172–191.

Sat	VT: 14,400 s	VT: 10,800 s	VT: 7200 s	VT: 3600 s	VT: 1800 s	VT: 900 s
E1	0.10%	0.10%	0.10%	0.79%	8.51%	33.44%
E14	0.10%	0.10%	0.10%	0.66%	5.78%	27.22%
E18	0.10%	0.10%	0.10%	1.67%	8.30%	29.58%

Looking at the tables, we can see that the use of a threshold equal to 1800 s for validity time implies a loss of epochs equal to about 16% for Milena (E14) and about 11% for Doresa (E18) in the second analyzed dataset. Considering the third dataset, the discarded epochs decrease, passing to 5.78% for Milena (E14) and 8.3% for Doresa (E18). If a validity time of 900 s is used, the discarded epochs are about the 30% of the total. This led us to prefer a validity time of 1800 s compared to one of 900, since a small improvement in the error corresponds to a high reduction in usable periods. Therefore, considering a validity time of 1800 s, we could obtain a $SISE_{orbit}$ comparable to that of the other satellites, but having to renounce the use of the two satellites in a number of epochs that is less than 8% in percentage terms. A validity time of 1800 s probably is better than one of 900 s because, by using the last one (as reported in Tables 5 and 6), a higher number of epochs will be lost.

Figures 13 and 14 show the radial, cross and along errors calculated with the validity time threshold set at 1800 s (Figures 13 and 14, left) and 10,800 s (Figures 13 and 14, right) of Milena (E14) and Doresa (E18) satellites, respectively, for the second dataset of DOYs 172–191. By observing the two figures, an improvement of all three components can be noted when the validity time threshold is set at 1800 s. The radial component of a satellite's ephemeris error is normally the smallest; however, it has the largest impact on the user's calculated position. Along-track and cross-track components are larger than the radial component by an order of magnitude but have little impact of the resultant user position error [17]. Therefore, the improvement obtained using a threshold of 1800 s as validity time of ephemeris should allow us to be able to use the satellites under analysis as an augmentation to the navigational solution.

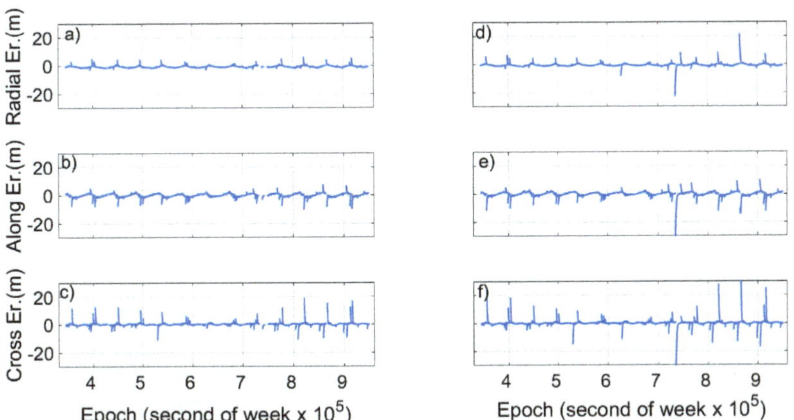

Figure 13. Comparison of radial, along and cross error for Milena (E14) obtained by using validity time threshold of: 1800 s (**left**); and 10,800 s (**right**). To have a better readability, the maximum value of errors is limited to 30 m. Errors that exceed this threshold relate to the cross component with a maximum error of 118.1 m (**f**). Figure refers to DOYs 172–191.

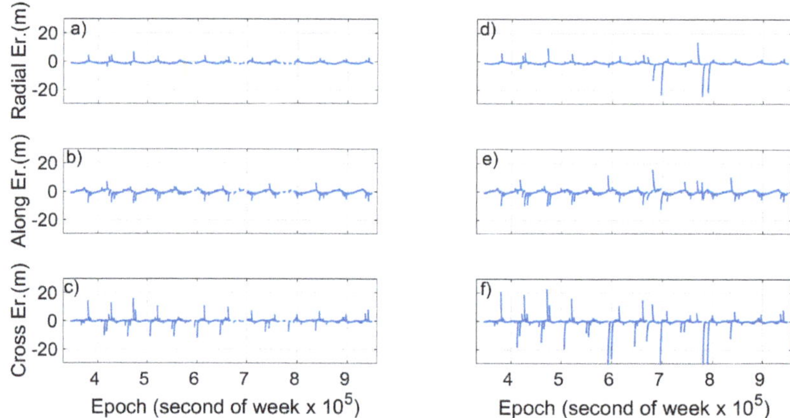

Figure 14. Comparison of radial, along and cross error for Doresa (E18) obtained by using validity time threshold of: 1800 s (**left**); and 10,800 s (**right**). To have a better readability, the maximum value of errors is limited to 30 m. Errors that exceed this threshold relate to the cross component with a maximum error of −173.1 m (f). Figure refers to DOYs 172–191.

4. Discussion

The position error between the broadcast orbit and the IGS precise orbit was determined for each Galileo satellite every second using six different validity time ephemeris thresholds (14,400 s, 10,800 s, 7200 s, 3600 s, 1800 s, and 900 s) for two periods: the first starts on 21 February 2018 (DOY 52) and ends on 12 March 2018 (DOY 71), the second starts on 21 June 2018 (DOY 172) and ends on 10 July 2018 (DOY 191) by using IGS products. For the first period (DOYs 52–71), two different types of transmitted ephemeris were used: those received by the receiver located in IGS YEL2 station located in Canada and the brdm ones. During the first period, 16 satellites were working while during the second the satellites were 19, thus more than 2.2×10^8 records were processed.

The analysis conducted show that Milena (E14) and Doresa (E18) satellites have a higher $SISE_{orbit}$ than the other satellites of the Galileo constellation for all three datasets analyzed. Using a synthetic indicator such as the RMS calculated on the entire dataset the Milena (E14) satellite shows a RMS $SISE_{orbit}$ equal to 3.56 m for the second dataset and of 1.12 m for the third dataset, while the Doresa (E18) satellite shows an RMS equal to 1.88 and 1.43 m for the same datasets. If the RMS calculation is done on a daily basis, the Milena (E14) and Doresa (E18) satellites have a higher RMS error with respect to the other satellites showing very high peaks in well-defined days (DOYs 61–65, 71, and 175–178).

Analyzing the temporal evolution of $SISE_{orbit}$, we can see how Milena (E14) and Doresa (E18) satellites show discontinuities and very high peaks compared with those of the other Galileo satellites. Despite these differences, the $SISE_{orbit}$ of Milena (E14) and Doresa (E18) satellites show the same periodicity of 13.9 h showed by the $SISE_{orbit}$ of the other satellites, comparable with that was found by Nicolini and Caporali [10] for the other Galileo satellite. A big difference between Milena (E14) and Doresa (E18) with respect to the other satellites regards the validity time of their ephemeris: their orbital error shows a much higher sensitivity than the other satellites to the validity time variations. In particular, the reduction in validity time has a positive effect: by observing the days with the higher RMS (61–65, 71, and 175–178), it can be seen how this value decreases when the validity time decreases. This improvement is also highlighted in the radial, along and across components of the orbital error. As shown in the Results Section, to obtain a $SISE_{orbit}$ comparable to that of the other satellites, it is necessary to decrease the validity time to 1800 s. This reduction involves a reduction of the epochs in which the satellites can be used. This reduction has been calculated in percentage terms: the epoch-loss is around 5% for Milena (E14) and 8% for Doresa (E18). This is a good result that must be verified appropriately in the position domain with single point positioning algorithms. Here, we want to recall that having two other satellites available will surely improve the achievable positioning accuracy. This aspect will be discussed in a forthcoming publication on which the authors are already working.

Author Contributions: Conceptualization and Methodology, U.R. and G.P.; Data Curation and Software, U.R.; Formal Analysis, U.R. and G.P.; Validation, U.R., G.B. and G.P.; Visualization and Writing-Original Draft Preparation, U.R., G.B. and G.P.; Writing—Review & Editing, U.R. and G.P.; Supervision, G.P.

Funding: This research has been supported by the Parthenope University of Naples with a grant within the call "Support for Individual Research for the 2015–17 Period". The above support is gratefully acknowledged.

Acknowledgments: The authors gratefully acknowledge IGS MultiGNSS Experiment (MGEX) for providing GNSS data and products.

Conflicts of Interest: The authors declare no conflict of interest.

References

1. Notice Advisory to Galileo Users (Nagu). Available online: https://www.gsc-europa.eu/sites/default/files/NOTICE_ADVISORY_TO_GALILEO_USERS_NAGU_2014014.txt (accessed on 23 November 2018).
2. Navarro-Reyes, D.; Castro, R.; Bosch, P.R. Galileo first FOC launch: Recovery mission design. In Proceedings of the 25th International Symposium on Space Flight Dynamics ISSFD, Munich, Germany, 19–23 October 2015.
3. Carlier, N.; Gülmüs, O. Spacecraft recovery operations conducted to the Galileo FOC-1 L3. In Proceedings of the 25th International Symposium on Space Flight Dynamics ISSFD, Munich, Germany, 19–23 October 2015.
4. Sosnica, K.; Prange, L.; Kazmierski, K.; Bury, G.; Drozdzewski, M.; Zajedel, R.; Hadas, T. Validation of Galileo orbits using SLR with a focus on satellites launched into incorrect orbital planes. *J. Geod.* **2018**, *92*, 131–148. [CrossRef]
5. Giorgi, G.; Lülf, M.; Günther, C.; Herrmann, S.; Kunst, D.; Finke, F.; Lämmerzah, C. Testing general relativity using Galileo satellite signals. In Proceedings of the 24th European Signal Processing Conference, Budapest, Hungary, 28 August–2 September 2016; pp. 1058–1062.
6. Paziewski, J.; Sieradzki, R.; Wielgosz, P. On the applicability of Galileo FOC satellites with incorrect highly eccentric orbits: An evaluation of instantaneous medium-range positioning. *Remote Sens.* **2018**, *10*, 208. [CrossRef]
7. Robustelli, U.; Pugliano, G. GNSS code multipath short time fourier transform analysis. *Navigation* **2018**, *65*, 353–362. [CrossRef]
8. Robustelli, U.; Pugliano, G. Code multipath analysis of Galileo FOC satellites by time-frequency representation. *Appl. Geomat.* **2019**, *11*, 69–80. doi:10.1007/s12518-018-0241-3. [CrossRef]
9. Robustelli, U.; Baiocchi, V.; Pugliano, G. Assessment of Dual Frequency GNSS Observations from a Xiaomi Mi 8 Android Smartphone and Positioning Performance Analysis. *Electronics* **2019**, *8*, 91. doi:10.3390/electronics8010091. [CrossRef]
10. Nicolini, L.; Caporali, A. Investigation on Reference Frames and Time Systems in Multi-GNSS. *Remote Sens.* **2018**, *10*, 80. doi:10.3390/rs10010080. [CrossRef]
11. Pugliano, G.; Robustelli, U.; Rossi, F.; Santamaria, R. A new method for specular and diffuse pseudorange multipath error extraction using wavelet analysis. *GPS Solut.* **2016**, *20*, 499–508. [CrossRef]
12. Misra, P.; Enge, P. Global Positioning System: Signals, measurements and performance second editions. In *Global Positioning System: Signals, Measurements And Performance Second Editions*, 3rd ed.; Ganga-Jamuna Press: Lincoln, MA, USA, 2006.
13. Kaplan, E.D.; Hegarty, C. *Understanding GPS/GNSS: Principles and Applications*, 3rd ed.; Artech House: Norwood, MA, USA, 2017.
14. Robustelli, U.; Benassai, G.; Pugliano, G. Accuracy evaluation of Doresa and Milena Galileo satellites broadcast ephemerides. In Proceedings of the 2nd IEEE International Workshop on Metrology for the Sea, Bari, Italy, 8–10 October 2018. doi:10.1109/MetroSea.2018.8657859.
15. ESA. European GNSS (Galileo) Open Service Signal In Space Operational Status Definition. Available online: https://www.gsc-europa.eu/system/files/galileo_documents/OS_SIS_OSD_V1.1.pdf (accessed on 23 November 2018).

16. Chen, L.; Bai, X.Z.; Liang, Y.G.; Li, K.B. Orbital Prediction Error Propagation of Space Objects. In *Orbital Data Applications for Space Objects—Conjunction Assessment and Situation Analysis*; Springer: Berlin, Germany, 2017.
17. Warren, D.L.M.; Raquet, J.F. Broadcast vs. precise GPS ephemerides: A historical perspective. *GPS Solut.* **2003**, *7*, 151–156. [CrossRef]
18. ESA. European Union. Galileo Open Service, Signal in Space Interface Control Document (OS SIS ICD). Available online: https://www.gsc-europa.eu/system/files/galileo_documents/Galileo-OS-SIS-ICD.pdf (accessed on 23 November 2018).
19. ESA. European GNSS (Galileo) Initial Services Open Service Quarterly Performance Report July-September 2018. Available online: https://www.gsc-europa.eu/system/files/galileo_documents/Galileo-IS-OS-Quarterly-Performance_Report-Q3-2018_0.pdf (accessed on 23 November 2018).
20. ESA. European GNSS (Galileo) Open Service Definition Document (OS-SDD). Available online: https://www.gsc-europa.eu/system/files/galileo_documents/Galileo-OS-SDD.pdf (accessed on 23 November 2018).
21. IGS News. Available online: http://www.igs.org/article/igs14-reference-frame-transition (accessed on 7 April 2019).
22. Cai, C.; Luo, X.; Liu, Z.; Xiao, Q. Galileo Signal and Positioning Performance Analysis Based on Four IOV Satellites. *J. Navig.* **2014**, *67*, 810–824. doi:10.1017/S037346331400023X. [CrossRef]
23. Montenbruck, O.; Steigenberger, P.; Hauschild, A. Broadcast versus precise ephemerides: A multi-GNSS perspective. *GPS Solut.* **2015**, *19*, 321–333. [CrossRef]
24. Montenbruck, O.; Schmid, R.; Mercier, F.; Steigenberger, P.; Fatkulin, C.N.R.; Kogure, S.; Ganeshan, A.S. GNSS satellite geometry and attitude models. *Adv. Space Res.* **2015**, *56*, 1015–1029. [CrossRef]
25. Dow, J.M.; Neilan, R.E.; Rizos, C. The International GNSS Service in a changing landscape of Global Navigation Satellite Systems. *J. Geod.* **2009**, *83*, 191–198. doi:10.1007/s00190-008-0300-3. [CrossRef]
26. Montenbruck, O.; Steigenberger, P.; Prange, L.; Deng, Z.; Zhao, Q.; Perosanz, F.; Romero, I.; Noll, C.; Stürze, A.; Weber, G.; et al. The Multi-GNSS Experiment (MGEX) of the International GNSS Service (IGS)—Achievements, prospects and challenges. *Adv. Space Res.* **2017**, *59*, 1671–1697. doi:10.1016/j.asr.2017.01.011. [CrossRef]
27. Rizos, C.; Montenbruck, O.; Weber, R.; Neilan, R.; Hugentobler, U. The IGS MGEX experiment as a milestone for a comprehensive multi-GNSS service. In Proceedings of the ION 2013 Pacific PNT Meeting, Honolulu, HI, USA, 22–25 April 2013; pp. 289–295.

Article

High Accuracy Buoyancy for Underwater Gliders: The Uncertainty in the Depth Control †

Enrico Petritoli, Fabio Leccese * and Marco Cagnetti

Science Department, Università degli Studi "Roma Tre", Via della Vasca Navale n. 84, 00146 Rome, Italy; e_petritoli@libero.it (E.P.); ing.marco.cagnetti@gmail.com (M.C.)

* Correspondence: fabio.leccese@uniroma3.it; Tel.: +39-06-5733-7347

† This paper is an extended version of our paper published in Petritoli, E.; Leccese, F; Cagnetti, M. A High Accuracy Buoyancy System Control for an Underwater Glider. In Proceedings of the second IEEE International Workshop on Metrology for the Sea (MetroSea), Bari, Italy, 8–10 October 2018.

Received: 8 March 2019; Accepted: 12 April 2019; Published: 17 April 2019

Abstract: This paper is a section of several preliminary studies of the Underwater Drones Group of the Università degli Studi "Roma Tre" Science Department: We describe the study philosophy, the theoretical technological considerations for sizing and the development of a technological demonstrator of a high accuracy buoyancy and depth control. We develop the main requirements and the boundary conditions that design the buoyancy system and develop the mathematical conditions that define the main parameters.

Keywords: uncertainty; buoyancy; depth control; accuracy; AUV; glider; autonomous; underwater; vehicle

---

1. Introduction

This paper is part of several preliminary studies by Underwater Drones Group (UDG) of the Science Department of the Università degli Studi "Roma Tre", which is developing an advanced Autonomous Underwater Vehicle (AUV) for the exploration of the sea at high depths. The final aim of the project is to create a platform for underwater scientific research that can accommodate a wide range of different payloads.

We will examine the buoyancy system and evaluate its sizing; then we will illustrate the technological solution we have come up with in order to realize the hydraulic system to be assembled in the Underwater Glider Mk. III (see Figure 1) [1–5].

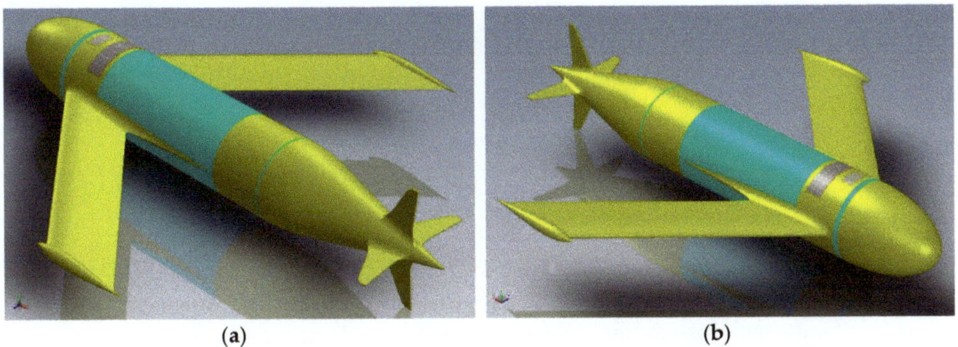

Figure 1. Perspective view of the Underwater Glider Mk. III. (**a**) Rear/port; (**b**) front/starboard.

1.1. The Underwater Glider

1.1.1. AUV Evolution

The exploration of the underwater world has always been one of mankind's dreams: submarines and bathyscaphes (for extreme depths) have been developed to study the "deep blue". Due to obvious dangers, the human exploration can take place only for very short periods and very limited areas: for these reasons, the exploration of the sea has immediately been drawn towards unmanned automatic systems [6–8].

An AUV is a vehicle that travels underwater without requiring input from an operator; this means that it must be equipped with a "brain" that regulates and coordinates its position, its depth and its speed: moreover, it is able to collect and store data from the payload. One of the first realizations was the Autonomous LAgrangian Circulation Explorer (ALACE) system, a buoy that was able to vary its buoyancy and therefore its depth. Although it possessed a great endurance, it only could be employed for great depths and in open sea—the consequences of these limitations are evident.

The next step was the use of Remote Operated Vehicles (ROVs). These, thanks to the constant development of electronic miniaturization, are extremely high performing vehicles for short-lasting marine operations, but the require the constant presence of a support vessel.

The need to get rid of the randomness of the currents has led to the natural development of the underwater glider concept [9–15].

1.1.2. The Underwater Glider

An underwater glider is a vehicle that, by changing its buoyancy, moves up and down in the ocean like a profiling float [16]. It uses hydrodynamic wings to convert vertical motion into horizontal motion, moving forward with very low power consumption [17–22]. While not as fast as conventional AUVs, the glider, using buoyancy-based propulsion, offers increased range and endurance compared to motor-driven vehicles and missions may extend to months and to several thousands of kilometres in range. An underwater glider follows an up-and-down, sawtooth-like mission profile providing data on temporal and spatial scales unavailable with previous types of AUVs [23–27].

1.1.3. The Mk. III Architecture

The Mk. III sub-glider has a cylindrical fuselage with a radome on the bow containing the customizable payload and, on the other end, the hydrodynamic fairing. The vehicle does not have moving surfaces: control is provided by the displacement of the battery package that varies the position of the centre of mass. The wings aerofoil is based on the Eppler E838 Hydrofoil. The aerofoil has the maximum thickness 18.4% at 37.2% chord and maximum camber 0% at 46.5% chord. The arrangement of the internal sectors is visible in Figure 2a,b. The buoyancy system is contained in the buoyancy control bay: it accommodates the buoyancy motor and the oil tank and provides longitudinal balance to the system by adjusting the level in the reservoir. The bladder is contained in the hydrodynamic fairing, in contact with the open water. The fairing is not a critical structural part—it has the task of not disturbing the hydrodynamic flow of the fuselage [28].

1.1.4. Conventions

We introduce, for clarity, the mathematical conventions and symbols that will be used in the subsequent discussion (see Figure 3a,b): where:

- α (or φ) is the angle between the x axis and the N axis.
- β (or θ) is the angle between the z axis and the Z axis.
- γ (or ψ) is the angle between the N axis and the X axis.

For the rotation matrix, we have:

$$\boldsymbol{R} = \begin{pmatrix} 0 & -\zeta_z & \zeta_y \\ \zeta_z & 0 & -\zeta_x \\ -\zeta_y & \zeta_x & 0 \end{pmatrix} \tag{1}$$

where ζ is the parameter vector [29].

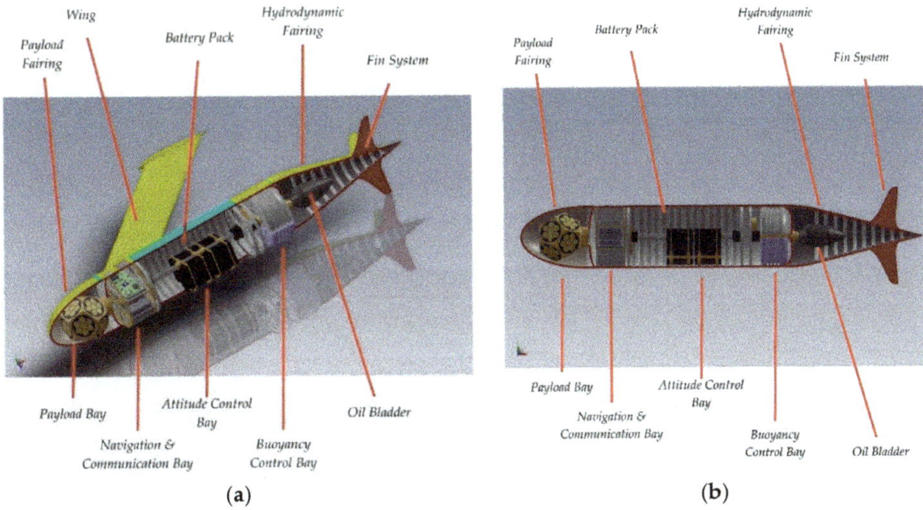

Figure 2. Underwater Glider Mk. III cutaway: (**a**) fuselage prospective section; (**b**) fuselage sagittal section.

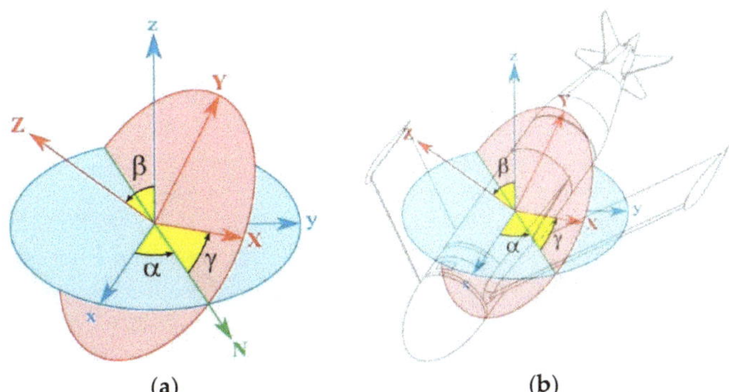

Figure 3. The Euler angles. (**a**) Body frame (blue) and reference frame (red); (**b**) The body frame referred to the drone.

2. Materials and Methods

2.1. The Buoyancy System

2.1.1. Basic Concepts

Gliders are controlled through hydrostatics (vertical forces) and manipulate hydrostatic balances in order to accomplish roll and pitch of the vehicle. Stability of the vehicle is a major critical factor:

a stable vehicle has the centre of gravity below the centre of buoyancy. In this configuration, the weight of the vehicle creates a restoring moment to add stability to the vehicle. Roll and pitch on the glider is accomplished by moving the battery pack. Figure 4 below displays a basic concept of a buoyancy system for the glider [30–36].

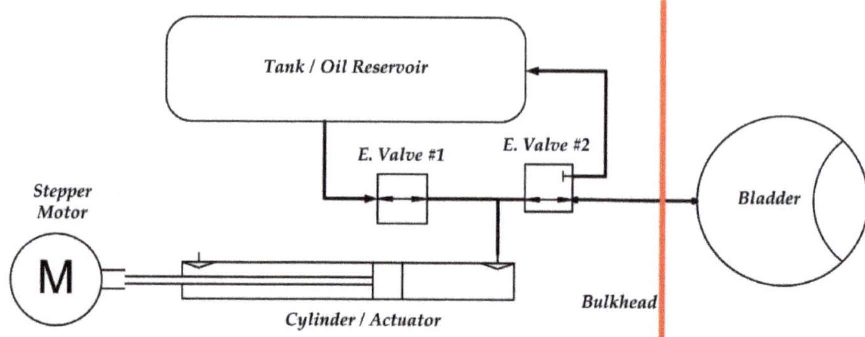

Figure 4. Basic scheme of the buoyancy system.

The system is extremely simple: while descending, hydraulic fluid moves from the external inflatable bladder, which produces a high pressure in the internal reservoir, which is at a low pressure through a valve: the decrease in volume of the bladder creates an increase in density, causing negative buoyancy [37–44].

While ascending, hydraulic fluid moves from internal accumulator to the external inflatable bladder through the pump. The increase in volume creates a decrease in density causing positive buoyancy. The seawater also flushes out the open hydrodynamic fairing of the vehicle, aiding it to rise to the surface. For neutral buoyancy, the vehicle must have a density equal to seawater [45–52].

2.1.2. The System Prototype

Our group has developed a technology demonstrator (see Figure 4) of the buoyancy system to validate the related technology and then to test it. To reduce the force required to actuate the oil piston, which pushes the oil in the bladder at high pressure, is necessary to reduce the piston surface (diameter) and increase the stroke: so, the buoyancy engine resembles a "shotgun". An open-loop stepper motor was used to drive the screw inside the actuator that, in turn, pushes the piston. Two solenoid valves regulate the flow of oil into the bladder [53–57].

The first problem was the occurrence of actuator buckling: under the push of the engine, the probability of a part bending is high, thus deforming the thread and jamming the mechanism. The problem was solved by constructing a rigid cage with four struts that support the piston's push load, leaving the screw only with the rolling friction load. In the early project development stages, the workgroup was oriented to use a centrifugal pump for all drives: this technology however did not allow us to create strong pressure differences; the need to use a more powerful engine was also highlighted because the prevalence was too low: this would have led to an excessive battery consumption. The second solution was to use volumetric pumps in order to obtain greater differences in pressure (even ones considerably higher than needed). Unfortunately, these would require too much power and are too heavy for our small vehicle [58–62].

At this stage of development, we have also thought to use the oil tank only as a passive fluid reservoir and trimmable counterweight of the payload and as an active actuator for longitudinal stability. The long travel of the piston (forward or backward) ensures the necessary variation of the bladder volume for manoeuvrability [63].

Now, the system prototype seems to work correctly and promises new developments: it is under in several fatigue cycle trials in order to investigate the parts more prone to failure as a sort of burn-in test. Our prototype (breadboard) is presented in Figure 5.

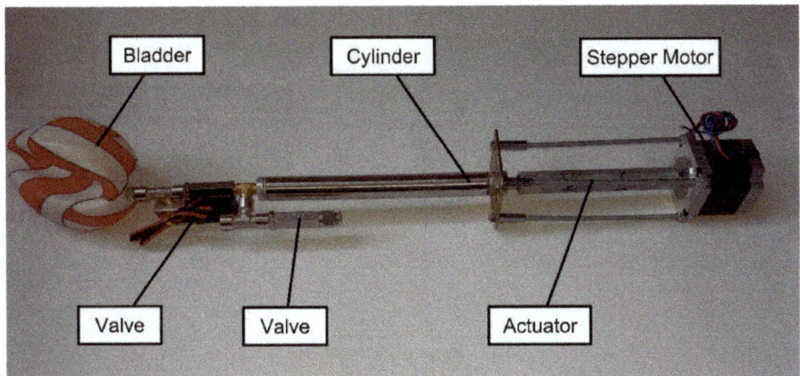

Figure 5. Buoyancy engine prototype.

2.2. Buoyancy

Archimedes' principle is the main concept underlying the buoyancy of underwater vehicles. When a vehicle is submerged in water, a buoyant force acts on the body vertically upward due to the pressure forces below the submerged body being greater than the pressure forces above. The buoyant force results in a value equal to the weight it displaces [64].

2.2.1. Static Buoyancy

Now, consider our drone (see Figure 6): it is in steady state.

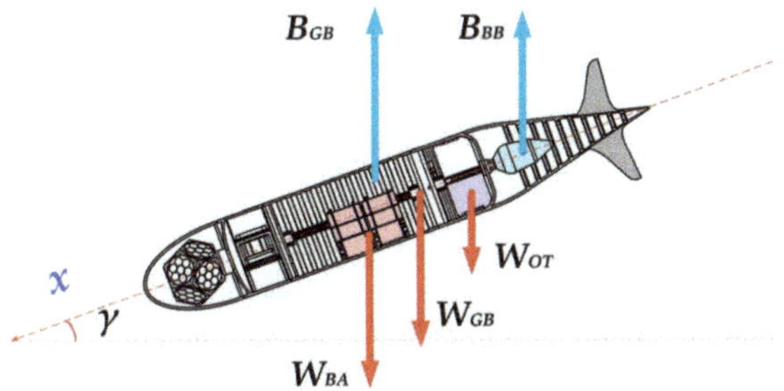

Figure 6. Drone in buoyancy Balance.

In these conditions, the total weight W_{TOT} of the glider is given by:

$$W_{TOT} = -W_{DW} + B_{GB} + B_{BB} \tag{2}$$

where:

W_{TOT} = Net total "weight" in the water.
W_{DW} = Dry Weight of the glider.
B_{BB} = Buoyancy of the oil bladder.
B_{GB} = Buoyancy of the naked glider.

The expression of dry weight is:

$$W_{DW} = W_{BA} + W_{OT} + W_{GB} \tag{3}$$

where:

W_{BA} = Weight of the battery pack.

W_{OT} = Weight of the oil tank.

W_{GB} = Weight of the naked glider (without oil tank and batteries). For "dry weight", we mean the weight of the vehicle out of the sea, without the hydrostatic force.

So Equation (2) becomes:

$$\sum F_z = W_{BA} + W_{OT} + W_{GB} + B_{GB} + B_{BB} = 0 \tag{4}$$

This series of equations will be useful later to establish the drone descent attitude.

2.2.2. Dynamic Balance on the Vertical Plane

Here, the drone dive (or emersion) is examined at constant speed: it is in steady-state gliding, the geometry of total forces is explained in Figure 7a. Figure 7b shows the Eppler E838 characteristic "*Cl vs. Alpha (=angle of attack α_{att})*".

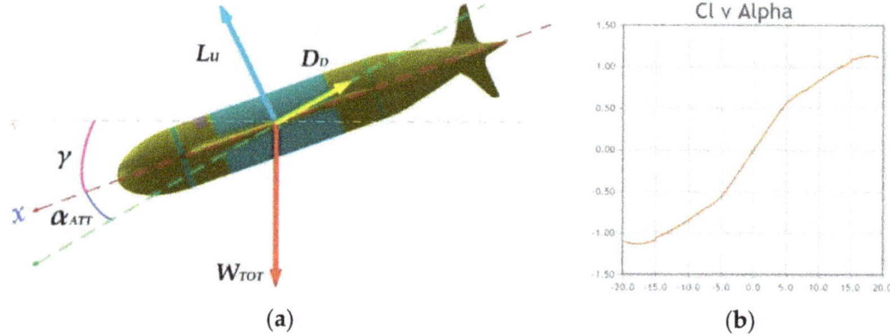

Figure 7. Dynamic balance of the forces (**a**) Drone geometrical balance of the forces; (**b**) Cl vs. Alpha (=angle of attack, α_{att} diagram for Eppler E838 aerofoil.

At equilibrium, for the dynamics on the vertical plane at constant speed we have:

$$\vec{W}_{TOT} + \vec{L} + \vec{D} = 0 \tag{5}$$

The expression for the lift is:

$$L = \frac{1}{2}\rho v^2 S C_L \tag{6}$$

According to the Eppler E838 characteristic "*Cl vs. Alpha*" (Figure 7b) when the angle of attack α_{att} = 0° (is null) the C_L is zero so that the lift force L is null. This shows that the drone cannot progress horizontally at constant speed (straight and level): the only mission profile allowed is a sawtooth curve [65].

2.2.3. Glider Trajectory

Because its motion is due to the difference between the forces of weight and buoyancy, the glider is unable to proceed straight and level, thus being forced to follow a dive/climb trajectory made smooth by the wings. Moreover, unlike gliders in air, AUVs can have ascending glide slopes if the net buoyancy is positive, producing a negative sink rate.

The buoyancy engine of the glider allows changing its net buoyancy into alternating positive and negative states, thereby imparting it with the ability to string together a succession of descending and ascending glide slopes referred to as a sawtooth glide.

The behaviour of the vehicle is considered in case of a simple glide slope (refer to Figure 7a):

$$\begin{aligned} Lift &= \mathbb{L} = qSC_L \\ Drag &= \mathbb{D} = qSC_D \\ Pitching\ moment &= \mathbb{m} = qScC_m \end{aligned} \tag{7}$$

where:

$q = \frac{1}{2}\rho v^2$: is the dynamic pressure.

S: is the characteristic area.

c: is the mean aerodynamic chord.

α_{att}: is the angle of attack.

For the other coefficients, we have:

$$\begin{aligned} C_L(\alpha) &= C_L^{\alpha}\cdot\alpha_{att} \\ C_D(\alpha) &= C_D^0 + C_D^{\alpha}\cdot\alpha_{att}^2 \\ C_m(\alpha) &= C_m^{\alpha}\cdot\alpha_{att} \end{aligned} \tag{8}$$

The coefficient of drag C_D is composed of two members: the first C_D^0 is insensitive to the angle of attack and is constant; the second one ($C_D^{\alpha}\cdot\alpha_{att}^2$) is instead a function of the square of the angle. Note that the zero lift coefficient $C_L^0 = 0$ because the wing profile that has been chosen for our project is symmetrical (type *Eppler 883*). Now is necessary to separate the contributions of the fuselage (body) and of the wings, for the three factors of lift, friction and pitching moment; so the expression is:

$$\begin{cases} \mathbb{L} = L^b + L^w \\ \mathbb{D} = D^b + D^w \\ \mathbb{m} = m^b + m^w \end{cases} \tag{9}$$

According to the Navier-Stokes (approximated) equations and the simplifications above cited, the previous system of equations becomes:

$$\begin{cases} \mathbb{L} = q\sqrt{V^3}\left\{C_L^{b_\alpha}\cdot\alpha_{att} + \frac{S_w}{\sqrt{V^3}}\cdot C_L^{w_\alpha}\cdot\alpha_{att}\right\} \\ \mathbb{D} = q\sqrt{V^3}\left\{\left[C_D^{b_0} + C_D^{b_\alpha}\cdot\alpha_{att}^2\right] + \frac{S_w}{\sqrt{V^3}}\left[C_D^{w_0} + C_D^{w_\alpha}\cdot\alpha_{att}^2\right]\right\} \\ \mathbb{m} = q\sqrt{V^3}\left\{C_m^{b_\alpha}\cdot\alpha_{att} - c_m\frac{l_{cb/ac_w}\cdot S_w}{c\cdot\sqrt{V^3}}\cdot C_L^{w_\alpha}\cdot\alpha_{att}\right\} \end{cases} \tag{10}$$

where:

S_w is the wing area.

l_{cb/ac_w} is the distance between the mean aerodynamic and the center of buoyancy.

c_m is a non-dimensional coefficient.

This parameter is necessary to know the exact attitude and therefore the α_{att} to obtain a constant descent profile [66].

2.2.4. Gliding Forces

In order to allow the vehicle to glide, it is necessary to create a differential buoyancy force and therefore the balance is not null: from the buoyant force expression, the change in volume needed for the buoyancy engine for a full dive is calculated as:

$$\sum F_{Z_{umbalanced}} = \Delta B_{BB} \neq 0 \tag{11}$$

In which $\Delta B_{BB} \neq 0$ is the buoyancy force due to the difference of volume of the bladder:

$$\sum F_{Zumbalanced} = \Delta B_{BB} = \frac{1}{2}\rho g \cdot \Delta V_{bladder} \tag{12}$$

where:

ρ = Seawater density (average 1.025 kg/l).
g = Gravity approx. to 9.81 m/s^2.
$\Delta V_{bladder}$ = Volume difference of the bladder.

The relationship between difference volume of the bladder and buoyancy force is:

$$\Delta V_{bladder} = \frac{2}{\rho g} \cdot \Delta B_{BB} \tag{13}$$

2.2.5. Restoring Moment on the Vertical Plan

The position of the center of mass is given by:

$$r_{CG} = \frac{\int r \cdot \rho(r) dV}{\int \rho(r) dV} \tag{14}$$

where:

$\rho(r)$: is the density.
dV: is the considered volume.
r: is the distance considered from the reference frame.

The gliders masses define above, is:

$$r_{CG} = \frac{\sum_i r_i \cdot W_i}{\sum_i W_i} = \frac{r_{BA} \cdot W_{BA} + r_{OT} \cdot W_{OT} + r_{GB} \cdot W_{GB}}{W_{BA} + W_{OT} + W_{GB}} \tag{15}$$

The vectorial balance of the forces becomes:

$$\begin{cases} \mathbb{F}_{gravitational} = W_{DW}\left(\boldsymbol{R^T} z\right) \\ \mathbb{F}_{buoyancy} = -V_{Dispalacement} \cdot g\left(\boldsymbol{R^T} z\right) \end{cases} \tag{16}$$

where

$V_{Dispalacement}$ is the volume of the drone (submerged).
$\boldsymbol{R^T}$ is the rotation function around the reference frame.
z is the upward direction.

When the geometrical centre of the body is offset from the CG frame, the resulting torque $\mathfrak{T}_G$ is given by:

$$\mathfrak{T}_G = r_{CG} \times W_{DW}\left(\boldsymbol{R^T} z\right) \tag{17}$$

The expression of r_{CG} is:

$$r_{CG} = \begin{pmatrix} x_{CG} \\ y_{CG} \\ z_{CG} \end{pmatrix} \tag{18}$$

The expression of $\hat{r}_{CG}$ is:

$$\hat{r}_{CG} = \begin{pmatrix} 0 & -z_{CG} & y_{CG} \\ z_{CG} & 0 & -x_{CG} \\ -y_{CG} & x_{CG} & 0 \end{pmatrix} \tag{19}$$

Therefore, we have:

$$\mathfrak{T}_G = W_{DW} \cdot \hat{r}_{CG}\left(\boldsymbol{R^T} z\right) \tag{20}$$

According to which, when the body frame is coincident with the centre of the figure:

$$W_{DW} \cdot r_{CG} = r_{GB} \cdot B_{GB} + r_{BB} \cdot B_{BB} \tag{21}$$

Moreover, the resulting balancing torque is

$$\mathfrak{T}_G = (r_{GB} \cdot B_{GB} + r_{BB} \cdot B_{BB}) \cdot g\left(\boldsymbol{R}^T z\right) \tag{22}$$

This parameter is necessary to know the exact torque and therefore the position to which the servomechanism to move the battery pack will have to obtain a constant descent profile [67].

2.2.6. Uncertainty in the Depth Control

In this part, the problem of uncertainty in controlling the depth of the drone is examined. In the case where the drone has a very low vertical speed, due to the high viscosity of the water, the vehicle stops (and therefore stabilizes itself in the vertical plan) in a few centimeters of water, so this is not a problem.

Consider the case in which the drone is gliding, at a constant speed ($V_z < 0$): at a depth of +5 m compared to the "target depth", the bladder swells up to make the drone assume a neutral buoyancy. From a dynamic point of view, our simulations show us that the drone will behave like a mass-spring-damper system, whose temporal behavior is described in Equation (23) and visible in Figure 8 (blue line).

$$\begin{cases} z_{CG}(t) = \mathbb{Z}e^{-\zeta\omega_n t} \cdot \cos\left(\sqrt{1-\zeta^2}\omega_n t - \Phi\right) \\ \omega_n = 2\pi f_n \end{cases} \tag{23}$$

The maximum precision in the measurement of depth is not given by the precision of the instrument, in this case a piezoelectric depth meter can be accurate up to 2.5 cm, but in the architecture of the drone. Because of all the possible attitudes of the vehicle, it is not known if the transducer is placed higher or lower than the centre of gravity: so, unless more than one transducer is installed (which is impractical) the only consideration is that the maximum distance at which the transducer should be placed from the centre of gravity is precisely 49.7 cm, so, the maximum range of precision obtainable from the instrument is:

$$\delta_z = 0.497 \text{ m} \tag{24}$$

As shown in Figure 8 (blue line) the natural dynamic stability of the vehicle, i.e., the oscillation within the error band, is obtained after a time $\mathbb{T}_{normal} = 9.78$ s which is not acceptable (see Equation (25)). It is not acceptable for two reasons: first of all because of the long damping time of the oscillation; secondly for the overshoot of the "target depth": the depth limit may have been placed not only by the type of mission but also by the nature (orography) of the seabed. Therefore, such behavior could lead to a collision of the vehicle with the seabed itself or a known obstacle, thus leading to damage:

$$\begin{cases} z_{CG}(t)_{normal} \leq \delta_z \\ \mathbb{T}_{normal} = 9.78 \text{ s} \end{cases} \tag{25}$$

The solution is to use the bladder as a "hydrostatic parachute" that changes the damping conditions of the system. In this case the simulation have evidenced that, if you place the bladder at maximum buoyancy (Figure 8—red line) shortly after reaching "+5" depth, the resulting behavior of the vehicle is visible in Figure 8 (azure line).

In this case the tolerance band is reached in $\mathbb{T}_{compensated} = 4.85$ s (see Equation (26)), half the previous time and there is no danger of overshooting the depth, thus keeping us always in safe conditions:

$$\begin{cases} z_{CG}(t)_{compensated} \leq \delta_z \\ \mathbb{T}_{compensated} = 4.85 \text{ s} \end{cases} \tag{26}$$

From a mathematical point of view, the system changes the damping factor, increasing it considerably in fact, $\zeta_{normal} < \zeta_{compensated}$. The model parameters are shown in Table 1 below.

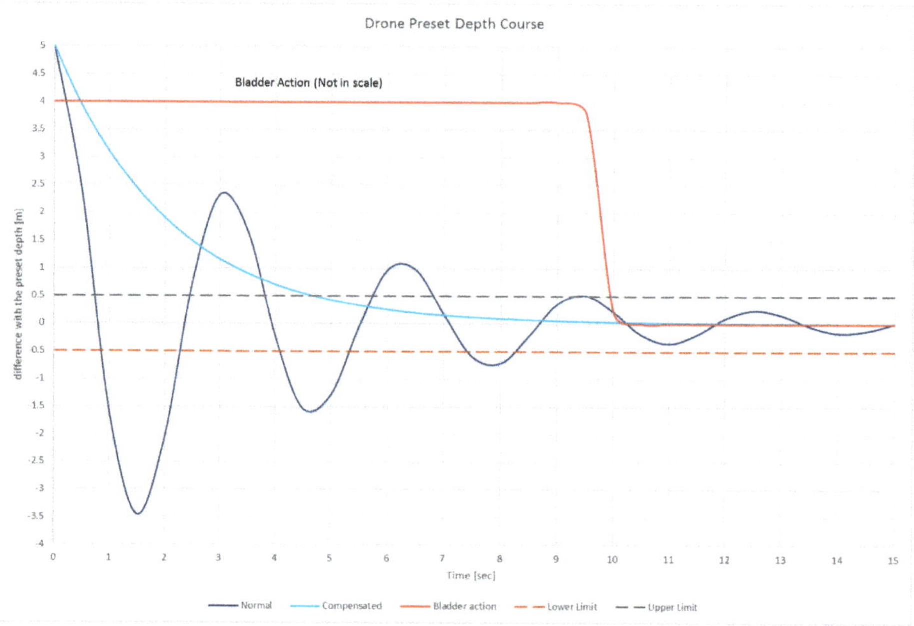

Figure 8. Drone depth dynamic behavior: in the blue line the normal (uncompensated) damping; the azure line the compensated behavior; dotted lines are the upper and lower limit. The red line (out of scale) is the bladder's "hydrostatic chute" action.

Table 1. Model parameters.

f_n	ζ_{normal}	$\zeta_{compensated}$	$\mathbb{Z}$
1	0.0794	0.15779	4.972

2.2.7. Dynamic Simulation

In the section, we simulate the behavior of the Eppler 838 aerofoil and the fuselage to evaluate, before a detailed and expensive but more sophisticated 3D simulation, if the proportions and dimensions of the drone fall within the range of measures evaluated as a requirement. For this purpose, the program "JavaFoil—Analysis of Airfoils" [68] is used. JavaFoil is a program which uses several traditional methods for aerofoil analysis. The backbone of the program consists of two methods:

- The evaluation of the potential flow. The analysis is done with a higher order panel method (linear varying vorticity distribution) and it calculates the local, inviscid flow velocity along the surface of the aerofoil for any desired angle of attack.
- The evaluation of the boundary layer. The analysis is steps along the upper and the lower surfaces of the aerofoil, starting at the stagnation point, solves a set of differential equations to find the various boundary layer parameters, according to the integral method.

The equations and criteria for transition and separation are based on the procedures described by Eppler. A standard compressibility correction according to Karman and Tsien has been implemented to take moderate Mach number effects into account. Usually this means Mach numbers between zero and 0.5 [69–72].

The simulation shows the critical parameters for the hydrodynamic behaviour of the model: the results are given in Table 2 below [73–78].

Table 2. Body and wing critical parameters [Speed = 0.52 kts, $R_e = 10^4$ and $\alpha_{att} = 3.3°$].

$C_L^{b_\alpha}$	$C_L^{w_\alpha}$	$C_D^{b_0}$	$C_D^{w_0}$	$C_D^{b_\alpha}$	$C_D^{w_\alpha}$	$C_m^{b_\alpha}$	c_m
0.3997	0.4085	0.01746	0.01676	0.00018	0.001515	0.005	0.4752

The trend of the flow lines with $\alpha_{att} = 3.3°$ around the body and the wing are visible in Figure 9a,b, below.

For completeness, in order to be able to express the induced resistance and the stall delay, the behavior of the aerofoil at different angles of attack is provided in Figure 10.

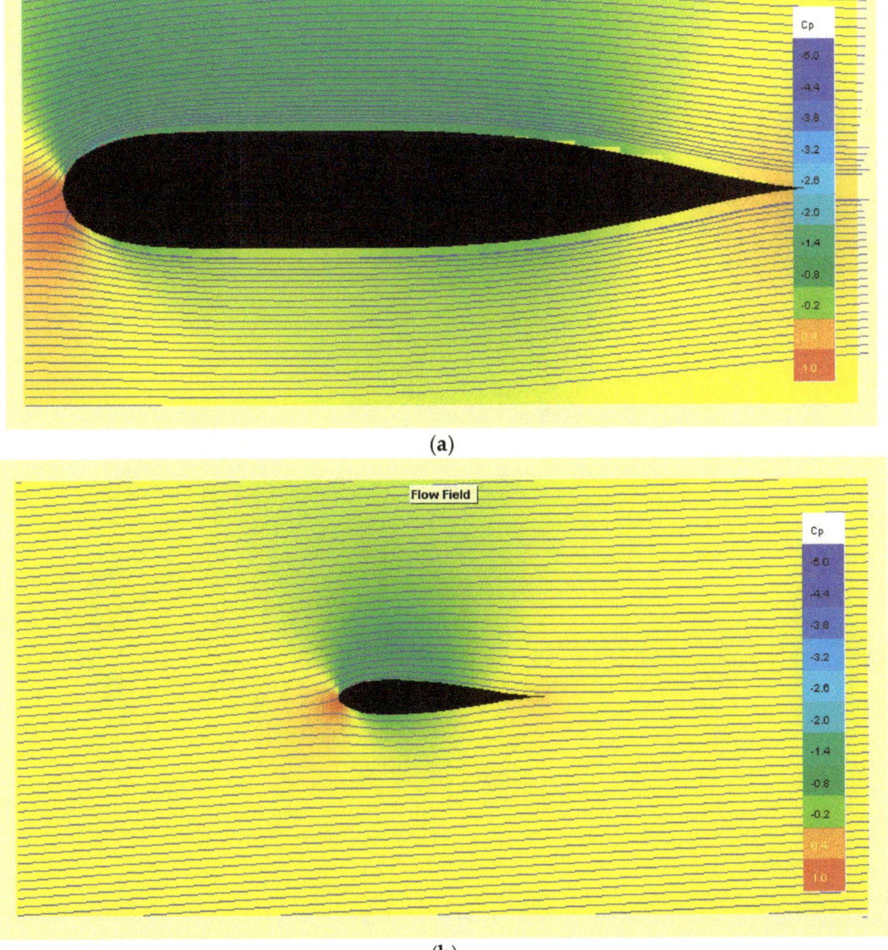

(a)

(b)

Figure 9. Flow field and pressure gradient of conditions: Speed = 0.52 kts, $R_e = 10^4$ and $\alpha_{att} = 3.3°$, (**a**) Body profile of the drone; (**b**) Eppler 838 Aerofoil.

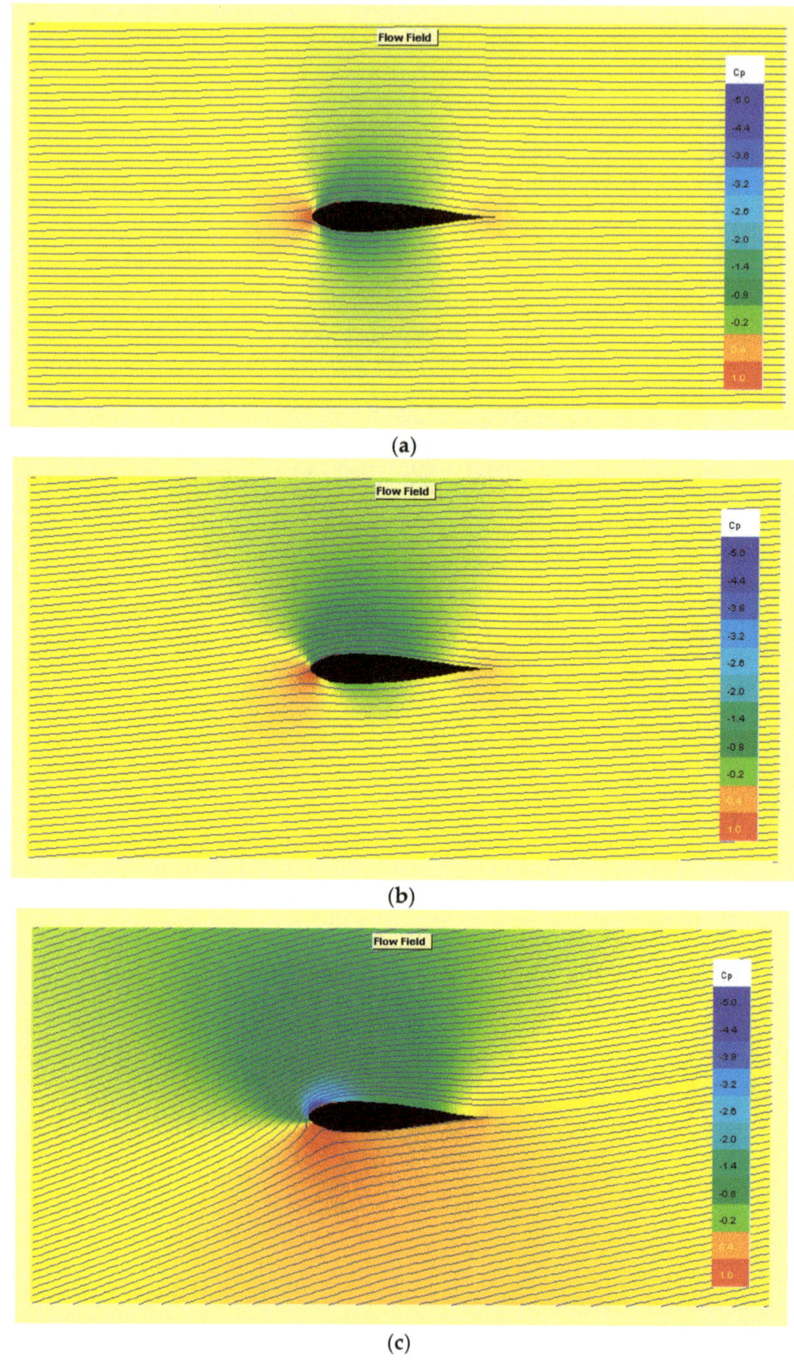

(a)

(b)

(c)

Figure 10. Flow field and pressure gradient of conditions (Speed = 0.52 kts, $R_e = 10^4$) are shown at the following α_{att}: (**a**) 0.5° (**b**) 4.0° (**c**) 15.0°.

3. Conclusions

This paper reports part of several preliminary studies of the Underwater Drones Group of the Università degli Studi "Roma Tre" Science Department and follows the route traced on several conference papers presented at the IEEE International Workshop on Metrology for the Sea (MetroSea); this part is dedicated to the design and engineering study of the part relating to the UAV buoyancy.

This paper highlights the large series of considerations and structural dimensions, going down in great detail, of the Underwater Glider Mk. III that is currently in an advanced development phase. The real novelties of this work are highlighted in the development due to two strong constraints that our group inserted during the design of this AUV: the first is to evaluate the project always under the most conservative (pessimistic) operating conditions; the second is to evaluate how any changes made to the subsystem in development (in our case the buoyancy s/s) is reflected (and forced) on all the other parts (or subsystems). All the results both from the partial simulations and from the construction and testing of subsystems will then be used in the operating vehicle.

The first section is dedicated to the design and engineering study of the part relating to the UAV buoyancy. In the first section, the architecture, the internal arrangement of the sub glider, the type of mission profile and the maximum requirements for the performances are broadly described. In the second part, the buoyancy system is described from an engineering-construction point of view: the solutions developed and implemented in a working prototype were illustrated.

The last part describes the mathematical requirements for sizing the vehicle. Firstly, the static requirements that are used to determine the mechanical and dimensional sizing of the buoyancy engine are examined. Then the dynamic stability (on the vertical plane) of the vehicle is analysed: this quantifies the forces involved during the "glide". The trajectory is analysed to decide the attitude and the angle of attack: the latter is necessary, in stationary conditions, to determine the work point of the profile or the position of the profile in the diagram "*Cl vs. Alpha*". In order to have a constant attitude, it is necessary to balance the moments in the vertical plane so that, once the wing profile is "started" and in progressive acceleration (i.e., while is nearly to reach the terminal velocity) the pitch up effects must be compensated by the movement of the centre of gravity. An evaluation of the uncertainty in the depth control is also provided.

Lastly, a simplified simulation is introduced in order to observe the hydrodynamic behaviour of the fuselage (limited to the profile) and of the aerofoil at different angles of attack, to highlight the stall characteristics.

Author Contributions: The authors have made an equal contribution in all the sections of this article.

Funding: This research received no external funding.

Conflicts of Interest: The authors declare no conflict of interest.

References

1. Wood, S. Autonomous underwater gliders. In *Underwater Vehicles*; Chapter 26; Inzartsev, A.V., Ed.; In-Tech: Vienna, Austria, 2009; pp. 499–524.
2. Kayton, M.; Fried, W.R. *Avionics Navigation Systems*; Wiley: New York, NY, USA, 2007; ISBN 9780471547952.
3. Jenkins, S.A.; Humphreys, D.E.; Sherman, J.; Osse, J.; Jones, C.; Leonard, N.; Graver, J.; Bachmayer, R.; Clem, T.; Carroll, P.; et al. *Underwater Glider System Study*; Technical Report No. 53; Scripps Institution of Oceanography: San Diego, CA, USA, May 2003.
4. Graver, J.G. Underwater Gliders: Dynamics, Control and Design. Ph.D. Thesis, Princeton University, Princeton, NJ, USA, 2005.
5. Kan, L.; Zhang, Y.; Fan, H.; Yang, W.; Chen, Z. MATLAB-Based simulation of buoyancy-driven underwater glider motion. *J. Ocean Univ. Chin.* **2007**, *7*, 113–118. [CrossRef]
6. Woolsey, C.A.; Leonard, N.E. Stabilizing underwater vehicle motion using internal rotors. *Automatica* **2002**, *38*, 2053. [CrossRef]

7. Jekeli, C. *Inertial Navigation Systems with Geodetic Applications*; Walter de Gruyter Berlin: New York, NY, USA, 2001.
8. Smith, R.N.; Chao, Y.; Jones, B.H.; Caron, D.A.; Li, P.; Sukhatme, G.S. Trajectory Design for Autonomous Underwater Vehicles based on Ocean Model Predictions for Feature Tracking. In *Field and Service Robotics*; Springer: Berlin/Heidelberg, Germany, 2010; pp. 263–273, 2009.
9. Mitchell, B.; Wilkening, E.; Mahmoudian, N. Low cost underwater gliders for littoral marine research. In Proceedings of the American Control Conference (ACC), Seattle, WA, USA, 17–19 June 2013; pp. 1412–1417.
10. Meyer, D. Glider Technology for Ocean Observations: A Review. *Ocean Sci. Discuss.* **2016**, *2016*, 1–26. [CrossRef]
11. Davis, R.E.; Eriksen, C.C.; Jones, C.P. Autonomous buoyancy-driven underwater gliders. In *Technology and Applications of Autonomous Underwater Vehicles*; Griffiths, G., Ed.; Taylor & Francis: London, UK, 2002; pp. 37–58.
12. Wang, C.; Zhang, Z.; Gu, J.; Liu, J.; Miao, T. Design and Hydrodynamic Performance Analysis of Underwater Glider Model. In Proceedings of the 2012 International Conference on Computer Distributed Control and Intelligent Environmental Monitoring (CDCIEM), Hunan, China, 5–6 March 2012; pp. 225–227. [CrossRef]
13. De Francesco, E.; De Francesco, E.; De Francesco, R.; Leccese, F.; Cagnetti, M. A proposal to update LSA databases for an operational availability based on autonomic logistic. In Proceedings of the 2nd IEEE International Workshop on Metrology for Aerospace, MetroAeroSpace, Benevento, Italy, 4–5 June 2015; pp. 38–43. [CrossRef]
14. Eriksen, C.C.; Osse, T.J.; Light, R.D.; Wen, T.; Lehman, T.W.; Sabin, P.L.; Ballard, J.W.; Chiodi, A.M. Seaglider: A longrange autonomous underwater vehicle for oceanographic research. *IEEE J Oceanic Eng.* **2001**, *26*, 424–436. [CrossRef]
15. Rudnick, D.L.; Eriksen, C.C.; Fratantoni, D.M.; Perry, M.J. Underwater Gliders for Ocean Research. *Mar. Technol. Soc. J.* **2004**, *38*, 1.
16. Austin, R. *Unmanned Aircraft Systems*; Wiley: Hoboken, NJ, USA, 2010.
17. Graver, J.G.; Liu, J.; Woolsey, C.; Leonard, N.E. Design and Analysis of an Underwater Vehicle for Controlled Gliding. In Proceedings of the Conference on Information Sciences and Systems (CISS), Princeton, NJ, USA, 2–5 July 1998.
18. Leccese, F.; Cagnetti, M.; Giarnetti, S.; Petritoli, E.; Luisetto, I.; Tuti, S.; Đurović-Pejčev, R.; Đorđević, T.; Tomašević, A.; Bursić, V.; et al. A Simple Takagi-Sugeno Fuzzy Modelling Case Study for an Underwater Glider Control System. In Proceedings of the 2018 IEEE International Workshop on Metrology for the Sea; Learning to Measure Sea Health Parameters (MetroSea), Bari, Italy, 8–10 October 2018; pp. 262–267. [CrossRef]
19. Teledyne Webb Research. Slocum Electric Glider. Available online: http://www.webbresearch.com/ (accessed on 1 February 2019).
20. Wood, S. State of Technology in Autonomous Underwater Gliders. *Mar. Technol. Soc. J.* **2013**, *47*, 5. [CrossRef]
21. Waldmann, C.; Kausche, A.; Iversen, M. MOTH-An underwater glider design study carried out as part of the HGF alliance ROBEX. In Proceedings of the 2014 IEEE/OES Autonomous Underwater Vehicles (AUV), Oxford, UK, 25–29 July 2014; pp. 1–3. [CrossRef]
22. Davis, R.E.; Webb, D.C.; Regier, A.; Dufour, J. The Autonomous Lagrangian Circulation Explorer (ALACE). *J. Atmos. Oceanic Technol.* **1992**, *9*, 264–285. [CrossRef]
23. Seo, D.C.; Gyungnam, J.; Choi, H.S. Pitching control simulation of an underwater glider using CFD analysis. In Proceedings of the Oceans—MTS/IEEE Kobe Techno-Ocean, Kobe, Japan, 8–11 April 2008; pp. 1–5.
24. Evans, C.D.; Riggins, R. The Design and Analysis of Integrated Navigation Systems Using Real INS and GPS Data. In Proceedings of the IEEE 1995 National Aerospace and Electronics Conference, Dayton, OH, USA, 22–26 May 1995.
25. Yu, J.; Zhang, F.; Zhang, A.; Jin, W.; Tian, Y. Motion Parameter Optimization and Sensor Scheduling for the Sea-Wing Underwater Glider. *IEEE J. Oceanic Eng.* **2013**, *38*, 243–254. [CrossRef]
26. Hussain, N.A.A.; Arshad, M.R.; Mohd-Mokhtar, R. Modeling and Identification of an Underwater Glider. In Proceedings of the 2010 International Symposium on Robotics and Intelligent Sensors (IRIS2010), Nagoya, Japan, 8–11 March 2010.
27. Bohenek, B.J. The Enhanced Performance of an Integrate Navigation System. In *A Highly Dynamic Environment*; Air Force Institute of Technology: Dayton, OH, USA, 1994.

28. Beard, R.W.; McLain, T.W. *Small Unmanned Aircraft—Theory and Practice*; Princeton University Press: Princeton, NJ, USA, 2012.
29. Petritoli, E.; Leccese, F.; Ciani, L. Reliability assessment of UAV systems. In Proceedings of the 4th IEEE International Workshop on Metrology for AeroSpace (MetroAeroSpace), Padua, Italy, 21–23 June 2017; pp. 266–270.
30. Watts, A.C.; Ambrosia, V.G.; Hinkley, E.A. Unmanned Aircraft Systems in Remote Sensing and Scientific Research: Classification and Considerations of Use. *Remote Sens.* **2012**, *4*, 1671–1692. [CrossRef]
31. Petritoli, E.; Leccese, F. Improvement of altitude precision in indoor and urban canyon navigation for small flying vehicles. In Proceedings of the 2nd IEEE International Workshop on Metrology for Aerospace (MetroAeroSpace 2015), Benevento, Italy, 4–5 June 2015; pp. 56–60. [CrossRef]
32. Andrade-Bustos, I.; Salgado-Jiménez, T.; García-Valdovinos, L.G.; Bandala-Sánchez, M. Stable Sliding PD Control for underwater gliders: Experimental results. In Proceedings of the Oceans 2016 MTS/IEEE Monterey, Monterey, CA, USA, 19–23 September 2016; pp. 1–7.
33. Pereira, A.; Heidarsson, H.; Caron, D.A.; Jones, B.H.; Sukhatme, G.S. A communication framework for the cost-effective operation of slocum gliders in coastal regions. In Proceedings of the 7th International Conference on Field and Service Robotics, Cambridge, MA, USA, 14–16 July 2009; pp. 1–10.
34. Skibski, C.E. Design of an Autonomous Underwater Glider focusing on External Wing Control Surfaces and Sensor Integration. Bachelor's Thesis, Florida Institute of Technology, Melbourne, FL, USA, 2009.
35. Isa, K.; Arshad, M.R. An analysis of homeostatic motion control system for a hybrid-driven underwater glider. In Proceedings of the 2013 IEEE/ASME International Conference on Advanced Intelligent Mechatronics (AIM), Wollongong, Australia, 9–12 July 2013; pp. 1570–1575.
36. Küchemann, D. *The Aerodynamic Design of Aircraft*; Pergamon international library of science, technology, engineering, and social studie; Pergamon Press: Oxford, UK, 1978.
37. Parthasarathy, G.; Sree, D.S.; Manasa, B.L. Design Mathematical Modeling and Analysis of Underwater Glider. *Int. J. Sci. Res.* **2015**, *4*, 711–714.
38. Song, D.L.; Yao, L.L.; Wang, Z.Y.; Han, L. Pitching Angle Control Method of Underwater Glider Based on Motion Compensation. In Proceedings of the 2015 International Conference on Computational Intelligence and Communication Networks (CICN), Jabalpur, India, 12–14 December 2015; pp. 1548–1551.
39. Smith, R.N.; Schwager, M.; Smith, S.L.; Rus, D.; Sukhatme, G.S. Persistent Ocean Monitoring with Underwater Gliders Towards Accurate Reconstruction of Dynamic Ocean Processes. In Proceedings of the 2011 IEEE International Conference Robotics and Automation (ICRA), Shanghai, China, 9–13 May 2011; pp. 1517–1524.
40. Techy, L.; Tomokiyo, R.; Quenzer, J.; Beauchamp, T.; Morgansen, K. *Full Scale Wind Tunnel Study of the Seaglider Underwater Glider*; UWAA Technical Report UWAATR-2010-0002; University of Washington: Seattle, WA, USA, September 2010.
41. Wilcox, J.S.; Bellingham, J.G.; Zhang, Y.; Baggeroer, A.B. Performance metrics for oceanographic surveys with autonomous underwater vehicles. *IEEE J Oceanic Eng.* **2001**, *26*, 711–725. [CrossRef]
42. Petritoli, E.; Giagnacovo, T.; Leccese, F. Lightweight GNSS/IRS integrated navigation system for UAV vehicles. In Proceedings of the 2014 IEEE International Workshop on Metrology for Aerospace (MetroAeroSpace), Benevento Italy, 29–30 June 2014; pp. 56–61. [CrossRef]
43. Webb, D.C.; Simonetti, P.J.; Jones, C.P. SLOCUM: An underwater glider propelled by environmental energy. *IEEE J Oceanic Eng.* **2001**, *26*, 447–452. [CrossRef]
44. Stommel, H. The Slocum Mission. *Oceanography* **1989**, *2*, 22–25. [CrossRef]
45. Sherman, J.; Davis, R.E.; Owens, W.B.; Valdes, J. The autonomous underwater glider "Spray". *IEEE J Oceanic Eng.* **2001**, *26*, 437–446. [CrossRef]
46. Titterton, D.; Weston, J. *Strapdown Inertial Navigation Technology*, 2nd ed.; The Institution of Electrical Engineers and The American Institute of Aeronautics and Astronautics: Reston, VA, USA, 2004.
47. Jun, B.H.; Park, J.Y.; Lee, F.Y.; Lee, P.M.; Lee, C.M.; Kim, K.; Lim, Y.K.; Oh, J.H. Development of the AUV 'ISiMI' and free running test in an ocean engineering basin. *Ocean Eng.* **2009**, *36*, 1. [CrossRef]
48. Leonard, N.E.; Graver, J.G. Model-Based feedback control of autonomous underwater gliders. *IEEE J. Ocean Eng.* **2001**, *26*, 4. [CrossRef]
49. Arima, M.; Ichihashi, N.; Ikebuchi, T. Motion characteristics of an underwater glider with independently controllable main wings. In Proceedings of the OCEANS 2008—MTS/IEEE Kobe Techno-Ocean, Kobe, Japan, 8–11 April 2008; pp. 1–7.

50. Roger, E.O.; Genderson, J.G.; Smith, W.S.; Denny, G.; Farley, P.J. Underwater acoustic glider. In Proceedings of the IEEE International Geoscience and Remote Sensing Symposium, Anchorage, AK, USA, 20–24 September 2004; pp. 2241–2244.
51. Woithe, H.C.; Kremer, U. A programming architecture for smart autonomous underwater vehicles. In Proceedings of the IEEE/RSJ International Conference on Intelligent Robots and Systems 2009—IROS 2009, St. Louis, MI, USA, 11–15 October 2009; pp. 4433–4438.
52. Woithe, H.C.; Tilkidjieva, D.; Kremer, U. *Towards a Resource-Aware Programming Architecture for Smart Autonomous Underwater Vehicles*; Technical Report, DCS-TR-637; Rutgers University: New Brunswick, NJ, USA, 2008.
53. Lennart, L. *System Identification Theory for the User*, 2nd ed.; Prentice Hall: Upper Saddle River, NJ, USA, 1999.
54. Petritoli, E.; Leccese, F. High accuracy attitude and navigation system for an autonomous underwater vehicle (AUV). *Acta IMEKO* **2018**, *7*, 3–9. [CrossRef]
55. Petritoli, E.; Leccese, F. A high accuracy navigation system for a tailless underwater glider. In Proceedings of the IMEKO TC19 Workshop on Metrology for the Sea, MetroSea 2017: Learning to Measure Sea Health Parameters, Naples, Italy, 11–13 October 2017; pp. 127–132.
56. Petritoli, E.; Leccese, F. A high accuracy attitude system for a tailless underwater glider. In Proceedings of the IMEKO TC19 Workshop on Metrology for the Sea, MetroSea 2017: Learning to Measure Sea Health Parameters, Naples, Italy, 11–13 October 2017; pp. 7–12.
57. DSIAC. Reliability of UAVs and Drones. Available online: https://www.dsiac.org/resources/journals/dsiac/spring-2017-volume-4-number-2/reliability-uavs-and-drones (accessed on 28 February 2019).
58. Etkin, B. *Dynamic of Flight*; John Wiley and Sons: Hoboken, NJ, USA, 1959.
59. Tan, X.; Kim, D.; Usher, N.; Laboy, D.; Jackson, J.; Kapetanovic, A.; Rapai, J.; Sabadus, B.; Zhou, X. An Autonomous Robotic Fish for Mobile Sensing. In Proceedings of the 2006 IEEE/RSJ International Conference on Intelligent Robots and Systems, Beijing, China, 9–15 October 2006; pp. 5424–5429.
60. Smith, R.N.; Pereira, A.; Chao, Y.; Li, P.P.; Caron, D.A.; Jones, B.H.; Sukhatme, G.S. Autonomous Underwater Vehicle trajectory design coupled with predictive ocean models: A case study. In Proceedings of the 2010 IEEE International Conference on Robotics and Automation (ICRA), Anchorage, AK, USA, 4–8 May 2010; pp. 4770–4777.
61. Petritoli, E.; Leccese, F.; Cagnetti, M. A High Accuracy Buoyancy System Control for an Underwater Glider. In Proceedings of the 2018 IEEE International Workshop on Metrology for the Sea; Learning to Measure Sea Health Parameters (MetroSea), Bari, Italy, 8–10 October 2018; pp. 257–261. [CrossRef]
62. Techy, L.; Kristi Morganseny, A.; Woolseyz, C.A. Long-baseline acoustic localization of the Seaglider underwater glider. In Proceedings of the American Control Conference (ACC) 2011, San Francisco, CA, USA, 29 June–1 July 2011; pp. 3990–3995.
63. Hussain, N.A.A.; Ali, S.S.A.; Saad, M.N.M.; Nordin, N. Underactuated nonlinear adaptive control approach using U-model for multivariable underwater glider control parameters. In Proceedings of the 2016 IEEE International Conference on Underwater System Technology: Theory and Applications (USYS), Penang, Malaysia, 13–14 December 2016; pp. 19–25. [CrossRef]
64. Yu, P.; Zhao, Y.; Wang, T.; Zhou, H.; Su, S.; Zhen, C. Steady-state spiral motion simulation and turning speed analysis of an underwater glider. In Proceedings of the 2017 4th International Conference on Information Cybernetics and Computational Social Systems (ICCSS), Dalian, China, 24–26 July 2017; pp. 372–377.
65. Bloch, T.; Krishnaprasad, P.S.; Leonard, N.; Murray, R. Jerrold Eldon Marsden [Obituaries]. *Control Syst. IEEE* **2011**, *31*, 105–108.
66. Williams, C.D. AUV systems research at the NRC-IOT: An update. In Proceedings of the Underwater Technology 2004—UT '04. 2004 International Symposium on, Taipei, Taiwan, 20–23 April 2004; pp. 59–73.
67. Jing, D.; Haifeng, W. System health management for unmanned aerial vehicle: conception, state-of-art, framework and challenge. In Proceedings of the Electronic Measurement & Instruments (ICEMI). In Proceedings of the 2013 IEEE 11th International Conference, Harbin, China, 16–19 August 2013; Volume 2, pp. 859–863.
68. Javafoil—Analysys of Airfolis. Available online: https://www.mh-aerotools.de/airfoils/javafoil.htm (accessed on 15 February 2019).
69. Vogeltanz, T. A Survey of Free Software for the Design, Analysis, Modelling, and Simulation of an Unmanned Aerial Vehicle. *Arch. Comput. Meth. Eng.* **2016**, *23*, 449. [CrossRef]

70. Vogeltanz, T.; Jašek, R. Free software for the modelling and simulation of a mini-UAV. In Proceedings of the Mathematics and Computers in Science and Industry, Varna, Bulgaria, 13–15 September 2014; pp. 210–215.
71. Jodeh, N.; Blue, P.; Waldron, A. Development of small unmanned aerial vehicle research platform: Modeling and simulating with flight test validation. In Proceedings of the AIAA Modeling and Simulation Technologies Conference and Exhibit, Keystone, CO, USA, 21–24 August 2006; p. 6261.
72. Mueller, T.J.; DeLaurier, J.D. Aerodynamics of small vehicles. *Annual review of fluid mechanics* **2003**, *35*, 89–111. [CrossRef]
73. Fabiani, P.; Fuertes, V.; Piquereau, A.; Mampey, R.; Teichteil-Königsbuch, F. Autonomous flight and navigation of VTOL UAVs: From autonomy demonstrations to out-of-sight flights. *Aerosp. Sci. Technol.* **2007**, *11*, 183–193. [CrossRef]
74. Wu, N.; Wu, C.; Ge, T.; Yang, D.; Yang, R. Pitch Channel Control of a REMUS AUV with Input Saturation and Coupling Disturbances. *Appl. Sci.* **2018**, *8*, 253. [CrossRef]
75. Hahn, A. Vehicle sketch pad: A parametric geometry modeller for conceptual aircraft design. In Proceedings of the 48th AIAA Aerospace Sciences Meeting Including the New Horizons Forum and Aerospace Exposition, Orlando, FL, USA, 4–7 January 2010; p. 657. [CrossRef]
76. Wang, T.; Wu, C.; Wang, J.; Ge, T. Modeling and Control of Negative-Buoyancy Tri-Tilt-Rotor Autonomous Underwater Vehicles Based on Immersion and Invariance Methodology. *Appl. Sci.* **2018**, *8*, 1150. [CrossRef]
77. Boussalis, H.; Valavanis, K.; Guillaume, D.; Pena, F.; Diaz, E.U.; & Alvarenga, J. Control of a simulated wing structure with multiple segmented control surfaces. In Proceedings of the 2013 21st Mediterranean Conference IEEE Control & Automation (MED), Crete, Greece, 25–28 June 2013; pp. 501–506.
78. Wang, T.; Wang, J.; Wu, C.; Zhao, M.; Ge, T. Disturbance-Rejection Control for the Hover and Transition Modes of a Negative-Buoyancy Quad Tilt-Rotor Autonomous Underwater Vehicle. *Appl. Sci.* **2018**, *8*, 2459. [CrossRef]

Article

Estimation of Wave Characteristics Based on Global Navigation Satellite System Data Installed on Board Sailboats

Paolo De Girolamo [1], Mattia Crespi [1], Alessandro Romano [1], Augusto Mazzoni [1], Marcello Di Risio [2,*], Davide Pasquali [2], Giorgio Bellotti [3], Myrta Castellino [1] and Paolo Sammarco [4]

1 "Sapienza", University of Rome, DICEA, Via Eudossiana, 18, 00184 Roma, Italy; paolo.degirolamo@uniroma1.it (P.D.G.); mattia.crespi@uniroma1.it (M.C.); alessandro.romano@uniroma1.it (A.R.); augusto.mazzoni@uniroma1.it (A.M.); myrta.castellino@uniroma1.it (M.C.)
2 Department of Civil, Construction-Architectural and Environmental Engineering (DICEAA)—Environmental and Maritime Hydraulic Laboratory (LIam), University of L'Aquila, 67100 L'Aquila, Italy; davide.pasquali@univaq.it
3 Engineering Department, Roma Tre University, Via V. Volterra, 62, 00146 Roma, Italy; giorgio.bellotti@uniroma3.it
4 Department of Civil Engineering and Computer Science , University of Rome "Tor Vergata", DICII, Via del Politecnico, 1, 00133 Roma, Italy; sammarco@ing.uniroma2.it
* Correspondence: marcello.dirisio@univaq.it

Received: 15 March 2019; Accepted: 14 May 2019; Published: 18 May 2019

Abstract: This paper illustrates a methodology to get a reliable estimation of the local wave properties, based on the reconstruction of the motion of a moving sailboat by means of GNSS receivers installed on board and an original kinematic positioning approach. The wave parameters reconstruction may be used for many useful practical purposes, e.g., to improve of autopilots, for real-time control systems of ships, to analyze and improve the performance of race sailboats, and to estimate the local properties of the waves. A Class 40 oceanic vessel (ECO40) left from the port of "Riva di Traiano" located close to Rome (Italy) on 19 October 2014 to perform a non-stop sailing alone around the world in energy and food self-sufficiency. The proposed system was installed on ECO40 and the proposed method was applied to estimate the wave properties during a storm in the Western Mediterranean Sea. The results compared against two sets of hindcast data and wave buoy records demonstrated the reliability of the method.

Keywords: GPS data analysis; off-shore wave climate; sailboat; ship motions; wave characteristics

---

1. Introduction

Estimation of ship motions and wave properties from a traveling ship is of utmost importance for many ocean engineering applications [1], for example: (i) to improve the performance of autopilots and control systems of ships [2] or in general of floating structures [3]; (ii) to provide real-time wave data for Global Forecasting Systems (GFS) numerical models and for the calibration/validation of wave forecasting/hindcasting models (e.g., [4]), etc.; and (iii) to analyze the performance of a race sailboat, i.e., to derive actual velocity polar curves in the presence of waves and to correct wind measurements carried out onboard. As far as the estimation of the wave properties is concerned, a possible approach is to derive the parameters of interest directly on the basis of the ship motion, taking into account the dynamic ship frequency response function (e.g., [1,5–7]).

In the last years, the measurement of directional wave spectra on the basis of moving ships kinematic has been extensively studied (e.g., [8,9]). Some approximated solutions of the main theoretical problem arising from the transformation of encounter frequencies measured onboard into the true wave frequencies have been derived. This problem is known as the "triple-valued function problem", which is due to the non-linear Doppler shift effect induced by frequency dispersive waves (e.g., [10,11]). Approximated solutions have been obtained by using both parametric methods, which assume the shape of the wave spectrum, and non-parametric (stochastic) methods, where the spectrum shape is not prescribed a priori [1,11,12].

Only in the last ten years, the GNSS technology has been applied on floating buoys for measuring directional wave spectra [13,14]. The advantages, as well as the disadvantages, of using GNSS buoys were pointed out by Herbers et al. [14]. On the other hand, onboard measurements of ship-motion is normally carried out by means of accelerometers used for standard pitch/roll directional wave buoys [10,11,15–17].

This paper describes the estimation of ship motion using GPS receivers, installed on a traveling sailboat for the evaluation of the local wave properties. It has to be stressed that the use of GNSS on an offshore traveling sailboat is challenging. Indeed, the distance of the moving GNSS receivers installed onboard (i.e., rover receivers) from reference static receivers of known coordinates (i.e., base stations) should not exceed few tens of kilometers. Indeed, Herbers et al. [14] used off-the-shelf GPS receivers able to provide positions with accuracy limited to a few meters, mentioning that sub-meter accuracy is achievable with post-processing methodology only. In this work, we propose to adopt the original methodology Kin-Vadase, presented and widely validated by Branzanti et al. [18], able to provide accuracies at decimeter level and at few millimeters/second for positions and velocities directly on-board, respectively (no need of external data, such as differential positioning or precise point positioning) and in real time. With respect to the work of De Girolamo et al. [19], this paper deals with a more detailed description of the methods and extends the results.

The paper is structured as follows. Section 2 illustrates the methodology to analyze the GNSS data and to use them in order to reconstruct the boat motion and, then, the wave parameters. Section 3 illustrates the real scale ECO40 sailboat used to test the proposed methodology during a non-stop sailing alone around the world. Section 4 discusses the results and demonstrates the reliability of the proposed method by comparing the estimated wave parameters against measured and hindcast data. Concluding remarks close the paper.

2. Methods

During offshore oceanic sailing, GPS measurements cannot be corrected using on land reference receivers, or a GNSS (Global Navigation Satellite System) network, since the distance between rovers and reference station should not be larger than few tens of kilometers. To solve this problem, in the present work, two GPS data processing techniques have been applied, by using a novel approach. The processed data provided by the GPS receivers have been used to compute the boat motion and to estimate the waves properties faced by the sailboat during its navigation around the world. This sections aims at detailing the proposed methods.

2.1. Analysis of Gnss Data

The GPS raw code and phase observations on both L1 and L2 frequencies were acquired by each GPS antenna/receiver system (in the following referred to as "receiver") with a sampling rate of 2 Hz. The raw observations were stored on a flash-card by each receiver. The data analysis was carried out in post-processing after the recovery of the flash-cards. The analysis carried out in the post processing described in the following, however, can be carried out in real-time onboard of the sailboat. The post-processing was carried out by employing two different techniques: the Variometric approach and the Moving Base Kinematic.

The Variometric Approach for Displacements Analysis Standalone Engine (VADASE) is an innovative GPS data processing approach proposed in the recent past [20,21]. The approach is based on timing single differences of carrier phase observations continuously collected using a standalone GPS receiver and on standard GPS broadcast products (orbits and clocks) that are available in real-time. Therefore, one receiver works in standalone mode and the epoch-by-epoch displacements (equivalent to velocities) are estimated. In this work, a kinematic extension of the Variometric approach (Kin-VADASE) developed specifically for the navigation field was used.

A second approach was used. Indeed, the most widely used technique in GPS kinematic positioning is based on the using of two devices: the rover moving receiver and a reference static receiver of known coordinates. To obtain high accuracy results, the distance between the rover and the reference receiver should not exceed few tens of kilometers. Thus, it is almost impossible to apply this technique in off-shore navigation, due to the lack of close reference receivers availability. Nevertheless, in this work, we present an innovative application of differential kinematic positioning applied to a reference moving (hereinafter referred to as Moving Base Kinematic, MBK). With respect to this reference moving receiver, it was possible to estimate, epoch by epoch, the positions of another receiver. This technique does not allow defining the absolute position of the receiver but only the relative one, which is very accurate, since the distance between each couple of receivers is known. The results of the GPS data post processing may then be used to reconstruct the six Degrees of Freedom (hereinafter referred to as 6DOFs) boat motion and to estimate the wave parameters encountered by sailing boats (i.e., significant wave height and mean wave direction). Actually, only heave, roll and pitch motions are described as they are needed to estimate the waves parameters.

2.2. Boat Motion Reconstruction

This subsection describes the method used to reconstruct the 6DOFs boat motion by using the two approaches described in Section 2.1. The two approaches have been used in a complementary way. In the following, the methods used to compute each degree of freedom of the boat from GPS measurements needed to estimate the wave parameters are described. The boat 6DOFs are referred in the following to the center of the triangle defined by the three GPS positions and not to the center of gravity of the boat. However, a simple translation may be applied to the results in order to express the motion of the boat with respect to its center of gravity, as usual. It has to be stressed, however, that this work aims at estimating the wave parameters and that the 6DOF boat motion is used to reach the goal.

2.2.1. Heave

To compute the boat heave, it is necessary to use a technique able to provide the absolute vertical position of the boat. Nevertheless, the up-velocity component provided by Kin-VADASE method is subjected to bias, mainly related to the number of visible satellites and to their position with respect to the GPS antennas location. A sensitivity analysis was performed by using different numerical techniques (i.e., first order, Simpson's rule, etc.) to minimize the integration error. It is worth noting that heave motion should exhibit a mean value very close to zero. Nevertheless, whatever the used numerical technique, the bias on the velocities gives rise to a vertical movement of the antenna due to low frequency drift related to the numerical integration. To eliminate the bias, a high pass filter was used. The high pass filter was applied to the vertical movement time history obtained by integrating over time the up-velocity component provided by a GPS antenna. The cut-off frequency was chosen large enough to avoid the low frequency drift related to the numerical integration, but small enough to analyze both the sea and the swell wave features.

2.2.2. Roll

For the computation of the boat roll motion, it is not necessary to use a technique able to provide the absolute position of the boat, as roll may be obtained by means of the relative vertical differences between two points located along an axis orthogonal to the main longitudinal axis of the boat.

The roll angle is not characterized by a zero mean value mainly because of two reasons. The first one is related to the wind acting on the sail and on the boat hull, which causes the heeling of the sailboat. It is a function of the wind speed and of the sails trim and configuration. The second one is due to the movable weight (i.e., ballast, not used sails, spare equipment, etc.), which are moved by the crew in order to reduce the heeling induced by the wind action. The heeling angle may be obtained by calculating the mean total roll angle value over a time longer than the waves encountered periods.

2.2.3. Pitch

The computation of the pitch motion, as for the roll, can be carried out by using the moving base kinematic (MBK) technique. For this parameter, two GPS receivers located along the main longitudinal axis of the sailboat should be used. Nevertheless, other antennas configurations could be used (see Section 3). In this case, it has to be stressed that the signals may be affected by the influence of the roll motion. To get the estimation of the pitch signal but the roll component, a mean signal (i.e., the actual pitch) can be computed (see Section 4).

2.3. Wave Parameters Estimation

The synthetic parameters of the waves faced by sailboats can be derived from the heave, roll and pitch motions. First, the transfer function between the boat and the sea surface movements has to be evaluated and applied to the time series estimated on the basis of GNSS data. Then, the duration of the time window to be analyzed needs to be defined. Herein, time windows of 30 min were considered. Indeed, this duration, which is typical for standard analysis of the free surface elevation signals as measured from conventional wave buoys, is considered to be long enough to properly describe, in a statistical sense, the considered sea state and short enough for considering the wave features to be stationary [22]. On the one hand, the significant wave height may be directly inferred from the heave motion by performing standard time and frequency domain analyses. On the other hand, the directional spectrum may be estimated by using the Direct Fourier Transform Method (i.e., by means of the DIWASP package, e.g., [23]) applied to the heave, pitch and roll signals without taking into account the Doppler shift effect due to the sailboat traveling. It has to be stressed that the influence of the waves upon the roll angle may be obtained by subtracting the heeling angle from the total measured roll angle, obtaining in this way a signal characterized by a mean value which is equal to zero.

3. The Experimental Sailboat ECO40

The considered sailboat is a small oceanic race boat length overall of 12.0 m) characterized by a very light displacement. She is a Class 40 oceanic vessel named ECO40, which left from the port of "Riva di Traiano" located close to Rome (Italy) on 19 October 2014 to perform a non-stop sailing alone around the world in energy and food self-sufficiency. The boat route goes through the Gibraltar Strait, then descends the Atlantic Ocean and sailing around the Antarctic, at a mean latitude of 50° S, from west to east, rounding the famous capes of the world: Cape of Good Hope, Cape Leeuwin and Cape Horn. Finally, it ascends the Atlantic Ocean, passing again the Strait of Gibraltar and coming back to homeport [24,25].

The boat was equipped with three high precision GPS receivers (rovers), provided by Leica Geosystem, for measuring the movements of the boat. An ad hoc survey was performed just after their installation in order to know the positions of the GPS antennas needed to infer the three-dimensional position and attitude of the boat as a rigid body.

The three geodetic class GPS antenna/receiver systems are able to acquire code and phase observations on both L1 and L2 frequencies. The GPS antennas were installed on board ECO40, as shown in Figure 1. Two antennas were mounted on the stern roll bar (they are indicated as GPS Stern-right and GPS Stern-left in the figure) and one antenna (GPS Bow) was installed along the main axis of the boat, close to the entrance and protected by a small fiberglass structure. The positions of the three GPS antenna were at the vertices of an isosceles triangle, as described in the Figure 1. The proposed method was applied to three couples of receivers deployed on board (Figure 1): GPS Bow–GPS Stern-right; GPS Bow–GPS Stern-left; and GPS Stern-right–GPS Stern-left.

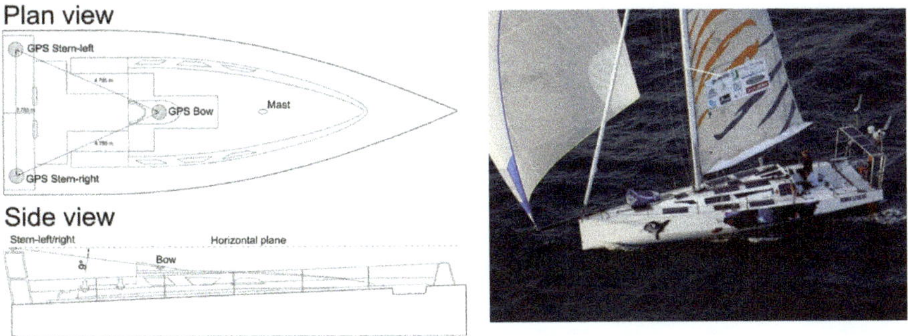

Figure 1. Plan view of the sailboat ECO40 and of the three GPS antennas (**Top Left**) and side view of the sailboat (**Bottom Left**); and (**Right**) a picture of the sailboat ECO40 during a test.

4. Results

When ECO40 left from the Italian Port Riva di Traiano (Italy) directed to Gibraltar Strait (on 19 October 2014), the weather forecasts suggested that, within the next 24/48 h, the first seasonal front of cold air was expected to induce Mistral winds with speed exceeding 40 knots, blowing from the Gulf of Lion [24]. ECO40 was able to reach the Asinara Island and to follow the route towards the Balearic Islands before the arrival of the main storm: the sailboat faced the storm running on the quarter. The route between the Asinara Island and the Balearic Islands is represented in Figure 2. The figure also provides information on the travel times.

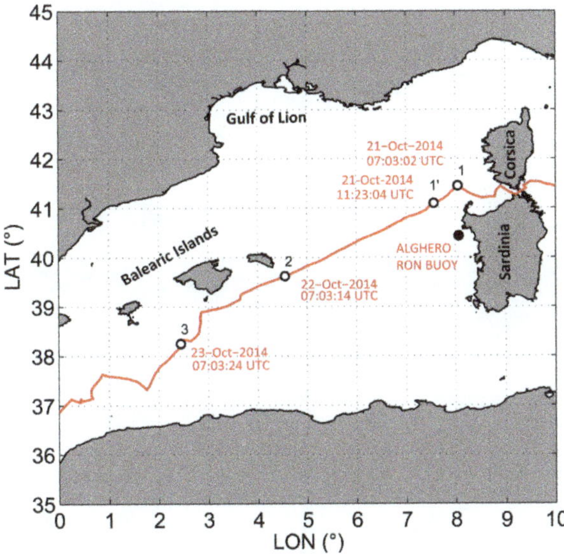

Figure 2. The route in the western Mediterranean Sea followed by ECO40. The empty dots refer to four points of interest along the route and the black dot identifies the position of the Alghero wave buoy.

This section aims at illustrating and discussing the results obtained by means of the proposed methodology described in Section 2.

The boat heave time series was obtained by computing the instantaneous mean values of the three filtered heave time series, as obtained by each of the three GPS antennas. Figure 3 shows an example of the calculated boat heave signal for a time window of four days.

The roll motion angle was computed by using the moving base kinematic (MBK) applied to the relative Up position of the couple of Stern-GPSs. Figure 4 shows the roll angle during four days of navigation of ECO40. As expected, the analysis revealed that the measured roll angle was not characterized by a zero mean value. The red line in Figure 4 (top) shows the heeling angle computed over a time of 10 min. It is possible to identify the route changes of the boat which take place when the heeling angle changes from positive (port) to negative values (starboard) or vice versa. The influence of the waves upon the roll angle obtained by subtracting the heeling angle to the total measured roll angle is also shown in Figure 4 (bottom).

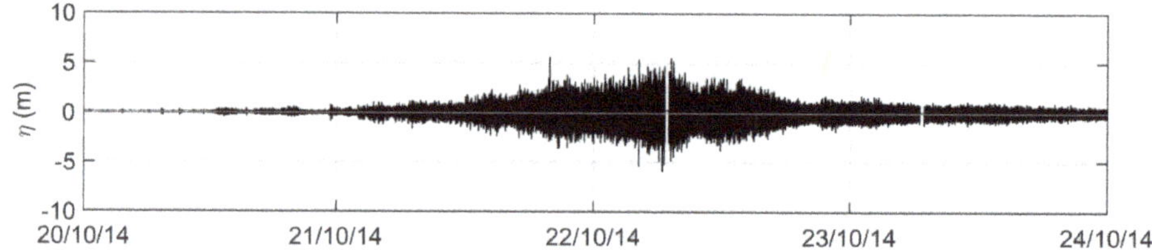

Figure 3. Example of the boat heave time series over a time window of four days, obtained by calculating the instant mean values of the three filtered heave time series as obtained by each GPS antenna.

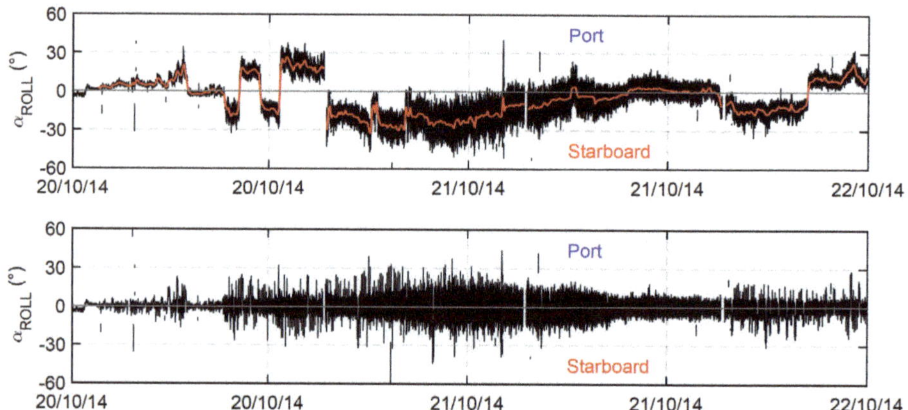

Figure 4. (**Top**) Example of the total measured roll angle time series (black line) over a time window of four4 days. The red line refers to the heeling angle calculated by averaging over 10 min. (**Bottom**) The roll angle obtained by subtracting the heeling angle to the total measured roll angle.

As far as the pith motion is considered, it has to be observed that two GPS receivers located along the main longitudinal axis of the sailboat were not available and therefore a direct estimation of the pitch could not be obtained. Two couples of GPS receivers were therefore considered: the bow–stern-right and the bow–stern-left. Figure 5 shows an example of the 2 Hz time series of the pitch angle obtained directly from the relative up position of each couple of GPS measurements. The time series were evaluated with respect to the local horizontal plane. Figure 5 (top) shows the time series of the pitch angle obtained by using the bow–stern-left GPS couple, while Figure 5 (middle) represents the same quantity as obtained by using the bow–stern-right GPS couple. The figure shows that both signals were contaminated by a component of the roll motion because the two axes passing for each couple of GPS (bow-stern right and bow-stern left) were not parallel to the main boat axis. Furthermore, the comparison of Figure 5 (top and middle) shows that the two signals are in phase opposition because of the changing of the reference GPS. To obtain the pitch signal depurated by the roll component, a mean signal obtained from the two couple bow–stern right and the bow–stern left was computed (Figure 5, bottom).

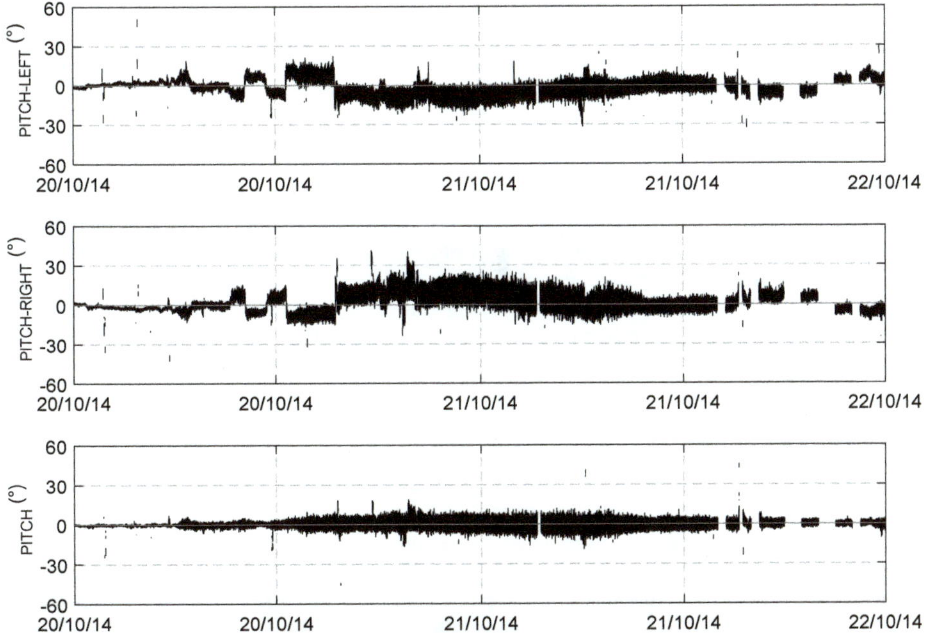

Figure 5. Time series of the angle obtained by using the bow–stern-left GPS couple (**Top**); and the same quantity as obtained by using the bow–stern-right GPS couple (**Middle**) over a time window of four days. (**Bottom**) Time series of the mean signal (pitch) obtained from the two couples bow–stern right and the bow–stern left over a time window of four days.

The features of the waves faced by the sailboat during the Gulf of Lion event were derived from the sailboat heave, roll and pitch motions. Since ECO40 is an ultra-light displacement boat (she has full loaded displacement of about 4700 kg, a length of about 12.0 m and a maximum breadth of 4.5 m), it was assumed that the sailboat follows the sea surface, thus assuming a unitary transfer function between the boat and sea surface movements (i.e., the response amplitude operator is the unit matrix).

Figure 6 shows the significant wave height time series of the storm obtained from the heave signal by using both the zero-crossing analysis ($H_{1/3}$) and the frequency spectrum analysis (H_{m0}) carried out in the encountered frequency domain. The maximum zero crossing wave height (H_{max}) is also represented. The maximum values obtained during the storm for $H_{1/3}$ and for H_{m0} were, respectively, 5.84 m and 5.42 m, while H_{max} was of 9.6 m. The vertical dashed lines refer to four points of interest along the route (as shown in Figure 2).

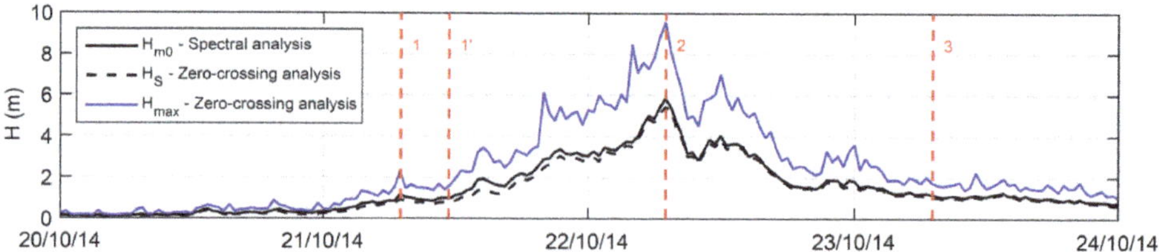

Figure 6. Significant wave height obtained from the heave by using zero crossing analysis (black dashed line) and frequency spectrum analysis (black line). The maximum wave height is reported as well (blue line). The vertical red dashed lines refer to the four points of interest along the route (as shown in Figure 2).

To assess the accuracy of the estimated wave parameters, Figure 7 shows the comparison between the significant wave height and mean wave direction as: (i) measured by the Alghero RON wave buoy; (ii) estimated by the ECMWF analysis; and (iii) estimated by DICCA for a grid point close to the buoy position. It has to be stressed that this study aimed at estimating the wave parameters based on GNSS data analysis (see Section 2). Therefore, only the synthetic wave parameters were compared to the results of other research works, being the original methodology Kin-VADASE widely validated (see Section 2, [18]) even if applied to a different environment.The Alghero wave buoy belongs to the "RON-Rete Ondametrica Nazionale" (Italian Wave Measurements Network [26]), managed by ISPRA up to 2014 and now dismissed after about 23 years; the buoy was located offshore the northwest coast of Sardinia and it was exposed to the waves generated by the storm at hand. The six-hourly hindcast wave data provided in analysis by ECMWF (European Centre for Medium-Range Weather Forecasts) and the the hourly wave hindcast data provided by DICCA (Department of Civil, Chemical and Environmental Engineering, University of Genoa, Italy), covering 36 (1979–2014) years over the Mediterranean Sea [27] were also considered. were interpolated in time and in space by means of bi-linear technique. Since the buoy was operating at a fixed position, the comparison between the buoy wave data and the ones measured onboard was performed just for the time window in which the two measurements can be reasonably compared. The analysis revealed that the synthetic parameters reconstructed on the basis of sailboat motion was in close agreement with the DICCA data. On the other hand, the ECMWF data seemed to underestimate the storm peak, as already outlined by previous work e.g., [4], likely due to the time and spatial resolution of the hindcast data. As far as the mean wave direction is concerned, the accuracy of the proposed method was reasonable.

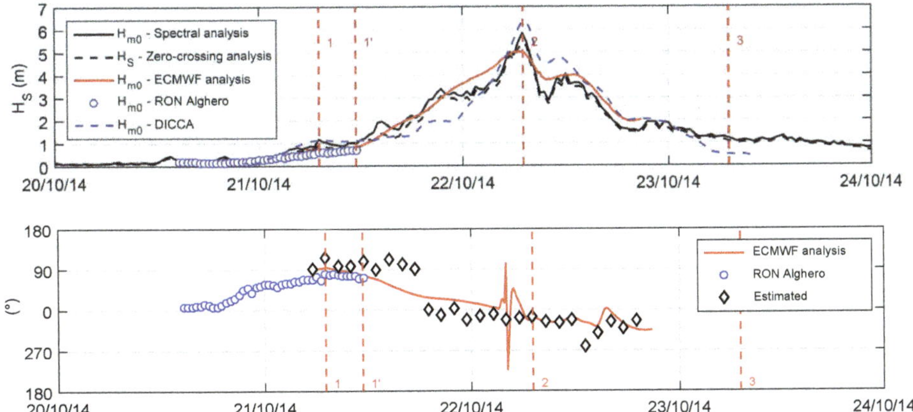

Figure 7. (**Top**) Comparison between the significant wave height measured onboard and the available wave data; and (**Bottom**) comparison between the mean spectral wave direction as estimated by means of the proposed method and the same quantity as obtained by the ECMWF data (red line) and Alghero wave buoy data (blue dots). The red dashed lines refer to four points of interest along the route (as shown in Figure 2).

5. Concluding Remarks

A method to reconstruct the movements of a sailboat in order to get a reliable estimate of wave parameters, using GPS receivers, is presented. It is based on the combined approach of the Kin-VADASE and Moving Base Kinematic techniques. The proposed method was applied to the records collected onboard the ECO40 sailboat during a storm in the Western Mediterranean Sea. Three GPS antennas were deployed on the sailboat. Their positions were measured by means of a survey performed when they were installed. Therefore, the three-dimensional locations of the GPS antennas suffice to infer the three-dimensional position and attitude of the boat as a rigid body, hence to reconstruct wave parameters. A fourth GPS antenna could be considered to improve the reliability of the estimation of the position and attitude of the boat.

Wave properties as derived on the basis of the estimated boat movements were compared to available wave data. It appears that the wave properties were calculated with a high degree of accuracy. Note that the main aim of this work was the estimation of wave parameters, hence the GNSS data analysis was not directly validated. Of course, the good agreement between estimated and available data may be viewed as an indirect validation of GNSS data analysis as well.

In general, the techniques proposed in this paper appear to be suitable to estimate the 6DOFs movements of sailboats navigating offshore, where onshore reference receivers or GNSS network cannot be used to correct GPS errors. The motion of the boat can provide quantitative information for many important practical applications, such as the improvement of autopilots, the analysis and the optimization of ship performances, and the estimation of local wave properties. Furthermore, it can be used to correct wind measurements carried out onboard sailboats.

Author Contributions: Conceptualization, P.D.G., M.C. (Mattia Crespi) and A.R.; Methodology, P.D.G., M.C. (Mattia Crespi), A.R., A.M., M.D.R., D.P., G.B., M.C. (Myrta Castellino) and P.S.; Software, A.R., A.M. and D.P.; Validation, P.D.G., M.C. (Mattia Crespi); Formal Analysis, P.D.G., M.C. (Mattia Crespi), A.R., A.M., M.D.R., D.P., G.B., M.C. (Myrta Castellino) and P.S.; Investigation, P.D.G., M.C. (Mattia Crespi), A.R., A.M., M.D.R., D.P., G.B., M.C. (Myrta Castellino) and P.S.; Resources, P.D.G. and M.C. (Mattia Crespi); Data Curation, A.R. and A.M.; Writing—Original Draft Preparation, P.D.G., M.C. (Mattia Crespi), A.R., A.M., M.D.R., D.P., G.B., M.C. (Myrta Castellino) and P.S.; Writing—Review & Editing, P.D.G., M.C. (Mattia Crespi), A.R., A.M., M.D.R., D.P., G.B., M.C. (Myrta Castellino) and P.S.; Visualization, A.R. and D.P.; Supervision, P.D.G and M.C. (Mattia Crespi).

Funding: This research received no external funding.

Acknowledgments: Pier Paolo Pecoraro (IDRO-GEOTEC) is acknowledged for his help and expertise during the installation of the instruments. The skipper Matteo Miceli is also acknowledged for his contribution. A special acknowledgement is due to Leica Geosystem for providing the high-precision GPS receivers and to Giovanni Besio for providing the DICCA wave data.

Conflicts of Interest: The authors declare no conflict of interest.

References

1. Iseki, T.; Ohtsu, K. Bayesian estimation of directional wave spectra based on ship motions. *Control. Eng. Pract.* **2000**, *8*, 215–219. [CrossRef]
2. Belleter, D.J.; Galeazzi, R.; Fossen, T.I. Experimental verification of a global exponential stable nonlinear wave encounter frequency estimator. *Ocean. Eng.* **2015**, *97*, 48–56. [CrossRef]
3. Tannuri, E.A.; Sparano, J.V.; Simos, A.N.; Da Cruz, J.J. Estimating directional wave spectrum based on stationary ship motion measurements. *Appl. Ocean. Res.* **2003**, *25*, 243–261. [CrossRef]
4. De Girolamo, P.; Di Risio, M.; Beltrami, G.; Bellotti, G.; Pasquali, D. The use of wave forecasts for maritime activities safety assessment. *Appl. Ocean. Res.* **2017**, *62*, 18–26. [CrossRef]
5. Hirayama, T. Real-time estimation of sea spectra based on motions of a running ship, 2nd report: Directional wave estimation. *J. Kansai Soc. Nav. Archit.* **1987**, *204*.
6. Iseki, T.; Ohtsu, K.; Fujino, M. A Study on Estimation of Directional Spectra Based on Ship Motions. *Jpn. Inst. Navig. J.* **1992**, *86*, 179–188. [CrossRef]
7. Webster, W.C.; Dillingham, J.T. Determination of directional seas from ship motions. In Proceedings of the Conference on Directional Wave Spectra Applications, Berkeley, CA, USA, 14–16 September 1981.
8. Roggenbuck, O.; Reinking, J. Sea Surface Heights Retrieval from Ship-Based Measurements Assisted by GNSS Signal Reflections. *Mar. Geod.* **2019**, *42*, 1–24. [CrossRef]
9. Roggenbuck, O.; Reinking, J.; Lambertus, T. Determination of Significant Wave Heights Using Damping Coefficients of Attenuated GNSS SNR Data from Static and Kinematic Observations. *Remote Sens.* **2019**, *11*, 409. [CrossRef]
10. Nielsen, U.D. Deriving the absolute wave spectrum from an encountered distribution of wave energy spectral densities. *Ocean Eng.* **2018**, *165*, 194–208. [CrossRef]
11. Nielsen, U.D.; Brodtkorb, A.H.; Sørensen, A.J. Sea state estimation using multiple ships simultaneously as sailing wave buoys. *Appl. Ocean Res.* **2019**, *83*, 65–76. [CrossRef]
12. Nielsen, U.D. Estimations of on-site directional wave spectra from measured ship responses. *Mar. Struct.* **2006**, *19*, 33–69. [CrossRef]
13. Kahma, K.; Hauser, D.; Krogstad, H.; Lehner, S.; Monbaliu, J.; Wyatt, L. *Measuring and Analysing the Directional Spectra of Ocean Waves*; COST 714; EUR 21367; COST Office: Brussels, Belgium, 2005.
14. Herbers, T.; Jessen, P.; Janssen, T.; Colbert, D.; MacMahan, J. Observing ocean surface waves with GPS-tracked buoys. *J. Atmos. Ocean. Technol.* **2012**, *29*, 944–959. [CrossRef]
15. Longuet-Higgins, M.; Cartwright, D.; Smith, N. Observations of the Directional Spectrum of Sea Waves Using the Motions of a Floating Buoy. *Deep Sea Res. Oceanogr. Abstr.* **1965**, *12*, 53.
16. Long, R.B. The statistical evaluation of directional spectrum estimates derived from pitch/roll buoy data. *J. Phys. Oceanogr.* **1980**, *10*, 944–952. [CrossRef]
17. Mitsuyasu, H.; Tasai, F.; Suhara, T.; Mizuno, S.; Ohkusu, M.; Honda, T.; Rikiishi, K. Observations of the directional spectrum of ocean WavesUsing a cloverleaf buoy. *J. Phys. Oceanogr.* **1975**, *5*, 750–760. [CrossRef]
18. Branzanti, M.; Colosimo, G.; Mazzoni, A. Variometric approach for real-time GNSS navigation: First demonstration of Kin-VADASE capabilities. *Adv. Space Res.* **2017**, *59*, 2750–2763. [CrossRef]
19. De Girolamo, P.; Crespi, M.; Romano, A.; Mazzoni, A.; Di Risio, M.; Pasquali, D.; Bellotti, G.; Castellino, M.; Sammarco, P. Wave characteristics estimation by GPS receivers installed on a sailboat travelling off-shore. In Proceedings of the 2018 IEEE International Workshop on Metrology for the Sea; Learning to Measure Sea Health Parameters (MetroSea), Bari, Italy, 8–10 October 2018; pp. 18–22. [CrossRef]
20. Colosimo, G.; Crespi, M.; Mazzoni, A. Real-time GPS seismology with a stand-alone receiver: A preliminary feasibility demonstration. *J. Geophys. Res. Solid Earth* **2011**, *116*. [CrossRef]
21. Colosimo, G.; Crespi, M.; Mazzoni, A.; Dautermann, T. Co-seismic displacement estimation: Improving tsunami early warning systems. *Gim Int.* **2011**, *25*, 19–23.

22. Holthuijsen, L.H. *Waves in Oceanic and Coastal Waters*; Cambridge University Press: Cambridge, UK, 2010.
23. Barber, N. The directional resolving power of an array of wave detectors. In *Proceedings of the Conference Ocean Wave Spectra*; Prentice-Hall, Inc.: Upper Saddle River, NJ, USA; pp. 137–150.
24. De Girolamo, P.; Romano, A.; Bellotti, G.; Pezzoli, A.; Boscolo, A.; Crespi, M.; Mazzoni, A.; Di Risio, M.; Pasquali, D.; Franco, L.; et al. Analysis of the 21/22 October 2014 storm experienced by the sailboat ECO40 in the Gulf of Lion. In Proceedings of the 3rd International Congress on Sport Sciences Research and Technology Support, icSPORTS, Lisbon, Portugal, 15–17 November 2015; pp. 290–298.
25. De Girolamo, P.; Romano, A.; Bellotti, G.; Pezzoli, A.; Castellino, M.; Crespi, M.; Mazzoni, A.; Di Risio, M.; Pasquali, D.; Franco, L.; et al. Met-ocean and heeling analysis during the violent 21/22 october 2014 storm faced by the sailboat ECO40 in the gulf of lion: Comparison between measured and numerical wind data. *Commun. Comput. Inf. Sci.* **2016**, *632*, 86–105. [CrossRef]
26. De Boni, M.; Cavaleri, L.; Rusconi, A. The Italian waves measurement network. In Proceedings of the 23rd International Conference on Coastal Engineering, Venice, Italy, 4–9 October 1992; pp. 1840–1850.
27. Mentaschi, L.; Besio, G.; Cassola, F.; Mazzino, A. Performance evaluation of Wavewatch III in the Mediterranean Sea. *Ocean Model.* **2015**, *90*, 82–94. [CrossRef]

Article

The AMERIGO Lander and the Automatic Benthic Chamber (CBA): Two New Instruments to Measure Benthic Fluxes of Dissolved Chemical Species †

Federico Spagnoli [1,*], Pierluigi Penna [1], Giordano Giuliani [1], Luca Masini [2] and Valter Martinotti [3]

1. Istituto per le Risorse Biologiche e le Biotecnologie Marine (IRBIM), Consiglio Nazionale delle Ricerche (CNR), Ancona 60125, Italy; pierluigi.penna@cnr.it (P.P.); giordano.giuliani@cnr.it (G.G.)
2. Institute for Microelectronics and Microsystems, Consiglio Nazionale delle Ricerche, Bologna 40129, Italy; masini@bo.imm.cnr.it
3. Dipartimento Sviluppo sostenibile e Fonti Energetiche, RSE SpA, Milano 20134, Italy; valter.martinotti@rse-web.it

* Correspondence: federico.spagnoli@cnr.it; Tel.: +39-071-2078847

† This paper is an extended version of paper published in Spagnoli, F.; Giuliani, G.; Martinotti, V.; Masini, L.; Penna, P. AMERIGO and CBA: A new lander and a new automatic benthic chamber for dissolved benthic flux measurements. In Proceedings of the 2018 IEEE International Workshop on Metrology for the Sea; Learning to Measure Sea Health Parameters (MetroSea), Bari, Italy, 8–10 October 2018.

Received: 28 April 2019; Accepted: 5 June 2019; Published: 10 June 2019

Abstract: Marine environments are currently subject to strong ecological pressure due to local and global anthropic stressors, such as pollutants and atmospheric inputs, which also cause ocean acidification and warming. These strains can result in biogeochemical cycle variations, environmental pollution, and changes in benthic-pelagic coupling processes. Two new devices, the Amerigo Lander and the Automatic Benthic Chamber (CBA), have been developed to measure the fluxes of dissolved chemical species between sediment and the water column, to assess the biogeochemical cycle and benthic-pelagic coupling alterations due to human activities. The Amerigo Lander can operate in shallow as well as deep water (up to 6000 m), whereas the CBA has been developed for the continental shelf (up to 200 m). The lander can also be used to deploy a range of instruments on the seafloor, to study the benthic ecosystems. The two devices have successfully been tested in a variety of research tasks and environmental impact assessments in shallow and deep waters. Their measured flux data show good agreement and are also consistent with previous data.

Keywords: lander; benthic chambers; benthic fluxes of dissolved chemical species; marine technology; marine instrumentation

---

1. Introduction

Marine environments are affected by strong ecosystem stressors that include direct human activities (e.g., marine traffic, offshore activities, mining, coastal works) and inputs (e.g., dumping of solid waste on the seafloor, anthropic inputs transported by rivers, ballast water discharge) and chemical and climate changes that act on a global scale (e.g., raised CO_2 levels and air temperature). These stressors affect marine chemistry and processes by inducing ocean acidification, global sea warming, and changes in hydrological and biogeochemical cycles [1]. Human activities also result in an increased supply of trophic substances, which, in some environmental settings such as shallow enclosed seas with low hydrodynamics, can lead to dystrophic crises [2–4]. These stressors also have the potential to alter the biogeochemical cycles of elements such as carbon, phosphorus, nitrogen, silicon, and metals, which can severely damage economic activities, such as fishing and tourism [4,5]. Moreover, the introduction and accumulation of heavy metals and organic substances (Polycyclic Aromatic Hydrocarbons (PAHs), pesticides, drugs) can induce strong pollution problems involving both the water column and sediment. These pollutants can modify or be incorporated in the food chains, damaging the ecosystem and heightening the risk for human health [5]. In particular, in the past few decades the marine biogeochemical cycle of carbon has undergone an acceleration as a consequence of increased atmospheric CO_2, which has resulted in reduced seawater pH and carbon sinking rates [6].

The alterations in the marine biogeochemical cycles of elements and pollution due to human activities can affect the water column, the bottom sediment, and the transfer processes at the sediment-water interface. The research into and the development of devices to enhance the study of marine biogeochemical and benthic-pelagic coupling processes is therefore very useful [7–9], also in Italy [10–13].

A variety of devices have been developed to study benthic-pelagic coupling processes. In the past few decades, benthic landers with different setups have been developed for a wide range of purposes and their technological features and models have extensively been reviewed [14–22]. Benthic landers are equipped with diverse sensors and devices according to the tasks they are deployed to perform, including microprofilers, planar optodes, and digital cameras to study sediment-water interface properties; eddy correlation systems to measure the fluxes of dissolved chemical species in extensive areas; video cameras to investigate the deep sea biota; and oceanographic sensors (O_2, pH, redox potential (Eh or ORP), optical turbidity, CTD, current meters, sediment traps) to study the water column [8,9,23–28]. In the past few years, other and much more complex and expensive devices have also been developed, such as landers for hadal environments [29–33], for multipurpose uses, and for transporting other mobile devices [34–37].

An important benthic-pelagic coupling process is the flux of dissolved chemical species at the sediment-water interface, generated by early diagenesis processes [38–40] or by volcanic benthic exhalation [41,42]. Such fluxes can strongly affect the chemistry of the water column, hence its ecology. This is especially true of shelf and coastal environments, where the high intensity of early diagenesis processes, due to a high reactive organic matter content in surface sediments, produces strong fluxes that affect shallow water columns [43].

The benthic fluxes of dissolved substances can be studied by onboard incubation or by in situ experiments, which are usually more reliable [44]. Fluxes are measured in benthic chambers handled by divers [45] in shallow waters, or mounted on benthic landers in deeper waters [44].

Landers equipped with one or more benthic chambers have been developed in the past [46–55].

Two new, low-cost, light, and easy to handle devices, the Amerigo Lander and the Automatic Benthic Chamber (CBA), which have recently been devised to study benthic ecosystems, particularly the fluxes of dissolved substance between sediment and the water column, are illustrated in this paper [56]. The two devices can also be employed jointly to improve the reliability of particular investigations.

The Amerigo Lander is basically a vector that can be deployed in shallow and deep bottoms and which returns to the surface at the end of its mission. It can carry a variety of instruments to study the water column and the benthic ecosystems by measuring various environmental parameters and processes.

The Amerigo Lander and the CBA have provided the Italian scientific community with new and highly innovative instruments to investigate the benthic ecosystems and their interactions with the water column, bridging a scientific and technological gap with North American [25,35,37,46–55], Northern European [23,24,26,36,49] and Asian [27–30,33,34,50] countries.

Their technical features make the two devices versatile and easy to use. The modularity of the lander, the various instruments that it can carry, and the much larger number of parameters that it can measure, compared with existing devices, make it a complete and original apparatus, suitable for operating both in shallows and at great depths. The most innovative features of the CBA are the larger area where measurements are acquired, which makes the fluxes more representative [23,24], and its easy and fast rigging, light weight, and maneuverability, which make it easy to use.

Both devices operate autonomously. Amerigo can work from shallow bottoms to depths of 6000 m, whereas the CBA has been designed for shallow water to shelf environments (up to about 200 m). Furthermore, the CBA can fit in the Amerigo, if required by research needs.

2. Amerigo Lander and Automatic Benthic Chamber: Technical Specifications and Equipment

2.1. *The Amerigo Lander*

The Amerigo Lander is essentially a carrier with a tripod structure that can host different types of instrumentation to measure biogeochemical and geophysical parameters and can collect water and sediment samples (Figure 1). It has been conceived as a simple, low-cost, and practical device that can be employed frequently in a variety of research tasks and environments, since it does not require huge resources, large vessels, or long missions.

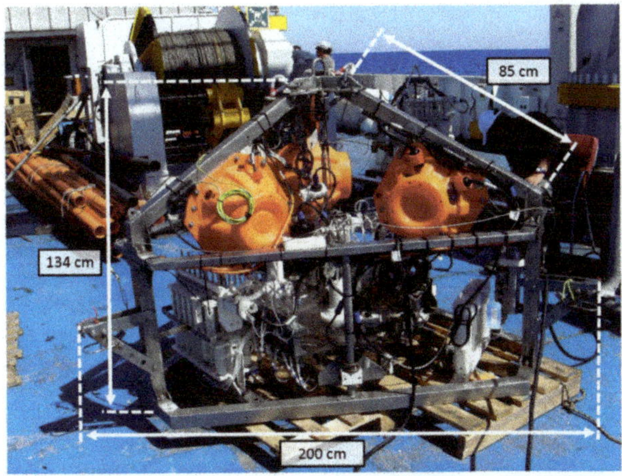

Figure 1. Photograph of the Amerigo Lander showing the tripod structure, the measurements, and the main instrumentation and devices.

Since it reaches the bottom through a speed-controlled free-fall and returns to the surface by positive buoyancy, after a timed release of the ballast weights, it requires no drive or propulsion systems, such as thrusters or cables, nor divers for positioning and recovery. It is also completely automatic, because all hosted instruments and mechanical devices are powered in situ and activated and managed by electronics and software, again without the need for divers or cables. Obviating the need for divers also allows for overcoming stringent safety issues, particularly Italian safety regulations, with considerable savings. The elimination of a cable connection to the vessel makes the lander easy to handle, because the support ship can move while the lander is working on the seafloor.

Another important advantage is its modular structure. In fact, the basic tripod structure can support a variety of components and instrumentation, which can be assembled and set to meet diverse research needs and environmental situations, ensuring flexible and simple operability. In particular, the lander's electronics and power supply have been developed to enable management of additional devices and operations and for increased deployment time. Its modular structure allows the Amerigo Lander to operate in shallow waters (lagoons, estuaries, continental shelves) and deep-bottoms (abyssal plains). Further savings have been obtained from the electronic housings and the release mechanisms.

The lander's basic structure consists of a stainless-steel tripod measuring 200 cm in width and 134 cm in height (Figure 1). All the electronic and mechanical devices required for reaching the seabed and returning to the surface and for hosting and managing the instruments operating on the seafloor are installed in the tripod (Figure 1).

The bottom is reached by a free fall without the need for thrusters or power, because the descent configuration envisages three ballast weights at the three ends and four buoys (Figure 2), which result in negative buoyancy.

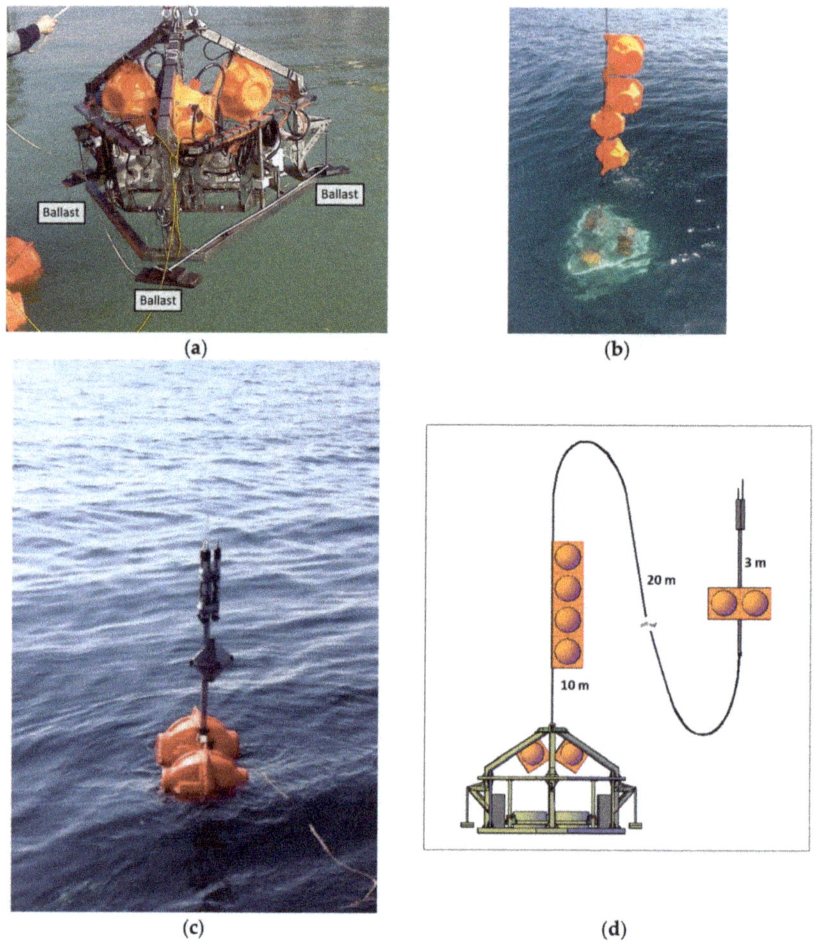

Figure 2. Amerigo Lander descent configuration. (**a**) The ballast weights mounted on the tripod; (**b**) the four buoys tethered to the tripod structure; (**c**) the recovery pole with the two buoys and the three localization devices; and (**d**) a schematic drawing of the Amerigo Lander during deployment on the seafloor.

The buoys are commercially available glass spheres (outer diameter 432 mm, thickness 14 mm, buoyancy 260 N), built for depths up to 6700 m (Nautilus Marine Service GMBH, VITROVEX Deep Sea Floatation Sphere), protected by a plastic shell (Nautilus Marine Service GMBH, SR330). The equilibrium between the buoys and the weights depends on the lander's weight in the water, its descent and landing speed, and ascent requirements.

In the present configuration, the lander weighs 294 kg in air and 131 kg in water and the 4 buoys have a buoyancy in seawater of 1040 kg, whereas the ballast weighs 45 kg (15 kg per weight). The buoys are tethered to the tripod with a rope 10 cm in thickness and 5 m in length (Nautilus Marine Service GMBH, EDDYROPE). Another rope, 10 cm in thickness, is tied to the recovery pole. The pole is fitted with 2 buoys the same size as the 4 buoyancy array buoys and with a 10 kg ballast weight, to support the recovery devices about 2 m above the sea surface (Figure 2d).

With this configuration the lander's initial descent speed is 0.78 m/s. The thrust of the 2 buoys tied to the recovery pole reduces the descent and landing speed to 0.52 m/s (Figure 3). With this set up, the lander can reach the bottom fast enough to avoid being shifted too much by lateral currents, and slowly enough to avoid an excessively strong impact.

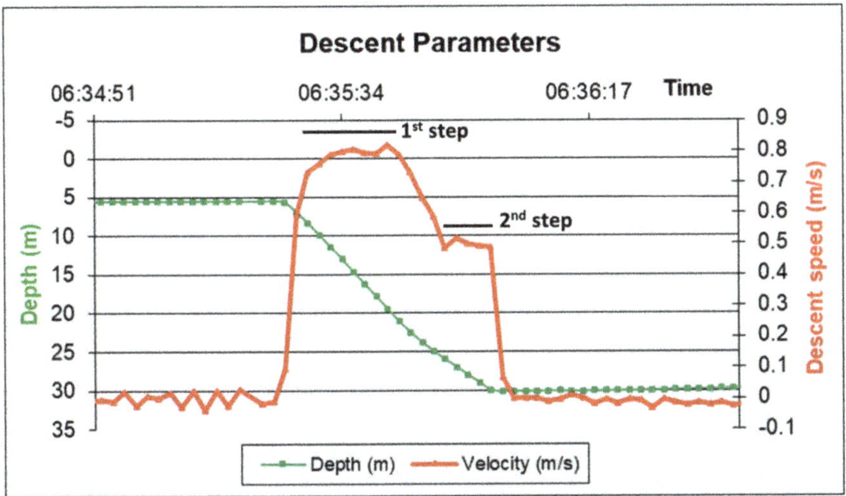

Figure 3. Amerigo Lander descent data. The data refer to the setup with three benthic chambers. Red line: Descent and landing speed. First step: Speed before activation of the first 2-buoy array; Second step: Speed after activation of the two buoys of the recovery pole. Green line: Depth profile.

The buoys are tethered to the tripod by a rope, not mounted directly on the tripod as in several other landers [26,47,50,51,54,55]. This solution has been adopted for two reasons, as follows: (i) As separation of the main tripod structure from the buoys makes the lander lighter and smaller, enhancing maneuverability in deployment and recovery operations, this solution requires smaller frames and less powerful winches; and (ii) this setup allows the lander to be transported in an ordinary van that can be driven with an ordinary license. Furthermore, the buoys and the recovery pole may not be needed in case of operation in shallow water, further enhancing maneuverability and reducing vessel size requirements. These features increase cost-effectiveness and the ease of organization.

After completion of the measurement and sampling operations, the lander returns to the surface autonomously. The release of the three ballast weights and the thrust of the buoy array results in positive buoyancy. In the present configuration (4 thrust buoys, 2 recovery pole buoys, 3 benthic chambers) its ascent speed is about 0.2 m/s (Figure 4).

At the end of the bottom operations, the ballast weights are released by a burn wire mechanism that unlocks the three lever hooks (Figure 5). This system is much less expensive than acoustic release [57] and further contributes to make the lander cost-effective and easy to use and to program.

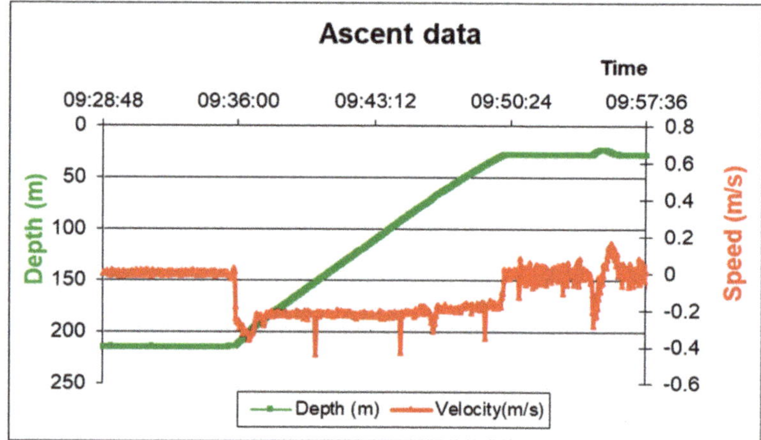

Figure 4. Amerigo Lander ascent data. The data refer to the setup with three benthic chambers. Red line: Ascent speed. Green line: Depth profile.

Figure 5. The three lever hooks in a locked (**a**) and unlocked (**b**) position.

In case of deployment in deep water, the Amerigo Lander may surface at a considerable distance from the dropping site, due to lateral sea currents during the ascent. To facilitate recovery, the lander is equipped with three redundant localization devices fitted on the top of the recovery pole, a GPS (Novatech ARGOS Beacons, Dartmouth, NS, Canada), a directional radio (Novatech Radio Beacon, Dartmouth, NS, Canada), and a flash (Novatech Xenon Flasher, Dartmouth, NS, Canada) for night recovery (Figure 6).

The Amerigo Lander is also equipped with instruments for monitoring and measuring the physical-chemical parameters throughout deployment, from descent to ascent. They include a CTD (SBE 37-SI MicroCAT, Sea-Bird Scientific, Bellevue, WA, USA) for continuous water column pressure, conductivity, and temperature recording, and a camera supporting an SD card from 4GB to 32GB (Telesub Lanterna, La Spezia, Italy) (Figure A1, Appendix A) for monitoring the lander's operation. In fact, both instruments monitor the lander's activities, particularly the beginning of descent, the descent speed, the landing, the functioning of the mechanical devices, the beginning and speed of the ascent, and the lander's surfacing.

Additionally, in this case, both the CTD and the video camera are commercially available to save construction costs. In particular, the video camera is a commercially available camera hosted in a pressure-resistant case.

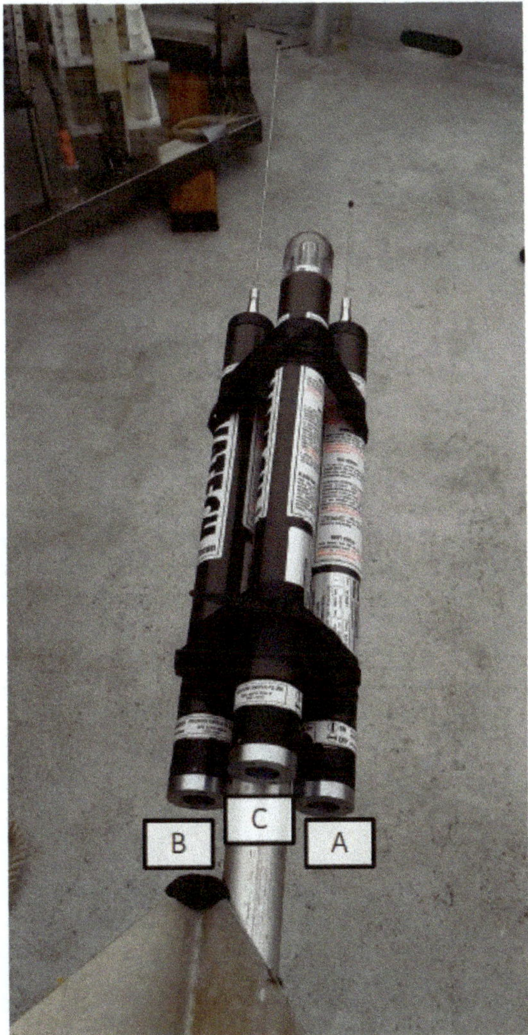

Figure 6. Amerigo Lander recovery system. (A) GPS localization (Novatech ARGOS Beacons); (B) directional radio localization system (Novatech Radio Beacon); and (C) flash for visual localization at night (Novatech Xenon Flasher).

2.2. *Electronics and Power Supply*

The Amerigo Lander's electronics and power supply are developed in-house (Figure 7, Figure A2 of Appendix A). The system is available on request. The burn wire system and all the mechanical and electronic devices, sensors, and probes are powered, turned on, turned off, and managed by the electronics and the batteries fitted in the tripod. The data collected in situ are also stored in the Amerigo electronics. The main hardware components are configured as illustrated in Tables A1 and A2, and Figure A2 of Appendix A. The philosophy of electronics is to be as open as possible, i.e., to enable fitting other sensors or devices by the availability of redundant on/off, serial, and analogic ports and the possibility of managing these sensors/devices and other operations by the software.

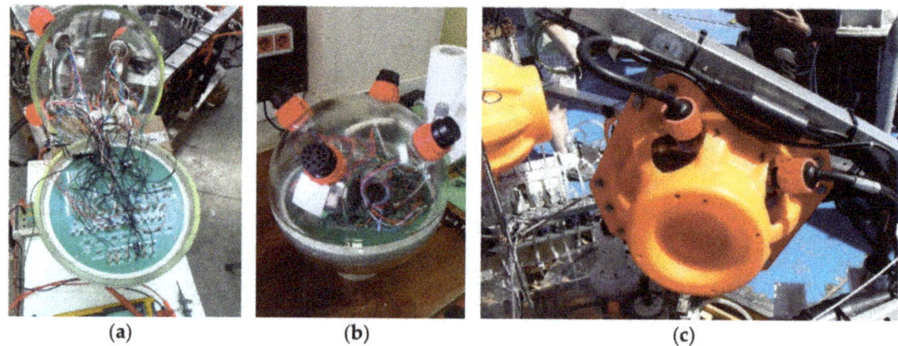

Figure 7. Amerigo Lander electronics: (a) Firmware testing phase and soldered cables inside the glass sphere; (b) electronics fitted in the glass sphere; (c) a glass sphere mounted on the tripod with the plastic case, marine connectors, and cables.

The Amerigo Lander is powered by two pairs of 12 V, 18 Ah, rechargeable lead batteries that are connected in parallel and fitted in a pressure-resistant case. The fact that they are commercially available and Pb-based involves lower managing and setup costs, compared to other metal-based batteries. The first pair is supported by the redundant second pair, which is only activated in case of exhaustion or failure of the first pair. This power supply supports about a 40 h operation of the lander in its present configuration. The Pb batteries can be increased or their type changed to support different configurations or to extend the operating time.

In the event of a failure of the general power supply or of the main electronics, a safety burn wire device powered by an independent 9 V battery and controlled by a dedicated electronic circuit is activated, after a predetermined time, to release the ballast for the final ascent. An additional safety system consists of a magnesium ring that is corroded by seawater. In case of the failure of all the electronic devices, its disappearance releases the ballast weights [58].

The electronics and the batteries are hosted in glass spheres built for depths of up to 7000 m (Nautilus Marine Service GMBH, VITROVEX Deep Sea Instrument Sphere; size: 13", outer/inner diameter: 330/306 mm, glass type: DURAN 8330), which are connected to the electronic devices, sensors, probes, and thrusters by marine connectors (Figure 7). The glass spheres are also commercially available and cost less than metal cylinders.

The two battery pairs are recharged on board by a cable that is removed before deployment. The cable is also used for serial port communication with the PC.

The main serial port (RS232–1) is devoted to communication with the PC. It allows for entering commands, changing the setup, downloading recorded data from the RAM flash memory or, in case of direct monitoring, it enables visualizing the data, the situation on the bottom, and the ongoing operations, as well as reporting malfunction alarms in real time (Figure A2 of Appendix A). The same port can be used to connect an acoustic modem for underwater communication.

The second serial port (RS232–2) is multiplexed in order to communicate with infinite RS-232 serial sensors, limited only by the hardware power connections (Figure A2 of Appendix A).

Different systems have been designed to protect the lander's electronics. The power supply to each electrical device (sensors, motors, batteries, burn wires) is protected by an electrical shunt that limits current drain. In case of failure of a device, the power supply to it is cut off to prevent a general electronic failure.

Furthermore, the open electronics and the excess of on/off, serial (by means of a multiplexed serial port), and analogic communication ports allows fitting many other electronic sensors, devices, and instruments, with respect to other system that are devised for standardization as well as useful use [28].

The following sensors are currently installed in the Amerigo Lander (Table 1) (Figure A2 of Appendix A):

Table 1. Technical specifications of the sensors installed in the Amerigo Lander.

Sensor Type	Company	Parameter	Resolution	Initial Accuracy	Maximum Depth (m)
SBE37-SI (CTD)	Sea Bird Electronics	Conductivity (S/m)	0.00001	0.0003	7000
		Temperature (°C)	0.002	0.0001	7000
		Depth (m)	0.002% FS	0.1% FS	7000
SBE5T (Pump)	Sea Bird Electronics				7000
Oxygen Optode 3830	AANDERAA	Oxygen concentration (µM)	<1 µM	<8 µM	6000
		Air saturation (%)	0.4%	<5%	6000
Seapoint Turbidity Meter	Seapoint	Turbidity (FTU)	10 mV/FTU (range 500 FTU)	<1%	6000
Mets	Franatech	Methane	2.44 µM (range 50 nM–10 µM)	<1%	4000
pH	AMT	pH	0.01	0.05	6000

An important question is the operation limit connected with temperature. The temperature operation limit of Amerigo coincides with the lower operation value between the sensors fitted in the benthic chambers and the polycarbonate (i.e., 40 °C (optode sensor); if the lander is used without sensors the temperature limit is <140 °C (polycarbonate)).

2.3. Burn Wire Device

The burn wire mechanism [51] consists of a metal wire coated with a plastic film that is corroded and then broken by an electric current at a bare point where the coating is interrupted (Figure A2, Appendix A). A simple 12 kg fishing wire coated with a thermo-shrinkable tube is a typical design. The plastic-coated wire usually keeps a lever hook locked. When the electric current runs through the wire, the bare point interacts with seawater, which triggers a reaction (1) that consumes the wire completely, leading to release of the lever hook, as follows:

$$M^0_{solid} = M^{2+}_{solute} + 2e^-, \tag{1}$$

where M^0 is the metal of the wire, $2e^-$ is the electric current, and M^{2+} is the metal in the solution. The electric circuit is closed by an electric mass on the metal tripod. By this method, any spring- or gravity-based mechanical device can be actioned by the release of a mechanical device or lever hook that can hold several tens of kilos, depending on the length of the lever (Figure 5). A metal wire, 0.4 mm in thickness with a resistance of a 12 Ω/m, exposed to an electric current of 200 mA in normal seawater (36 PSU) is usually consumed and broken in about 20 s. As mentioned above, the burn wire system is much less expensive than any acoustic release system.

2.4. Current Configuration of the Amerigo Lander

In its current configuration, the Amerigo Lander is equipped with instruments and sensors for measuring benthic fluxes of dissolved chemical species and for monitoring physical-chemical parameters in the near-bottom sea water column. In particular, the former measurements at the sediment-water interface are performed using three benthic chambers, two water sampling systems, and some sensors fitted in the chambers (Figures 8–12).

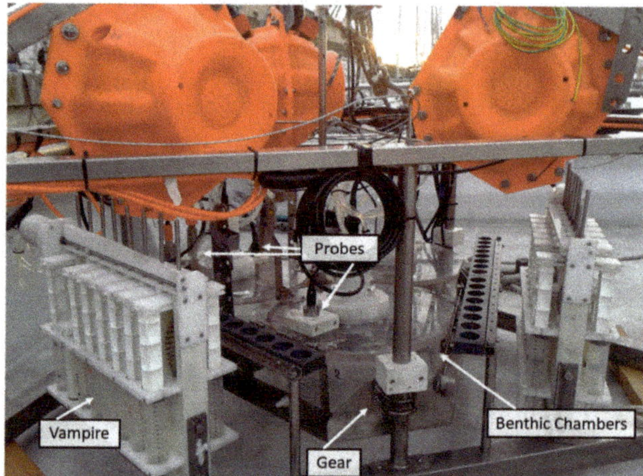

Figure 8. Amerigo Lander: The three benthic chambers, the two water sampling devices, the three probes in a chamber, and the chassis on which the chambers are mounted.

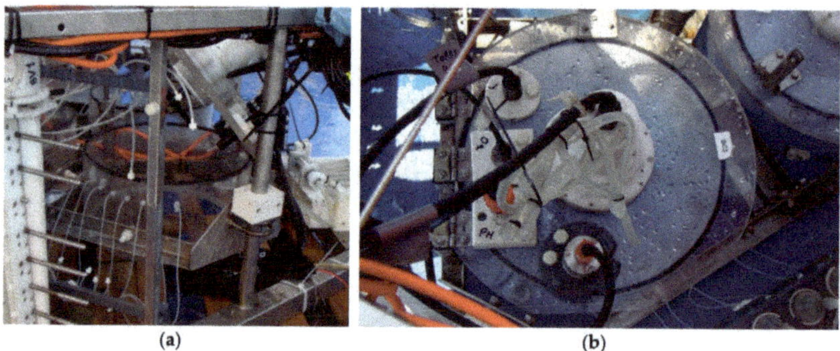

(a) (b)

Figure 9. Amerigo Lander. (**a**) Lateral view of a benthic chamber mounted on the chassis with the top lid open, the tubes on the lateral wall to collect water samples, and the rotating paddle. (**b**) Top view of a benthic chamber with the closed top lid fitted with the oxygen, pH, methane, and turbidity sensor.

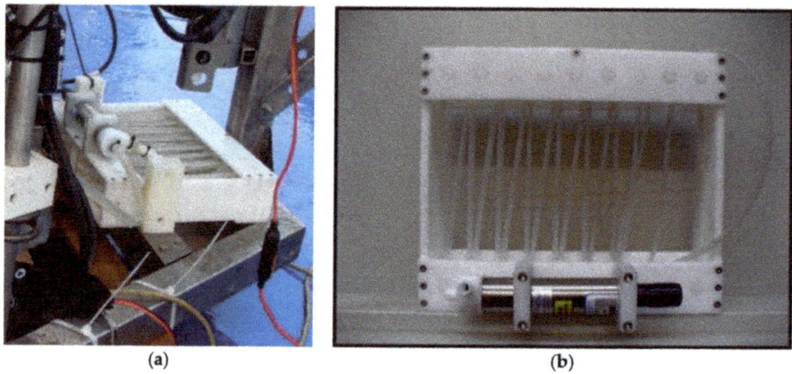

(a) (b)

Figure 10. The OxyStat oxygen replacement device. (**a**) The pump and the tube connected to the interior of the chamber; (**b**) view from above.

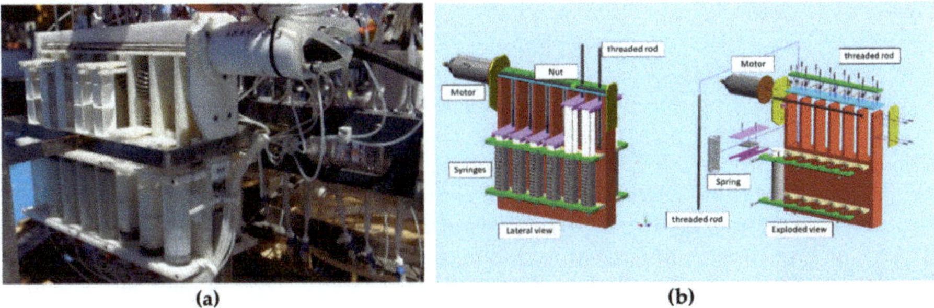

(a) **(b)**

Figure 11. The VAMPIRE device allows drawing water into the chamber or expelling it as well as injecting a tracer inside the chamber. (**a**) Photograph of the VAMPIRE device installed on the Amerigo Lander; (**b**) schematic drawing of the VAMPIRE device.

Figure 12. The Delrin case hosting the electric motors with the silicone tube pressure-compensation system.

The benthic chambers are polycarbonate cylinders with a movable polycarbonate top lid (Figures 8 and 9a). The cylinders measure 37 cm (inner diameter) by 20 cm (height) and have a countersink in the bottom to facilitate penetration in sediment [46].

The three benthic chambers are mounted on a chassis (Figures 8 and 9a) that is released by a burn wire mechanism a few minutes after the tripod has landed on the seabed. The chambers are mounted 5 cm over the plate of the structure so that they can penetrate into the sediment for 5 cm, while the remaining 15 cm remain above the sediment. A few minutes after deployment of the chambers on the seabed, the lid of each chamber is unhooked, again by a burn wire device, thus closing the chamber and holding a known volume of water (approximately 17 L) overlying a known area of sediment [46]. The chassis and lid release time can be programmed by the lander's software to adapt them to the research task and the type of sea bottom.

The chemical and physical-chemical parameters in the chambers and some solute are measured by the following sensors fitted in each chamber during deployment: An oxygen sensor (AANDERAA, Oxygen Optode 3830, Aanderaa Data Instruments AS, Bergen, Norway), a turbidity sensor (Seapoint Turbidity Meter, Seapoint Sensors, Inc., Exeter, NH, USA), a methane sensor (ASD-Sensortechnik GmbH, METS methane sensor, Franatech Gmbh, Lüneburg, Germany) and a pH (AMT, pH-combined sensor, AMT Analysenmesstechnik GmbH, Rostock, Germany) sensor (Table 1) mounted on the lid (Figure 9b). The power-on/power-off and measurement intervals of each sensor can be set by the lander's software according to research requirements.

The chambers also contain an OxyStat device, which allows for replacing the oxygen consumed in the chamber [59]. The device is connected to the chamber by a water pump (SEABIRD SBE5T, Sea-Bird Scientific, Bellevue, WA, USA) and a silicone tube (Figure 10) and is controlled by the lander's software and the oxygen probe inside the chamber. In practice, the software receives the chamber oxygen concentration data and when its level falls below a given threshold, the software turns the OxyStat pump on. The oxygen-poor water in the chamber is pumped into the 15 m long gas-permeable silicone tube, it adsorbs oxygen from surrounding seawater, and is pumped back into the benthic chamber. Restoration of the oxygen level to the predetermined threshold results in the pump being turned off. The minimum and maximum oxygen concentrations can be set by the lander's software before the mission or calculated on the basis of the initial concentration measured in the chamber.

Each benthic chamber is connected (Figure 9a) by silicone tubes (inner diameter, 1.5 mm, outer diameter 3 mm) to a water sampling device (VAMPIRE) to collect water or to introduce tracers into the chambers (Figure 11). The VAMPIRE consists of a Delrin frame hosting 8 pairs of syringes, each pair capable of collecting/injecting a maximum volume of 280 mL of water or tracer. If one syringe of the pair is not connected to the inside, seawater outside the chamber can be collected while the other syringe draws water inside the chamber. Each couple of syringes is activated by a nut moving on a rotating stainless-steel rod. The nut moves the levers that release the stainless-steel springs, which actuate the syringe pair in suction or injection mode. The rod is controlled by an electric motor (CBF Motors SRL, CRB35GM, CBF motors srl, Lissone, Italy) which is powered on and off by the lander's software. Its timing and the activation of water sampling or tracer injection can also be set by the lander's software.

In research tasks involving analysis of dissolved gases, a set of glass ampoules can be added before the syringes of the VAMPIRE device (Figure A3, Appendix A) to store the water samples in a gas-impermeable vessel until analysis.

Each benthic chamber is also equipped with a stirring system, consisting of a rotating paddle mounted on the chamber lid (Figure 9a). The paddle is actioned by the coupling of an electric motor (CBF Motors SRL, CRB35GM, CBF motors srl, Lissone, Italy) to a permanent Neodymium magnet (Supermagnete, magnetic disk diameter 30 mm, height 15 mm, Neodymium, N42, nickel-plated, Webcraft GmbH, Gottmadingen, Deutschland). The paddle turns at a speed of 4–6 rpm, reproducing the hydrodynamics near the seabed, which is responsible of the formation of the benthic boundary diffusion layer at the sediment-water interface and, consequently, of the intensity of the benthic fluxes of dissolved chemical species. The motors of the rotating paddles are also activated by the lander's software some minutes after closing of the lid.

Whereas the cases housing the electronics and the batteries are pressure-resistant, those housing the motors actuating the stainless-steel rod of the VAMPIRE device and the rotating paddles of the benthic chambers are pressure-compensated. These cases are Delrin cylinders with two silicone tubes (Figure 12) filled with a non-conductive liquid (commercially available Vaseline oil). If any air bubbles remain in the case, compression of the silicone tubes offsets the pressure difference between inside and outside, avoiding a collapse of the case.

At the end of each mission, i.e., after completion of the water sampling and sensor measurements in the benthic chambers, the three ballast weights are released by activation of the burn wire device, which induces positive buoyancy. The lander returns to the surface, where it is localized by means of the three positioning devices, and finally recovered on board. Missions typically last 8 to 36 h, depending on the tasks to be performed or the intensity of the benthic fluxes of dissolved substances, and are limited by the power supply, which, at present, supports the systems for about 40 h (see Section 2.2). However, as noted above, the open philosophy of the electronics allows the increasing of this time.

After the recovery operations, the water samples taken by the syringes are collected from the VAMPIRE, divided and treated in an inert atmosphere (Figure A4, Appendix A) for immediate (on board) or subsequent (laboratory) chemical analysis of the solutes to be determined in the benthic fluxes.

The data collected by the sensors are downloaded into a computer.

The results of the chemical analyses and the data collected by the chemical sensor are then used to calculate the benthic fluxes of dissolved chemical species (see Section 3).

2.5. Typical Mission of the Amerigo Lander

The Amerigo Lander hardware is all managed by wizard software, developed in-house. The software helps the operator in setting all the parameters that are required for the lander's function, to activate and test the motors, to monitor all the parameters in real time, to simulate a measurement mission, and to plan the activities for scheduling a mission. The software allows for the downloading and processing of the data collected during the mission and stored in situ. Finally, the software has been developed to fit further sensors and devices and to manage operations that are not currently scheduled.

A typical mission of the Amerigo Lander consists of six sequential phases that need to be correctly planned for the success of the mission, as follows (Figure 13):

1) After the on board programming and checks, such as control of the mooring line and testing of motors, sensors, communication, and security equipment, the lander is immersed into the sea, where it is held at a depth of 5 m, until activation of a dedicated burn wire mechanism releases a small buoy that confirms the functioning of the whole electronic system;
2) Following the buoy check, the lander is released for its free fall to the bottom;
3) After the lander has reached the seabed, a short interval is envisaged to allow the settling of the sediment resuspension, due to the impact of the tripod on the bottom, to settle;
4) Activation of a burn wire mechanism releases the chamber chassis, enabling its settling on the bottom and penetration into sediment for the first 5 cm; all the sensors in the chamber (oxygen, methane, turbidity) and CTD are sampled; the chambers are still open;
5) After another interval, to allow settling of the sediment resuspension due to the impact of the chamber on the bottom and to enable sensor readings, activation of another burn wire releases the benthic chamber lids;
6) Activation of the stirring paddles allows for mixing the seawater in the chambers;
7) The 8 pairs of syringes of the 2 VAMPIRES are activated by user-programmable times;
8) Finally, the ballast weights are released by the last burn wire and the lander floats back to the surface by virtue of its positive buoyancy.

Figure 13. The Amerigo Lander photographed in some operational phases: (**a**) In the water before release; (**b**) on the seabed (view from above); (**c**) on the seabed (lateral view); and (**d**) on board after recovery.

2.6. Other Possible Configurations of the Amerigo Lander

The basic structure of the lander is the tripod, which is designed to land on the seabed by gravity, counteracted by the positive thrust of the buoys. It then performs its scheduled operations on the seabed and finally returns to the surface by virtue of positive buoyancy, after the release of the ballast weights. All the on-board instruments and devices are built to operate at depths up to 6000 m.

This setup makes the lander a vector that can host different types of instrumentation, such as sensors and probes for monitoring chemical and physical-chemical environmental parameters (pH, Eh, conductivity, temperature, salinity (calculated), oxygen, methane, pCO_2, H_2S) in the water column during its descent, its permanence on the bottom, and its ascent.

The lander has also been designed to host instruments such as a microprofiler, to study sediment-water interface properties, a penetrometer, to measure the mechanical properties of surface sediments, a gravimeter, to measure seismicity on the seafloor, a corer, to collect sediment cores for early diagenesis, pollution, stratigraphy, or other studies, and passive samplers of water column and sediment solutes, to study pollution and environmental processes and to determine background values. In any case, the open architecture of the electronics allows for fitting other instruments and performing other operations.

Finally, it is a modular device, consisting of the buoy array, the recovery pole, the ballast weights, and the removable and replaceable instruments that perform different tasks in different environments, from very shallow waters, like lagoons and salt marshes, to shelf areas and abyssal planes.

2.7. The Automatic Benthic Chamber

The CBA (Figure 14) has been developed as an alternative to the Amerigo Lander for missions involving measurements in shallow and transitional waters or work that needs to be carried out quickly and economically. This is made possible by the fact that the CBA does not require expensive maintenance, it is practical and fast to fit out, it is light and easy to maneuver, and it is deployed and recovered simply with a rope, which means that it can also be managed by small vessels.

Figure 14. The Automatic Benthic Chamber during (**a**) deployment and (**b**) operational on the seabed with various instruments: Multiparameter probe; VAMPIRE; electronics case; and battery pack cases.

The CBA can also be mounted on the Amerigo Lander, instead of the three benthic chambers, when measurements of benthic fluxes of dissolved chemical species are to be performed over a wider area.

The present CBA is an automated device based on earlier manual benthic chambers managed by divers [60,61]. It is a Plexiglas cylinder open on the bottom and closed on top, which confines a known volume of water (approximately 100 L) overlying a known sediment area (3116 cm^2) (Figure 14b). Its inner diameter is 63 cm and its height is 30 cm, of which 5 cm penetrate into the sediment and 25 cm remain above it, due to a lateral horizontal fin (Figure 14). The CBA is fitted with two valves on its top side, to let out the water entering the chamber during descent and landing (Figure 14). Like the Amerigo Lander, the CBA is equipped with an internal stirring system that reproduces the hydrodynamics near the seabed, which is responsible for the formation of the benthic boundary diffusion layer and for the intensity of the dissolved fluxes in the benthic chamber. The stirring system consists of a four-arm rotating paddle fitted on top of the inner side of the chamber. The paddle is actioned by the coupling of an electric motor (CBF Motors SRL, CRB35GM, CBF motors srl, Lissone, Italy) with a Neodymium magnet (Supermagnete, Magnetic disk 30 mm in diameter, 15 mm in height, Neodymium, N42, nickel-plated, Webcraft GmbH, Gottmadingen, Deutschland) and turns at a speed of 4–6 rpm. In the CBA, this motor is activated immediately before deployment by connecting directly the batteries to the motor.

The CBA is also equipped with a multiparameter probe (Hydrolab MS5, OTT HydroMet, Kempten, Germany) for continuous monitoring of temperature, pH, conductivity, dissolved oxygen, Eh, and salinity (calculated) in the chamber (Figure 14). Like the Amerigo Lander, it is also fitted with the VAMPIRE system for collecting water samples inside and outside the chamber and for injecting tracers inside the chamber at programmable times. The motor of the VAMPIRE is activated by simple, easily programmable, and commercially available electronics (Idec MicroSmart FC6A PLC,

IDEC Corporation, Sunnyvale, CA, USA). The cases housing the electronics, the battery packs, and the motors driving the VAMPIRE and the stirring paddle are made in Delrin and are built to withstand hydrostatic pressure up to a depth of about 200 m. Additionally, in the CBA, the VAMPIRE motor and the electronics cases are pressure-compensated by a silicone tube system filled with a non-conductive liquid (simple Vaseline oil), which affords resistance to high water pressures.

The power supply of the CBA consists of three battery packs (NI-MH size D, 12 V, 8 Ah, Torricella SRL, Milano, Italy) housed in cylindrical Delrin cases (Figure 14). Two packs are connected directly to the electronics that supply and manage the VAMPIRE motor and one pack is connected directly to the rotating paddle, while the multiparameter probe has its own power supply system.

As regards the planning of CBA operations, the syringe sampling time is set by programming the electronics, while the probe measurement time is programmed by the software of the probe itself.

With regard to deployment and recovery, the CBA is deployed on the seabed and recovered by a rope which, during measurement activities, is attached to and marked by a buoy and a light.

The CBA, both in the standalone configuration and installed in the Amerigo Lander, is a low-cost device that does not require divers or connection cables to the support ship, thus saving the steep cost of divers and the technical problems posed by the connection cable. Further savings are afforded by the fact that the electronics and the batteries are commercially available, hence the low-cost.

Like the Amerigo Lander, the CBA has a temperature operating limit which coincides with the lowest operating value of the sensors fitted in the benthic chamber and the polycarbonate, which is 50 °C (Hydrolab MS5 Multiprobe). However, if the CBA is used without the multiparameter probe the value is <140 °C (polycarbonate).

3. Benthic Flux Calculation

The benthic chambers of the Amerigo Lander and the CBA have been designed to measure the release/adsorption of dissolved substances at the sediment-water interface. The principle of their measurement with benthic chambers involves establishing the concentration differences of a solute over time in a known volume, confined over a known area of sediment [62].

Basically, the benthic fluxes of dissolved substances in each benthic chamber of the lander and of the CBA are calculated (2) by dividing the concentration of each solute, measured in the samples collected in the chamber by the syringes—typically nutrients such as ammonium, nitrites, nitrates, phosphates and silica, carbonate species (DIC, alkalinity, pCO_2), trace elements (heavy metals), and organic pollutants—or recorded by the sensors (oxygen, methane, and pH), at the time of the collection or measurement, multiplied by the volume of the benthic chamber and divided by its base area [44].

$$Di = \frac{\partial C_i}{\partial t} V/A, \tag{2}$$

where Di is the flux of solute i, C_i is the concentration of chemical i, t is the time of sample collection or sensor measurement, V and A are the real volume and the area of each benthic chamber.

In practice, the benthic fluxes of each solute are computed as Equation (3) by multiplying the slope of the line, calculated by a least square fit with time (days) on the x-axis and the concentration at different times, multiplied by the height of the benthic chamber, on the y-axis (Figure 15).

$$D_i = y_i, \tag{3}$$

where y_i is the slope of the time vs. the concentration line of Figure 15.

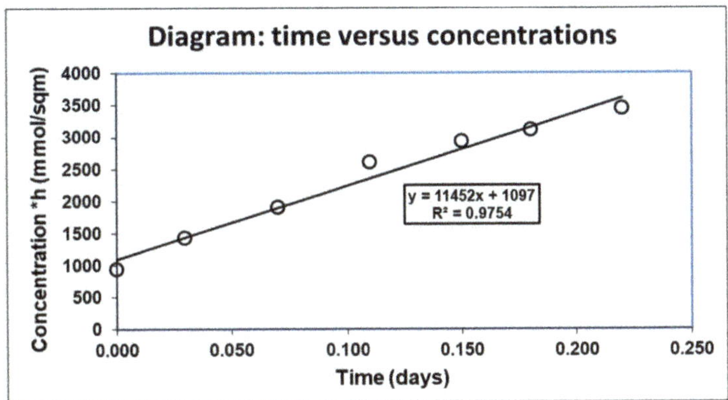

Figure 15. Diagram illustrating the time of water sample collection in the benthic chamber against its concentration and the correlation line.

During deployment, the real volume of each chamber is determined by injecting a solution of a non-reactive solute (tracer), e.g., CsCl, BrCl or deionized water, at a known concentration into the chambers [44] and subsequently measuring its concentration in the water samples collected in the syringes as Equation (4).

$$V_2 = \frac{V_1 * C_1}{C_2}, \tag{4}$$

where V_2 is the real volume of the benthic chamber, C_1 is the tracer concentration in the syringe, V_1 is the volume of the tracer injected into the chamber, and C_2 is the tracer concentration in the chamber after the injection.

Theoretically, the tracer concentration in the chamber after the initial injection should be constant. If this does not happen, there are three possible explanations, as follows: The benthic chamber is not well placed on the bottom, there are leaks, or an irrigation process is under way in the bottom sediment.

4. Discussion

The Amerigo Lander [63–71] and the CBA [41,42,70–77] have successfully been tested and used in measurement and research activities carried out in the framework of international and national projects and in environmental investigations into the impacts of human activities on marine (e.g., harbor sediment dredging) or land environments (e.g., quality of drinking water).

The Amerigo Lander has been tested and employed in shallow, medium, and deep-sea environments, whereas the CBA has been used up to a depth of 140 m.

As a discussion of the data collected by the two devices, we report (Figure 16) the trend of the dissolved oxygen concentrations, measured in the benthic chambers of the lander (by the AANDERAA optode oxygen sensors, Aanderaa Data Instruments AS, Bergen, Norway) and the CBA (by the Hydrolab MS5 oxygen sensor, OTT HydroMet, Kempten, Germany), at the same deployment time and site [71], and on a pelitic and organic matter-rich bottom in front of the Po River Estuary [78,79]. All the benthic chambers of the two devices recorded similar continuously decreasing values, due to mineralization of the high content in fresh reactive organic matter, deposited in front of the Po River Estuary.

The benthic fluxes of solutes, whose concentrations were determined in the water samples collected by the VAMPIRE syringes, also showed reliable data. In fact, very similar values were determined for the fluxes of dissolved inorganic carbon (DIC) (Figure 17) measured by the three benthic chambers of the Amerigo Lander and by the CBA, deployed at the same and site, i.e., on pelitic and fresh organic matter-rich bottom sediments. Furthermore, these DIC flux values are very similar to those measured in earlier studies using different benthic chamber devices [80–82] at the same site and in the same season (Figure 17).

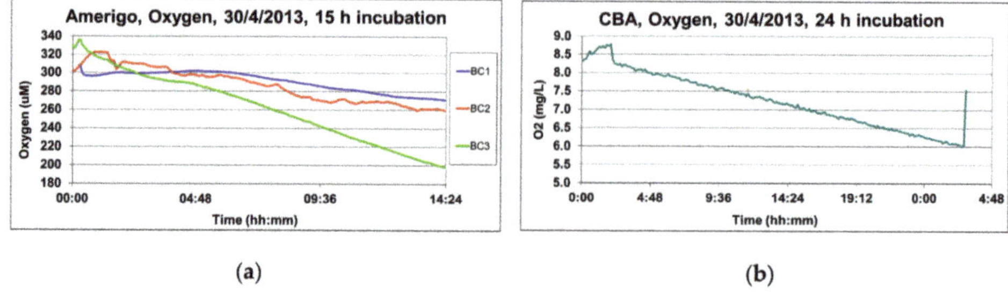

(a) (b)

Figure 16. (**a**) Oxygen values recorded in the three benthic chambers of the Amerigo Lander (BC1, BC2, BC3) during deployment; (**b**) oxygen values recorded in the CBA during deployment of the two devices at the same time and site, a mud bottom sediment rich in fresh organic matter (Po River Prodelta).

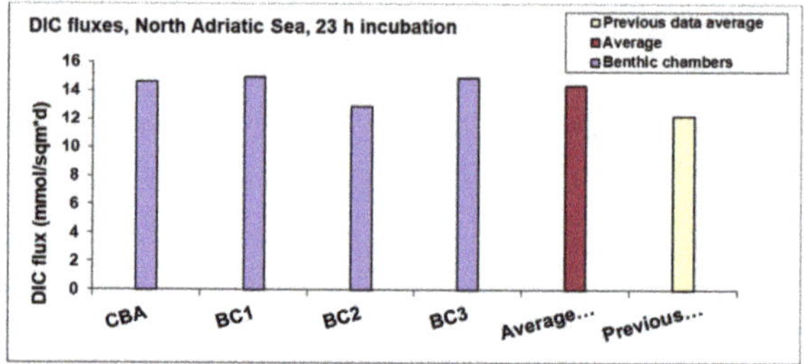

Figure 17. DIC fluxes measured by the CBA, by the Amerigo Lander (BC1, BC2, BC3), and in previous investigations at the same site, a mud bottom sediment rich in fresh organic matter (Po River Prodelta).

On the whole, the oxygen and DIC data reported above (and other solute flux data that are not shown but are available from the authors) demonstrate that the Amerigo Lander and the CBA provide very similar information on benthic fluxes of dissolved substances and that these data are comparable with flux information recorded in previous work conducted at the same site. These first data, therefore, provide very good support for the correct functioning of both our devices.

The Amerigo Lander and the CBA record oxygen concentrations in the different deployment phases. In particular, the oxygen sensors can monitor the oxygen concentrations inside the chambers, the oxygen fluxes at the sediment-water interface can be calculated, and, furthermore, the oxygen trend can be used to check the closing of the benthic chamber.

The trends of the oxygen concentrations recorded in the CBA and in two benthic chambers of the Amerigo Lander (BC1 and BC3) is shown in Figure 18. The data refer to the same time and station, on pelitic bottom sediment in front of the Po River Estuary.

The declining oxygen concentration in the CBA and in the Ox3 chamber, due to benthic respiration or sediment-water interface fluxes, can be appreciated in Figure 18a. Notably, the two peaks in the oxygen concentration trend in the Ox1 chamber of the lander demonstrate that the lid opened twice.

In Figure 18b, the oxygen concentrations were multiplied by the height of the benthic chambers calculated by the dilution of the Cs tracer (4). The oxygen flux was then obtained by the slope of the regression line between time (days) and concentration (μmol/m^2) (black line, Figure 18b).

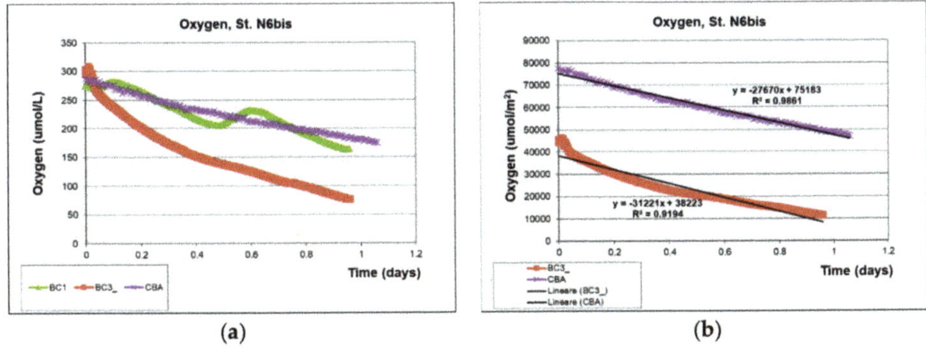

(a) (b)

Figure 18. (**a**) Oxygen concentrations measured in the CBA and in two benthic chambers (BC1 and BC3) of the Amerigo Lander at the same site, a mud bottom sediment rich in fresh organic matter (Po River Prodelta); (**b**) multiplication of the oxygen concentrations by the height of the benthic chamber of the CBA and in BC3 against time (days) allows calculating the benthic fluxes (slope of the regression line (black line of Figure 18b)).

The fluxes calculated by the slope of the regression line are shown in Figure 19 and Table 2. As demonstrated by the examination of Figure 18, the flux measured in the CBA is almost constant over the 24 h incubation. For this reason, only the total flux was calculated (Figure 19 and Table 2). In contrast, the oxygen concentration trend in BC3 shows a decreasing flux that can be divided into early (with higher values) and later (with lower values). This is due to the small size of the chamber of the lander, which is more responsive to changes in the environmental conditions inside the chamber, like the reduction in fresh reactive organic matter and oxygen consumption.

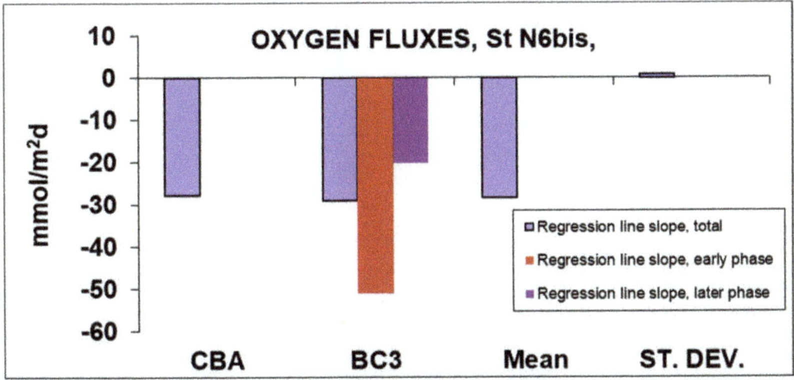

Figure 19. Oxygen fluxes calculated based on the data collected in the CBA and in the BC3 chamber of the Amerigo Lander. Deployment in front of the Po River Prodelta.

The CBA has also been deployed in a volcanic environment, at multiple sites on the seabed around the volcanic complex of Panarea, to measure the dissolved fluxes of DIC and metals released from the bottom in the gas vent area [39,43,44,54]. Figure 20 and Table 3 demonstrate the marked difference in DIC fluxes on the bottom between sites affected by vent fluxes (GEOCAL14CBA1, GEOCAL14CBA2, PANA14CBA1, PEG1, SP) and sites devoid of fluxes of dissolved substances at the sediment-water interface (PANA13CBA1, PIANA, 1, 2, 3), due to a surface layer of iron oxyhydroxide [39]. In addition, Figure 20 and Table 3 show very different DIC fluxes on the seafloor around the Panarea volcanic area, which is involved by vent fluxes, and the average DIC benthic fluxes measured in front of the Po River Estuary.

Table 2. Values of the oxygen fluxes at the sediment-water interface, calculated by the regression of all points (total), the phase with higher slope (early phase) and the lower slope (later phase).

.	Regression Line Slope, Total	Regression Line Slope, Early Phase	Regression Line Slope, Later Phase
Oxygen		mmol/m2*days	
CBA	−27.670	-	-
BC3	−28.903	−51.087	−20.144
Mean	−28.287	-	-
ST. DEV.	0.872	-	-

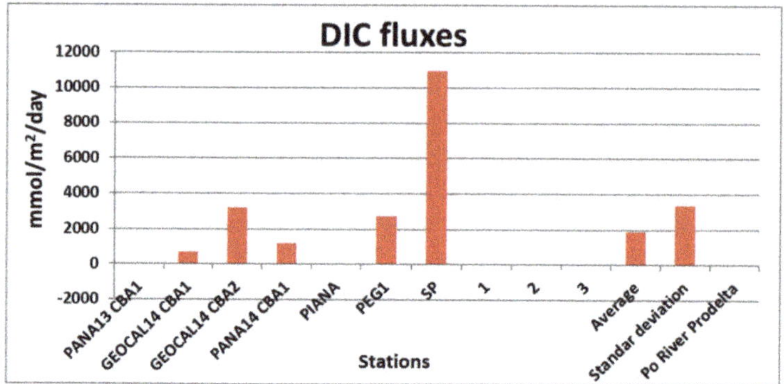

Figure 20. DIC fluxes measured by the CBA in the submarine volcanic area of Panarea and in front of the Po River Prodelta.

Table 3. Values of the DIC fluxes at the sediment-water interface measured by the CBA in the submarine volcanic area of Panarea and in front of the Po River Prodelta.

Cruise	Stations	Flux (mmol/m^2·d)
PANA13	PANA13 CBA1	60.60
GEOCAL14	GEOCAL14 CBA1	689.30
GEOCAL14	GEOCAL14 CBA2	3223.90
PANA14	PANA14 CBA1	1212.70
PANA15	PIANA	−17.99
PANA15	PEG1	2750.60
PANA15	SP	10978.00
PANA15B	1	61.54
PANA15B	2	−4.41
PANA15B	3	−19.29
	-	-
Average	-	1893.49
Standard deviation	-	3408.76
Po River Prodelta average	-	29.00

5. Conclusions

The Amerigo Lander and the CBA, two new instruments for measuring the benthic fluxes of dissolved substances, built by the authors, are presented herein. Both devices are autonomous and can operate in shelf (CBA) and deep-sea (Amerigo Lander) environments. The Amerigo Lander can also be used for other investigations of shallow and deep benthic ecosystems, because it can carry several different instruments. Both devices have been successfully tested and employed in international and national research projects and in environmental investigations of anthropic impacts by local authorities. These tests and activities have demonstrated the sound performance of the Amerigo Lander and the CBA, as also reflected by the DIC and oxygen data reported above. The CBA has also proved suitable for deployment in a volcanic area affected by gas and fluid vents, for which very few data are available due to technical measurement difficulties. In case of use in volcanic environments, the temperature of the solutes released from the bottom should be carefully monitored because of the temperature operation limits of the sensors (40–50 °C) or of the polycarbonate (140 °C).

Notably, these new instruments mark an important advancement in the Italian marine technology community, providing the means to compete for international research and applicative projects at the same level as foreign institutions.

Author Contributions: The research article has been realized with the following individual contributions. Conceptualization, F.S. and V.M.; methodology, F.S., P.P., G.G., V.M. and L.M.; software, P.P. and F.S.; validation, F.S. and P.P.; formal analysis, F.S. and V.M.; investigation, F.S., P.P. and G.G.; resources, F.S. and V.M.; data curation, F.S., V.M. and P.P.; writing—original draft preparation, F.S. and P.P.; writing—review and editing, F.S. and P.P.; visualization, F.S.; supervision, F.S.; project administration, F.S. and V.M.; funding acquisition, F.S. and V.M.

Funding: Research was funded by the contribution of "Fondo di Ricerca per il Sistema Elettrico nell'ambito, Accordo di Programma RSE S.p.A.—Ministero dello Sviluppo Economico—D.G. Nucleare, Energie rinnovabili ed efficienza energetica".

Acknowledgments: We should thank various people that have contributed in some way to the realization of the Amerigo Lander and of the CBA. In particular, we would to thank Giovanni Ciceri, Enrico Capodarca, Fabio Latini, and Massimo Cocciaretto for the fundamental help in the design phase; Patrizia Giordano, Laura Borgognoni, Massimo Leonetti, Franco Lanini, and Claudio Vannini, as well as the crew and the captains of the N/O Urania Emanuele Gentile and Vincenzo Lubrano Lavadera, and also the dive group of the Black Angels for their contributions in the experimental phases. A special thanks and a particular memory go to the mourned Giovanni Bortoluzzi for the infinite energy that transmitted to all of us.

Conflicts of Interest: The authors declare no conflict of interest. The funders had no role in the design of the study; in the collection, analyses, or interpretation of data; in the writing of the manuscript, or in the decision to publish the results.

Appendix A

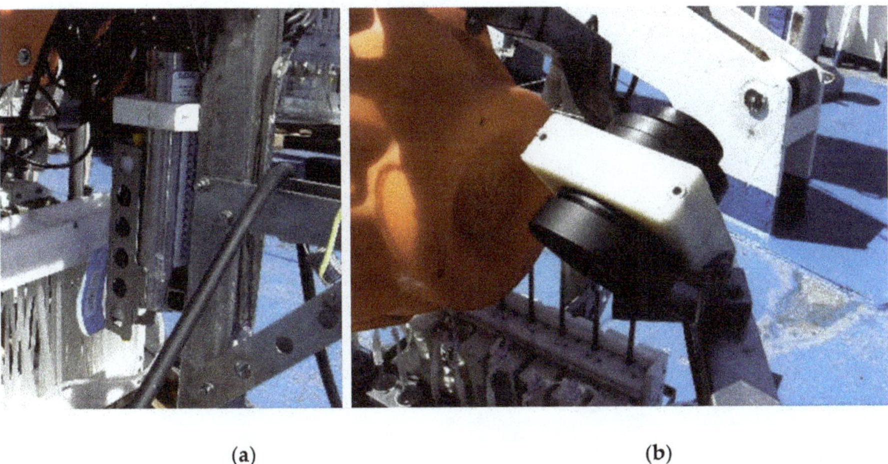

(a) (b)

Figure A1. The Amerigo Lander monitoring instrumentation. (**a**) The SBE CTD mounted on the Amerigo tripod; (**b**) the Telesub Lanterna camera installed on the Amerigo Lander.

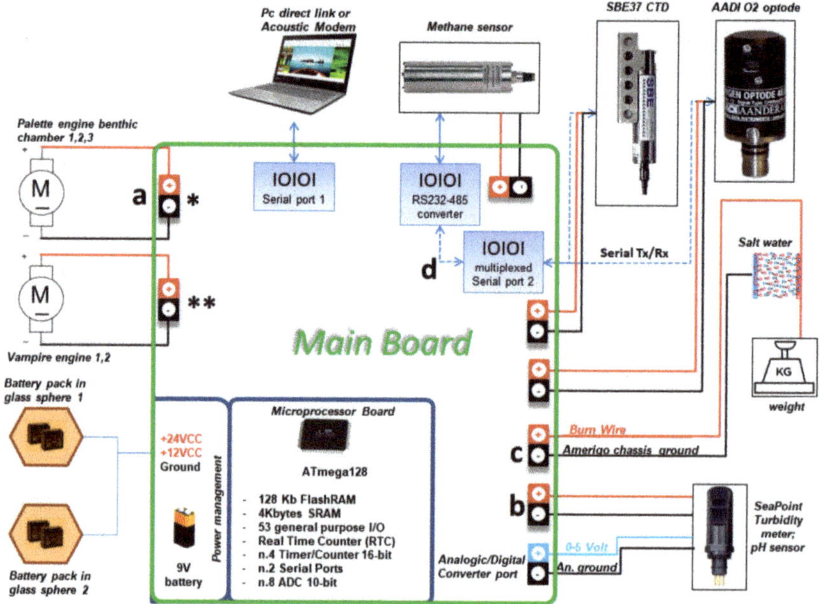

Figure A2. Electronics of the Amerigo Lander: Schematic drawing of the electronics diagram architecture. * Multiplied by three (palettes); ** multiplied by two (VAMPIRE motors); a: dedicated POWER-ON/OFF port (H bridge with electronic shunt); b: other "n" analogic devices; c: other "n" burn wire systems; d: other "n" digital sensors; blue dashed line: serial Tx/Rx port.

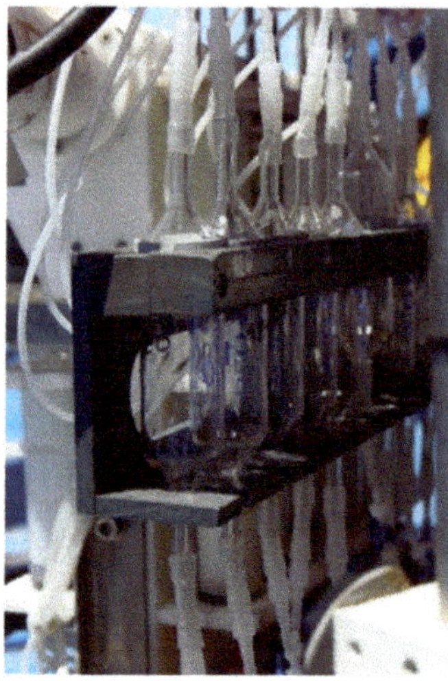

Figure A3. The set of gas-impermeable glass ampoules for water sampling for the analysis of dissolved gases.

Figure A4. Filtering, subdivision into aliquots, and treatment of the water samples collected by the syringes of the Amerigo Lander is conducted in a Nitrogen glove-box.

Table A1. Main configuration of the Amerigo Lander electronic.

Type	Description
Microprocessor	ATmega128
	RAM flash 128 Kb
	RAM 4Kb
	e2prom 4Kb
	Internal clock
	n. 3 timer 16 bit
	n. 8 ADC channels 10 bit
	n. 2 TTL serial ports
External circuit	serial flash RAM 4Mb
	n. 2 RS232 serial ports
	n. 3 RS485 ports
	n. 8 analogic inputs
	n. 2 digital inputsn.
	20 on/off ports (H-bridge with electronic shunt)

Table A2. Configurations of the general inputs and outputs of the electronic of the Amerigo Lander.

Name	Type	Description
OUT 1	ON/OFF	Power terminals for CTD sensor
OUT 2	ON/OFF	Power terminals for O2 sensor number 1, chamber n.1
OUT 2	ON/OFF	Power terminals for O2 sensor number 2, chamber n.2
OUT 4	ON/OFF	Power terminals for O2 sensor number 3, chamber n.3
OUT 5	ON/OFF	Power terminals for CH4 sensor number 1, chamber n.1
OUT 6	ON/OFF	Power terminals for CH4 sensor number 2, chamber n.2
OUT 7	ON/OFF	Power terminals for CH4 sensor number 3, chamber n.3
OUT 8	ON/OFF	Power terminals for analogic sensors (n.2 turbidity and n.1pH)
OUT 9	ON/OFF	Power terminals - Engine palette chamber n.1 (mixing water)
OUT 10	ON/OFF	Power terminals - Engine palette chamber n.2 (mixing water)
OUT 11	ON/OFF	Power terminals -Engine palette chamber n.3 (mixing water)
OUT 12	ON/OFF	Engine VAMPIRONE number 1
OUT 13	ON/OFF	Engine VAMPIRONE number 2
OUT 14	ON/OFF	Burn wire n.1 Buoy electronic check
OUT 15	ON/OFF	Burn wire n.2 Benthic chamber release
OUT 16	ON/OFF	Burn wire n.3 Benthic chamber cover release
OUT 17	ON/OFF	Burn wire n.4 Benthic chamber cover release
OUT 18	ON/OFF	Burn wire n.5 weights release
OUT 19	ON/OFF	Power terminals 24 Volts for microprofiler
OUT 20	ON/OFF	OxyStat pump
RS232 1	RS232	Main serial port - PC communication
RS232 2-1	RS232	CTD
RS232 2-2	RS232	serial communication with O2 sensor n. 1, chamber n.1
RS232 2-3	RS232	serial communication with O2 sensor n. 2, chamber n.2
RS232 2-4	RS232	serial communication with O2 sensor n. 3, chamber n.3
RS232 2-5	RS232	serial communication with pH sensor n. 1, chamber n.1
RS232 2-6	RS232	serial communication with pH sensor n. 2, chamber n.2
RS232 2-7	RS232	serial communication with pH sensor n. 3, chamber n.3
RS485 1	RS485	serial communication with CH4 sensor n. 1, chamber n.1
RS485 2	RS485	serial communication with CH4 sensor n. 2, chamber n.2
RS485 3	RS485	serial communication with CH4 sensor n. 3, chamber n.3
an 1	analogical in	Analogical (0-5V) Turbidity sensor n. 1, chamber n.1
an 2	analogical in	Analogical (0-5V) Turbidity sensor n. 2, chamber n.2
an 3	analogical in	Analogical (0-5V) Turbidity sensor n. 3, chamber n.3
an 4	analogical in	NC-future purpose (pCO_2)
an 5	analogical in	NC-future purpose (Eh)
an 6	analogical in	NC-future purpose (H_2S)
an 7	analogical in	NC-future purpose (other sensor)
an 8	analogical in	NC-future purpose (other sensor)
DI 1	digital IN	magnetic sensor (future purpose, VAMPIRONE position)
DI 2	digital IN	magnetic sensor (future purpose, VAMPIRONE position)

References

1. Douvere, F. The importance of marine spatial planning in advancing ecosystem-based sea use management. *Mar. Policy* **2008**, *32*, 762–771. [CrossRef]
2. Pérez-Albaladejo, E.; Rizzi, J.; Fernandes, D.; Lille-Langøy, R.; Goksøyr, A.; Oros, A.; Spagnoli, F.; Porte, C. Assessment of the environmental quality of coastal sediments by using a combination of in vitro bioassays. *Mar. Pollut. Bull.* **2016**, *108*, 53–61. [CrossRef] [PubMed]
3. Catalano, G.; Azzaro, M.; Bastianini, M.; Bellucci, L.G.; Bernardi Aubry, F.; Bianchi, F.; Burca, M.; Cantoni, C.; Caruso, G.; Casotti, R.; et al. The carbon budget in the northern Adriatic Sea, a winter case study. *J. Geophys. Res. Biogeosci.* **2014**, *119*, 1399–1417. [CrossRef]
4. Legendre, L. *Marine Biogeochemical Cycles*; John Wiley & Sons, Inc.: Hoboken, NJ, USA, 2014; pp. 145–187.

5. Pörtner, H.-O.; Karl, D.M.; Boyd, P.W.; Cheung, W.W.L.; Lluch-Cota, S.E.; Nojiri, Y.; Schmidt, D.N.; Zavialov, P.O. Ocean systems. In *Climate Change 2014: Impacts, Adaptation, and Vulnerability. Part A: Global and Sectoral Aspects. Contribution of Working Group II to the Fifth Assessment Report of the Intergovernmental Panel on Climate Change*; Field, C.B., Barros, V.R., Dokken, D.J., Mach, K.J., Mastrandrea, M.D., Bilir, T.E., Chatterjee, M., Ebi, K.L., Estrada, Y.O., Genova, R.C., et al., Eds.; Cambridge University Press: Cambridge, UK, 2014; pp. 411–484.
6. Gattuso, J.P.; Magnan, A.; Billé, R.; Cheung, W.W.; Howes, E.L.; Joos, F.; Allemand, D.; Bopp, L.; Cooley, S.R.; Eakin, C.; et al. Contrasting futures for ocean and society from different anthropogenic CO_2 emissions scenarios. *Science* **2015**, *349*, aac4722. [CrossRef] [PubMed]
7. Spagnoli, F.; Bergamini, M.C. Water-solid exchanges of nutrients and trace elements during early diagenesis and resuspension of anoxic shelf sediments. *Waterair Soil Pollut.* **1997**, *99*, 541–556. [CrossRef]
8. Apitz, S.E.; Bell, E.; Gilbert, F.; Hall, P.; Kershaw, P.; Nickell, L.; Parker, R.; Rabouille, C.; Shimmield, G.; Solan, M.; et al. Coastal Ocean Benthic Observatories (COBO): Integrated tools for the in situ observation and study of benthic ecosystem biogeochemical processes. In Proceedings of the 230th National Meeting of the American-Chemical-Society, Division of Geochemistry, in-situ Methods and Investigations in Environmental Science, Washington, DC, USA, 28 August–1 September 2005.
9. Apitz, S.E.; Bell, E.; Breuer, E.; Damgaard, L.; Gilbert, F.; Glud, R.; Hall, P.; Kershaw, P.; Lansard, B.; Nickell, L.; et al. Integrating new technologies for the study of benthic ecosystem response to human activity: Towards a Coastal Ocean Benthic Observatory (COBO). In Proceedings of the XVIII Congresso dell'Associazione Italiana di Oceanologia e Limnologia (AIOL), Napoli, Italy, 3–7 July 2006; Volume 19, pp. 73–78.
10. Spagnoli, F.; Marcaccio, M.; Frascari, F. Early diagenesis processes and benthic fluxes in different depositional environments of the Northern and Central Adriatic Sea. In Proceedings of the XVIII Congresso dell'Associazione Italiana di Oceanologia e Limnologia (AIOL), Napoli, Italy, 3–7 July 2006; Volume 19, pp. 483–487.
11. Spagnoli, F.; Bartholini, G.; Dinelli, E.; Marini, M.; Giordano, P. Early diagenesis of carbon and nutrients in sediments of the Gulf of Manfredonia (Southern Adriatic Sea). In *Marine research@CNR*; Brugnoli, E., Cavarretta, G., Mazzola, S., Trincardi, F., Ravaioli, M., Santoleri, R., Eds.; Consiglio Nazionale delle Ricerche, Dipartimento Terra & Ambiente: Roma, Italy, 2011; Volume DTA/06-2011, pp. 459–474.
12. Spagnoli, F.; Bartholini, G.; Marini, M.; Giordano, P. Biogeochemical processes in sediments of the Manfredonia Gulf (Southern Adriatic Sea): Early diagenesis of carbon and nutrient and benthic exchange. *Biogeosci. Discuss.* **2004**, *1*, 803–823. [CrossRef]
13. Spagnoli, F.; Bartholini, G.; Marini, M.; Giordano, P.; McCorkle, D.; Fiesoletti, F.; Specchiulli, A. Early diagenesis and benthic fluxes in Manfredonia Gulf (Southern Adriatic Sea). *Geochim. Cosmochim. Acta* **2004**, *68*, 348.
14. Berelson, W.M.; Hammond, D.E.; Smith, K.L.; Jahnke, R.A.; Devol, A.H.; Hinga, K.R.; Sayles, F. In situ benthic flux measurement devices-bottom lander technology. *Mar. Technol. Soc. J.* **1987**, *21*, 26–32.
15. Buchholtz-ten Brink, M.R.; Gust, G.; Chavis, D. Calibration and performance of a stirred benthic chamber. *Deep Sea Res. Part A Oceanogr. Res. Pap.* **1989**, *36*, 1083–1101. [CrossRef]
16. Glud, R.N.; Forster, S.; Huettel, M. Influence of radial pressure gradients on solute exchange in stirred benthic chambers. *Mar. Ecol. Prog. Ser.* **1996**, *141*, 303–311. [CrossRef]
17. Greinert, J.; Linke, P.; Sweetman, A. *Integrated Modular Systems for Monitoring of Ecosystem Functions in Deep-Sea Habitats with Relevance for Mining*; MIDAS: Itasca, IL, USA, 2015.
18. Parker, W.; Doyle, K.; Parker, E.; Kershaw, P.; Malcolm, S.; Lomas, P.; Kershaw, P. Benthic interface studies with landers. Consideration of lander/interface interactions and their design implications. *J. Exp. Mar. Boil. Ecol.* **2003**, *285*, 179–190. [CrossRef]
19. Priede, I.G.; Addison, S.; Bradley, S.; Bagley, P.M.; Gray, P.; Yau, C.; Witte, U. Autonomous deep-ocean lander vehicles; modular approaches to design and operation. In Proceedings of the IEEE Oceanic Engineering Society (OCEANS'98), Nice, France, 28 September–1 October 1998; Volume 3, pp. 1238–1244.
20. Tengberg, A.; De Bovee, F.; Hall, P.; Berelson, W.; Chadwick, D.; Ciceri, G.; Crassous, P.; Devol, A.; Emerson, S.; Gage, J.; et al. Benthic chamber and profiling landers in oceanography—A review of design, technical solutions and functioning. *Prog. Oceanogr.* **1995**, *35*, 253–294. [CrossRef]
21. Tengberg, A.; Stahl, H.; Gust, G.; Müller, V.; Arning, U.; Andersson, H.; Hall, P. Intercalibration of benthic flux chambers I. Accuracy of flux measurements and influence of chamber hydrodynamics. *Prog. Oceanogr.* **2004**, *60*, 1–28. [CrossRef]

22. Viollier, E.; Rabouille, C.; Apitz, S.; Breuer, E.; Chaillou, G.; Dedieu, K.; Furukawa, Y.; Grenz, C.; Hall, P.; Janssen, F.; et al. Benthic biogeochemistry: State of the art technologies and guidelines for the future of in situ survey. *J. Exp. Mar. Boil. Ecol.* **2003**, *285*, 5–31. [CrossRef]
23. Berg, P.; Glud, R.N.; Hume, A.; Ståhl, H.; Oguri, K.; Meyer, V.; Kitazato, H. Eddy correlation measurements of oxygen uptake in deep ocean sediments. *Limnol. Oceanogr. Methods* **2009**, *7*, 576–584. [CrossRef]
24. Berg, P.; Long, M.H.; Huettel, M.; Rheuban, J.E.; McGlathery, K.J.; Foreman, K.H.; Giblin, A.E.; Marino, R.; Howarth, R.W. Eddy correlation measurements of oxygen fluxes in permeable sediments exposed to varying current flow and light. *Limnol. Oceanogr.* **2013**, *58*, 1329–1343. [CrossRef]
25. Fones, G.R.; Davison, W.; Holby, O.; Thamdrup, B.; Jorgensen, B.B. High-resolution metal gradients measured by in situ DGT/DET deployment in Black Sea sediments using an autonomous benthic lander. *Limnol. Oceanogr.* **2001**, *46*, 982–988. [CrossRef]
26. Greeff, O.; Glud, R.N.; Gundersen, J.; Holby, O.; Jørgensen, B.B. A benthic lander for tracer studies in the sea bed: In situ measurements of sulfate reduction. *Cont. Shelf Res.* **1998**, *18*, 1581–1594. [CrossRef]
27. Miwa, T.; Iino, Y.; Tsuchiya, T.; Matsuura, M.; Takahashi, H.; Katsuragawa, M.; Yamamoto, H. Underwater observatory lander for the seafloor ecosystem monitoring using a video system. In Proceedings of the 2016 Techno-Ocean (Techno-Ocean), Kobe, Japan, 6–8 October 2016; pp. 333–336.
28. Best, M.M.; Favali, P.; Beranzoli, L.; Blandin, J.; Çağatay, N.M.; Cannat, M.; de Stigter, H. The EMSO-ERIC Pan-European Consortium: Data benefits and lessons learned as the legal entity forms. *Mar. Technol. Soc. J.* **2016**, *50*, 8–15. [CrossRef]
29. Chen, J.; Zhang, Q.; Zhang, A.; He, L.; Chen, Q. Sea trial and free-fall hydrodynamic research of a 7000-meter lander. In Proceedings of the OCEANS 2015-MTS/IEEE, Washington, DC, USA, 19–22 October 2015; pp. 1–5.
30. Chen, J.; Zhang, Q.; Zhang, A.; Tang, Y. 7000M lander design for hadal research. In Proceedings of the 2014 Oceans, St. John's, NL, Canada, 14–19 September 2014; pp. 1–4.
31. Hardy, K.; Cameron, J.; Herbst, L.; Bulman, T.; Pausch, S. Hadal landers: The Deepsea Challenge Ocean Trench Free Vehicles. In Proceedings of the 2013 OCEANS, San Diego, CA, USA, 23–27 September 2013; pp. 1–10.
32. Jamieson, A.J.; Fujii, T.; Solan, M.; Priede, I.G. HADEEP: Free-Falling Landers to the Deepest Places on Earth. *Mar. Technol. Soc. J.* **2009**, *43*, 151–160. [CrossRef]
33. Murashima, T.; Nakajoh, H.; Takami, H.; Yamauchi, N.; Miura, A.; Ishizuka, T. 11,000 m class free fall mooring system. In Proceedings of the Oceans 2009-Europe, Bremen, Germany, 11–14 May 2009; pp. 1–5.
34. Choi, J.K.; Fukuba, T.; Yamamoto, H.; Furushima, Y.; Miwa, T.; Kawaguchi, K. Pinpoint and Safe Installation of a Standalone Seafloor Observatory. In Proceedings of the 2018 OCEANS-MTS/IEEE Kobe Techno-Oceans (OTO), Kobe, Japan, 28–31 May 2018; pp. 1–4.
35. Pargett, D.M.; Jensen, S.D.; Roman, B.A.; Preston, C.M.; Ussler, W.; Girguis, P.R.; Scholin, C.A. Deep water instrument for microbial identification, quantification, and archiving. In Proceedings of the 2013 OCEANS, San Diego, CA, USA, 23–27 September 2013; pp. 1–6.
36. Wenzhöfer, F.; Wulff, T.; Floegel, S.; Sommer, S.; Waldmann, C. ROBEX-Innovative robotic technologies for ocean observations, a deep-sea demonstration mission. In Proceedings of the OCEANS 2016 MTS/IEEE, Monterey, CA, USA, 19–23 September 2016; pp. 1–8.
37. Williams, A.J. Expendable benthic lander (XBL). In Proceedings of the 2008 IEEE/OES US/EU-Baltic International Symposium, Tallinn, Estonia, 27–29 May 2008; pp. 1–8.
38. Berner, A. *Early Diagenesis: A Theoretical Approach*; Princeton University Press: Princeton, NJ, USA, 1980.
39. Hammond, D.E.; Fuller, C.; Harmon, D.; Hartman, B.; Korosec, M.; Miller, L.G.; Hager, S.W. Benthic fluxes in San Francisco bay. In *Temporal Dynamics of an Estuary: San Francisco Bay*; Springer: Dordrecht, The Netherlands, 1985; pp. 69–90.
40. Hammond, D.E.; McManus, J.; Berelson, W.M.; Kilgore, T.E.; Pope, R.H. Early diagenesis of organic material in equatorial Pacific sediments: Stoichiometry and kinetics. *Deep Sea Res.* **1996**, *43*, 1365–1412. [CrossRef]
41. Esposito, V.; Andaloro, F.; Canese, S.; Bortoluzzi, G.; Bo, M.; Di Bella, M.; Italiano, F.; Sabatino, G.; Battaglia, P.; Consoli, P.; et al. Exceptional discovery of a shallow-water hydrothermal site in the SW area of Basiluzzo islet (Aeolian archipelago, South Tyrrhenian Sea): An environment to preserve. *PLoS ONE* **2018**, *13*, e0190710. [CrossRef] [PubMed]

42. Spagnoli, F.; Andaloro, F.; Canese, S.; Capaccioni, B.; Esposito, V.; Giordano, P.; Romeo, T.; Bortoluzzi, G. Nuove recenti conoscenze sul sistema idrotermale del complesso vulcanico dell'Isola di Panarea (Arcipelago delle Eolie, Mar Tirreno Meridionale) New recent insights of the hydrothermal system of the Panarea Island (Aeolian Archipelago, South Tyrrhenian Sea). *Mem. Descr. Carta Geol.* **2019**, *105*, 85–90.
43. Berelson, W.; Hammond, D.; Johnson, K.; Johnson, K. Benthic fluxes and the cycling of biogenic silica and carbon in two southern California borderland basins. *Geochim. Cosmochim. Acta* **1987**, *51*, 1345–1363. [CrossRef]
44. Cummins, K.M.; McManus, J.; Smith, G.; Spagnoli, F.; Hammond, D.E.; Cummins, K.M.; Berelson, W.M. Methods for measuring benthic nutrient flux on the California Margin: Comparing shipboard core incubations to in situ lander results. *Limnol. Oceanogr. Methods* **2004**, *2*, 146–159.
45. Hammond, D.; Giordani, P.; Berelson, W.; Poletti, R. Diagenesis of carbon and nutrients and benthic exchange in sediments of the Northern Adriatic Sea. *Mar. Chem.* **1999**, *66*, 53–79. [CrossRef]
46. Berelson, W.; Hammond, D. The calibration of a new free-vehicle benthic flux chamber for use in the deep sea. *Deep. Sea Res. Part A Oceanogr. Res. Pap.* **1986**, *33*, 1439–1454. [CrossRef]
47. Black, K.S.; Fones, G.R.; Peppe, O.C.; Kennedy, H.A.; Bentaleb, I. An autonomous benthic lander: Preliminary observations from the UK BENBO thematic programme. *Cont. Shelf Res.* **2001**, *21*, 859–877. [CrossRef]
48. Devol, A.H. Verification of flux measurements made with in situ benthic chambers. *Deep. Sea Res. Part A Oceanogr. Res. Pap.* **1987**, *34*, 1007–1026. [CrossRef]
49. Ferrón, S.; Alonso-Pérez, F.; Castro, C.G.; Ortega, T.; Pérez, F.F.; Ríos, A.F.; Forja, J.M. Hydrodynamic characterization and performance of an autonomous benthic chamber for use in coastal systems. *Limnol. Oceanogr. Methods* **2008**, *6*, 558–571. [CrossRef]
50. Ishida, H.; Watanabe, Y.; Fukuhara, T.; Kaneko, S.; Furusawa, K.; Shirayama, Y. In situ Enclosure Experiment Using a Benthic Chamber System to Assess the Effect of High Concentration of CO_2 on Deep-Sea Benthic Communities. *J. Oceanogr.* **2005**, *61*, 835–843. [CrossRef]
51. Jahnke, R.; Christiansen, M. A free-vehicle benthic chamber instrument for sea floor studies. *Deep. Sea Res. Part A Oceanogr. Res. Pap.* **1989**, *36*, 625–637. [CrossRef]
52. Lee, J.S.; An, S.-U.; Park, Y.-G.; Kim, E.; Kim, D.; Kwon, J.N.; Kang, D.-J.; Noh, J.-H. Rates of total oxygen uptake of sediments and benthic nutrient fluxes measured using an in situ autonomous benthic chamber in the sediment of the slope off the southwestern part of Ulleung Basin, East Sea. *Ocean Sci. J.* **2015**, *50*, 581–588. [CrossRef]
53. Sayles, F.; Dickinson, W. The ROLAI^{2}D lander: A benthic lander for the study of exchange across the sediment-water interface. *Deep. Sea Res. Part A Oceanogr. Res. Pap.* **1991**, *38*, 505–529. [CrossRef]
54. Smith, K.L.; Clifford, C.H.; Eliason, A.H.; Walden, B.; Rowe, G.T.; Teal, J.M. A free vehicle for measuring benthic community metabolism1. *Limnol. Oceanogr.* **1976**, *21*, 164–170. [CrossRef]
55. Smith, K.L., Jr.; White, G.A.; Laver, M.B. Oxygen uptake and nutrient exchange of sediments measured in situ using a free vehicle grab respirometer. *Deep Sea Res. Part A Oceanogr. Res. Pap.* **1979**, *26*, 337–346. [CrossRef]
56. Spagnoli, F.; Giuliani, G.; Martinotti, V.; Masini, L.; Penna, P. AMERIGO and CBA: A new lander and a new automatic benthic chamber for dissolved benthic flux measurements. In Proceedings of the 2018 IEEE International Workshop on Metrology for the Sea, (MetroSea 2018), Bari, Italy, 8–10 October 2018; ISBN 978-1-5386-7643-1.
57. Morris, R.; Hardy, K. Selecting an acoustic release for a mooring or lander. In Proceedings of the OCEANS 2017, Anchorage, AK, USA, 18–21 September 2017; pp. 1–5.
58. Phleger, C.F.; Soutar, A. Free vehicles and deep-sea biology. *Am. Zool.* **1971**, *11*, 409–418. [CrossRef]
59. Morse, J.W.; Boland, G.; Rowe, G.T. A 'gilled' benthic chamber for extended measurement of sediment-water fluxes. *Mar. Chem.* **1999**, *66*, 225–230. [CrossRef]
60. Giordano, P.; Spagnoli, F.; Marcaccio, M.; Marini, M.; Frascari, F.; Modica, A.; Rivas, G. Il Mar Piccolo di Taranto: Osservazioni preliminari sul ciclo dei nutrienti all'interfaccia acqua—Sedimento. *Atti Della Assoc. Ital. Oceanol. Limnol.* **2004**, *17*, 59–70.
61. Masini, L.; Spagnoli, F.; Marcaccio, M.; Frascari, F. *Prototipo di Camera Bentica Automatica per Bacini Acquatici Continentali e Marini*; Consiglio Nazionale delle Ricerche, Istituto di Geologia Marina: Bologna, Italy, 2001.
62. Berelson, W.; Hammond, D.; O'Neill, D.; Xu, X.-M.; Chin, C.; Zukin, J. Benthic fluxes and pore water studies from sediments of the central equatorial north Pacific: Nutrient diagenesis. *Geochim. Cosmochim. Acta* **1990**, *54*, 3001–3012. [CrossRef]

63. Spagnoli, F. *AMERIGO: A Deep Sea Lander. CNR Marine Research Activities and Technologies, Thematics: Technologies*; Consiglio Nazionale delle Ricerche, Dipartimento Terra & Ambiente: Rome, Italy, 2010.
64. Spagnoli, F.; Andresini, A.; Borgognoni, L.; Bortoluzzi, G.; Campanelli, A.; Canonico, C.; Ferrante, V.; Giuliani, G.; Giordano, P.; Greco, M.; et al. *Campagna Oceanografica CASE4*; Rapporto Finale di Crociera; ISMAR-CNR: Ancona, Italy, 2012.
65. Spagnoli, F.; Allende Ccori, J.; Andresini, A.; Borgognoni, L.; Ferrante, V.; Giordano, P.; Pignotti, E.; Zuzolo, M.G. *Campagna Oceanografica PER1*; Rapporto Finale di Crociera; CNR-ISMAR: Ancona, Italy, 2013; p. 33.
66. Spagnoli, F.; Ciceri, G.; Giuliani, G.; Martinotti, V.; Penna, P. AMERIGO: A new benthic lander for dissolved flux measurements at sediment-water-interface. In Proceedings of the Goldschmidt 2013, Florence, Italy, 25–30 August 2013; Volume 77, p. 2242.
67. Spagnoli, F.; Giuliani, G.; Penna, P.; Martinotti, V. AMERIGO, a lander for benthic flux and chemical and physical parameter measurements and for the sampling of water and sediment at the sediment-water interface. In *Le Tecnologie del CNR per il Mare*; Consiglio Nazionale delle Ricerche CNR: Roma, Italy, 2013.
68. Spagnoli, F.; Bartholini, G.; Capaccioni, B.; Giordano, P. Benthic Nutrient Fluxes in Central and Southern Adriatic and Ionian Seas. In Proceedings of the 40th CIESM Congress, Marseille, France, 28 October–1 November 2013.
69. Spagnoli, F.; Borgognoni, L.; Campanelli, A.; Ciceri, G.; Martinotti, V.; Giordano, P.; Giuliani, G.; Penna, P. A new biogeochemical method for the monitoring of possible seeps in marine CCS fields. In Proceedings of the Geoitalia 2013—Le Geoscienze per la Società, IX edizione del Forum Italiano di Scienze della Terra, Pisa, Italy, 16–18 Setptember 2013.
70. Spagnoli, F.; Ciceri, G.; Giordano, P.; Martinotti, V.; Politi, M. Experimental biogeochemical approach to the monitoring of baseline levels of CO_2 fluxes at the sediment-water interface for CCS purpose. In Proceedings of the Geoitalia 2011, VIII Forum Italiano di Scienze della Terra, Torino, Italy, 19–23 September 2011.
71. Spagnoli, F.; Kaberi, H.; Giordano, P.; Zeri, C.; Borgognoni, L.; Bortoluzzi, G.; Campanelli, A.; Ferrante, V.; Giuliani, G.; Martinotti, V.; et al. Benthic fluxes of dissolved heavy metals in polluted sediments of the Adriatic Sea. In Proceedings of the Scientific Conference: Integrated Marine Research in the Mediterranean and the Black Sea, Bruxelles, Belgium, 7–9 December 2015; pp. 301–302, ISBN 978-960-9798-25-9.
72. Spagnoli, F.; Bartholini, G.; Giordano, P. Benthic fluxes and early diagenesis processes in Adriatic Sea. Sessione poster. In Proceedings of the Goldschmidt 2013, Florence, Italy, 25–30 August 2013.
73. Bortoluzzi, G.; Spagnoli, F.; Aliani, S.; Romeo, T.; Canese, S.; Esposito, V.; Grassi, M.; Masetti, G.; Dialti, L.; Cocchi, L.; et al. New geological, geophysical and biological insights on the hydrothermal system of the Panarea—Basiluzzo Volcanic complex (Aeolian Islands, Tyrrhenian Sea). In Proceedings of the SGI-SIMP, Milan, Italy, 10–12 September 2014.
74. Spagnoli, F.; Borgognoni, L.; Acri, F.; Caccamo, G.; De Marco, R.; Leonetti, M. *Indagini del Fondale del Porto di Fano a Mezzo di Camera Bentica Valutazione Degli Effetti del Dragaggio sui Sedimenti Marini. Fase Post-Operam*; Technical Report; Istituto di Scienze Marine del CNR: Ancona, Italy, 2016.
75. Spagnoli, F.; Giordano, P.; Borgognoni, L.; Acri, F. *Studi di Carattere Ambientale Volti a Valutare le Dinamiche e gli Effetti Ambientali dei Sedimenti Marini Provenienti da Escavi Portuali. Valutazione degli Effetti del Dragaggio sui Sedimenti Marini. Fase Post-Operam*; Technical Report; Istituto di Scienze Marine del CNR: Ancona, Italy, 2016.
76. Giordano, P.; Spagnoli, F.; Langone, L.; Miserocchi, S.; Ferrante, V.; Vasumini, I.; Graziani, G.; Savelli, F.; Dalpasso, E.; De Marco, M. *Rapporto delle Attività di Campagna, Campagna RID16*; Istituto di Scienze Marine: Venice, Italy, 2016.
77. Rovere, M.; Argnani, A.; Mercorella, A.; Spagnoli, F.; Frapiccini, E.; Pellegrini, C.; Ciccone, F.; Zaniboni, F.; Pagnone, G.; Paparo, M.A.; et al. *Accordo Operativo 2016 tra Consiglio Nazionale delle Ricerche—Istituto di Scienze Marine (ISMAR CNR) e Ministero dello Sviluppo Economico, Direzione Generale per la Sicurezza Anche Ambientale delle Attività Minerarie ed Energetiche—Ufficio Nazionale Minerario per gli Idrocarburi e le Georisorse (DGS-UNMIG)*; Istituto di Scienze Marine di Bologna: Bologna, Italy, 2018.
78. Frascari, F.; Spagnoli, F.; Marcaccio, M.; Giordano, P. Anomalous Po River flood event effects on sediments and the water column of the northwestern Adriatic Sea. *Clim. Res.* **2006**, *31*, 151–165. [CrossRef]
79. Spagnoli, F.; Dinelli, E.; Giordano, P.; Marcaccio, M.; Zaffagnini, F.; Frascari, F. Sedimentological, biogeochemical and mineralogical facies of Northern and Central Western Adriatic Sea. *J. Mar. Syst.* **2014**, *139*, 183–203. [CrossRef]

80. Spagnoli, F.; Dell'Anno, A.; De Marco, A.; Dinelli, E.; Fabiano, M.; Gadaleta, M.V.; Ianni, C.; LoIacono, F.; Manini, E.; Marini, M.; et al. Biogeochemistry, grain size and mineralogy of the central and southern Adriatic Sea sediments: A review. *Chem. Ecol.* **2010**, *26*, 19–44. [CrossRef]
81. Spagnoli, F.; Frascari, F.; Marcaccio, M.; Bergamini, M.C. Early diagenesis and nutrient benthic fluxes in the Adriatic Sea. *Geophys. Res. Abstr.* **2003**, *5*, 12257.
82. Capet, A.; Lazzari, P.; Spagnoli, F.; Bolzon, G.; Solidoro, C. Benthic contributions to Adriatic and Mediterranean biogeochemical cycles. *EGU Gen. Assem. Conf. Abstr.* **2017**, *19*, 17596.

Article

Towards Non-Invasive Methods to Assess Population Structure and Biomass in Vulnerable Sea Pen Fields †

Giovanni Chimienti [1,*], Attilio Di Nisio [2], Anna M. L. Lanzolla [2], Gregorio Andria [2], Angelo Tursi [1] and Francesco Mastrototaro [1]

1 Department of Biology and CoNISMa LRU, University of Bari Aldo Moro, 70125 Bari, Italy; angelo.tursi@uniba.it (A.T.); francesco.mastrototaro@uniba.it (F.M.)

2 Department of Electrical & Information Engineering (DEI), Polytechnic University of Bari, 70125 Bari, Italy; attilio.dinisio@poliba.it (A.D.N.); anna.lanzolla@poliba.it (A.M.L.L.); gregorio.andria@poliba.it (G.A.)

* Correspondence: giovanni.chimienti@uniba.it; Tel.: +39-080-544-3344

† This paper is extended version of the conference paper Biometric relationships in the red sea pen Pennatula rubra (Cnidaria: Pennatulacea).

Received: 8 April 2019; Accepted: 13 May 2019; Published: 15 May 2019

Abstract: Colonies of the endangered red sea pen *Pennatula rubra* (Cnidaria: Pennatulacea) sampled by trawling in the northwestern Mediterranean Sea were analyzed. Biometric parameters, such as total length, peduncle length, number of polyp leaves, fresh weight, and dry weight, were measured and related to each other by means of regression analysis. Ad hoc models for future inferencing of colonies size and biomass through visual techniques were individuated in order to allow a non-invasive study of the population structure and dynamics of *P. rubra*.

Keywords: corals; Pennatulacea; *Pennatula*; model; Mediterranean Sea; biometry; ROV; trawling; fishery; VME

1. Introduction

The mesophotic and aphotic zones of the Mediterranean Sea are inhabited by a variety of benthic organisms, some of which are able to create biogenic habitats due to their three-dimensionality and their aggregative behavior. Among them, corals play a crucial role as habitat formers, being the main builders of peculiar coral frameworks, in the case of stony corals [1], or coral forests, in the case of arborescent corals [2,3]. These habitats are featured by a high sensitivity to human pressures, particularly on trawlable grounds, where their abundance and their associated community significantly decrease [4]. Among soft-bottom octocorals, pennatulaceans can form extensive populations, known as sea pen fields, providing relevant structure in flat, low-relief muddy habitats where there is little physical habitat complexity. These fields create essential biogenic habitat for suprabenthic and benthic invertebrates, as well as an important feeding and nursery area for a rich demersal fish fauna [5–8] representing Essential Fish Habitats and Vulnerable Marine Ecosystems (VMEs) [9] worthy of protection.

Sea pen fields are often difficult to be found on muddy bottoms using indirect methods because they cannot be detected using the common habitat mapping geophysical techniques. The most effective way to identify these fields, through the analysis of data coming from commercial fishing bycatch or from experimental trawl fishing surveys, on a large scale still remains trawling. After the identification of a sea pen field, the visual techniques used onboard oceanographic cruises, such as Remotely Operated Vehicles (ROVs), allow to carry out more detailed studies and/or monitoring of VMEs on a relatively small area [10]. Non-contact and non-destructive imaging techniques, based on properly developed object segmentation and detection algorithms, have been proved to be a viable alternative to contact measurement and diagnostic techniques in a large variety of sectors, ranging from industrial quality control to characterization of devices, to medical imaging and clinical applications.

Hence, it is foreseeable the development of automatic vision methods for identification, counting, and measurement of the sea [11–20].

Despite the recognized ecological importance of pennatulaceans, little is known about their biology and ecology, and their vulnerability to human pressures has been assessed under a precautionary approach [21]. Destructive sampling is still needed to estimate the biomass of a sea pen field, as well as to collect information on colonies' size, population structure, and dynamics, in order to assess the main features of the population and to establish its need for protection. Nowadays, advances in the design, lowering of costs, and increased performance of unmanned vehicles, including ROVs and drones, pave the way to the extended exploration of marine environments [22–28]. Underwater imagery techniques, such as ROVs and towed cameras, are now allowing a better understanding of the sea pen numerical importance on a relatively small scale, but these methods are still not effective in understanding biomass and size structure [10]. Hence, the development of a non-invasive technique for determining the weight of colonies is desirable to obtain quantitative data of biomass (both fresh and dry) from ROV imaging, useful to support conservation measures. Moreover, the possibility to estimate the colonies' length with the same approach would also enhance the size structure assessment of sea pens population and their monitoring, consistently with their need of protection and their fragility to the trawled sampling gears (e.g., trawl nets and dredges). Finding of proper biometric correlations could avoid the need of further sampling for the study of these vulnerable populations in the near future [29], allowing non-invasive methods, and representing a valid alternative to destructive sampling.

The present study modeled the biometric measurements collected from a population of the red sea pen *Pennatula rubra* (Ellis, 1761) sampled by trawling. This species, endemic of the Mediterranean Sea, belongs to the suborder Subsessiliflorae because of the presence of polyps disposed in pinnately arranged leaves. It represents one of the most important field-forming sea pens of the Mediterranean continental shelf [30,31], reported as vulnerable in the Red List of the International Union for the Conservation of Nature (IUCN) [32] among the seventeen threatened coral species of the basin. The colonies of *P. rubra* live with the peduncle (i.e., the basal part of the colony) into the sediment, the rachis representing most of the visible portion of the colony (Figure 1). For this reason, the total length cannot be directly measured through visual methods. In this study, we found reliable biometric relationships using the number of polyp leaves as a proxy to estimate the total length and the biomass of the colonies, enabling future in situ assessments of population structure and biomass. This would also avoid the need for sampling for the study of the population dynamic of *P. rubra*, representing a necessary knowledge base for a non-invasive study of the wild populations.

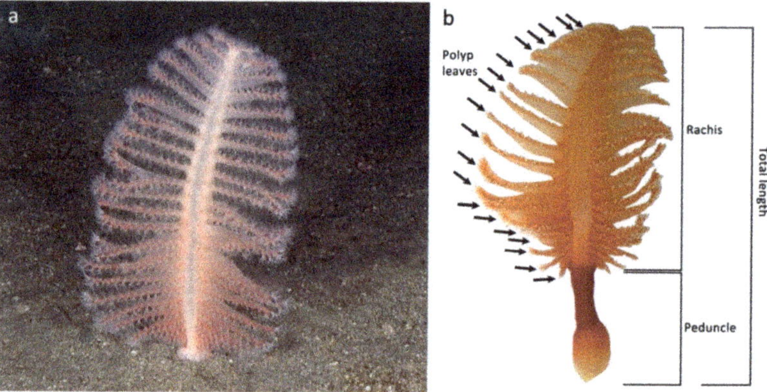

Figure 1. The colony of *Pennatula rubra*. (**a**) In vivo appearance of the species; (**b**) indication of polyp leaves (black arrows), rachis, peduncle, and total length.

2. Materials and Methods

Colonies of *P. rubra* were sampled using an experimental trawl net, with a stretched mesh of 20 mm in the codend, in the frame of the MEDITS (Mediterranean International Bottom Trawl Survey) project [33]. Sampling was carried out during 2013 northwest Punta Alice (Ionian Sea, southern Italy; start: 39°35.05′N–16°52.26′E; end: 39°34.05′N–16°53.63′E) at 61–65 m depth [31], onboard the *Pasquale e Cristina* fishing vessel. A SCANMAR acoustic system (Scanmar AS, Åsgårdstrand, Norway) [34] was used to measure the horizontal and vertical openings of the net in order to estimate the swept area. The colonies of *P. rubra* sampled were preserved on board at −20°C.

A total of 168 colonies, sampled over an area of 41,000 m^2, were analyzed. The following biometric parameters were measured for each colony: length of the peduncle, the total length of the colony (considering both rachis and peduncle), fresh weight, and number of polyp leaves (Figure 1). Length measures were carried out using a manual caliber with 1 mm resolution, and fresh weight was measured using a DENVER MXX-212 electronic balance (Denver Instrument GmbH, Goettingen, Germany; 0.01 g resolution, 0.04 g worst-case uncertainty).

Measurements were carried out in the laboratory after thawing, considering that the freezing process causes the complete contraction of the colonies. On the contrary, living colonies can contain a highly variable quantity of seawater driving their contraction and considerably changing their size and fresh weight [31]. Then, a suitable procedure to obtain dry weight measurement was performed for a reduced number of sampled colonies. In particular, 54 colonies were selected having different fresh weight values to obtain a statistically significant population. Each colony was identified by a unique ID and fresh weighted, and then the colonies were dried in an oven at 40 °C for 96 h. Dry weight was measured for each colony. A detailed study highlighting the relationship among all the biometric parameters of *P. rubra* was performed, with the aim to develop suitable models for the colonies' size and fresh weight based on the number of polyp leaves. The model for the estimation of size was developed to obtain both the rachis length and the total length, the former being visible with imagery techniques (Figure 1), while the latter used in direct measurements from samples. Moreover, the relationship between fresh and dry weight of the colonies was also assessed. Finally, data obtained from an ROV survey carried out on the same population sampled by trawling [10] were used to compare the distribution of the number of polyp leaves obtained from both the methods (i.e., visual vs. sampling). In particular, polyp leaves were counted for a total of 207 colonies observed in vivo, whose position, contraction, and ROV framing allowed to clearly distinguish the polyp leaves of at least one side of the colony.

Although it is not very common, the number of polyp leaves from the two sides of the same colony can be different due to mechanical damage or predation events. For this reason, both right and left polyp leaves of each sampled colony were counted. When the number of leaves was different between the two sides of the same colony, the mean value was calculated and used for the models. Whereas, the maximum number of polyp leaves per colony side was preferred over the mean value for the size estimation model, considering the direct link between the length of the colony and the number of polyp leaves.

3. Results and Discussion

The number of polyp leaves proved to be a reliable proxy to estimate size and biomass of *P. rubra* colonies using non-invasive approaches based on ROV imaging, as previously highlighted with regression analysis [10,29]. In particular, the analysis of colonies sampled through experimental trawl fishing surveys allowed to identify suitable relationships, which were able to estimate the size and biomass of sea pens using models based on the number of polyp leaves, that can be measured from images.

3.1. Model for Size Estimation

In the first step, the relationship between rachis length and the number of polyp leaves was investigated. The experimental results highlighted a good linear behavior (correlation coefficient = 0.84) between number n of polyp leaves and the estimated rachis length as reported below in Equations (1) and (2) and shown in Figure 2.

$$l_r = l_t - l_p \tag{1}$$

$$\hat{l}_r(n) = 2.6 \cdot n - 1.1 \tag{2}$$

where l_t is the total length of the colony, l_p is the length of the peduncle, n is the number of polyp leaves, and l_r and $\hat{l}_r$ are rachis length and its estimation by means of linear regression, respectively.

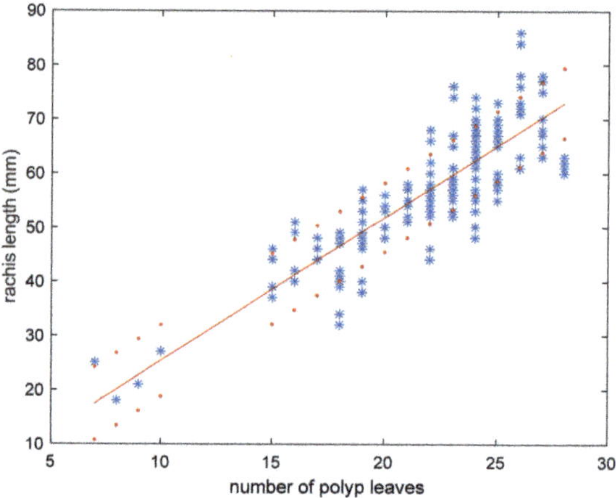

Figure 2. Rachis length vs. the number of polyp leaves in *Pennatula rubra*; the red line is the linear regression, and red dot lines include 95% of data.

To quantify the accuracy of the proposed model, the root mean square relative error e_{l_r} was calculated by using Equation (3), then a value of 11.8% for this parameter was obtained. This value is due to intraspecific variability, and it can be accepted for size-frequency distribution analysis in soft coral populations.

$$e_{l_r} = \sqrt{\frac{1}{M}\sum_{i}^{M}\left(\frac{\hat{l}_r(n_i) - l_{r_i}}{l_{r_i}}\right)^2} \tag{3}$$

In Equation (3), M is the number of samples, n_i and l_{r_i} are the values of polyp leaves number and rachis length of the i-th sample, respectively. Experimental values are equally distributed around the regression line (Figure 2).

Figure 2 shows a significant variation in colony length for each value of the number of polyp leaves; therefore the dispersion of obtained data was analyzed. Figure 3 shows the distribution of l_r values grouped for each n, where mean and standard deviation were represented with a box plot. The values with $n < 15$ were not included in the dispersion analysis because they did not provide length variation in correspondence of the same values of n. In general, colonies with a high number of polyp leaves provide more dispersion, with a maximum standard deviation value of 8.2 mm. This can be due to intraspecific variability, particularly evident in older colonies whose number of polyp leaves is higher.

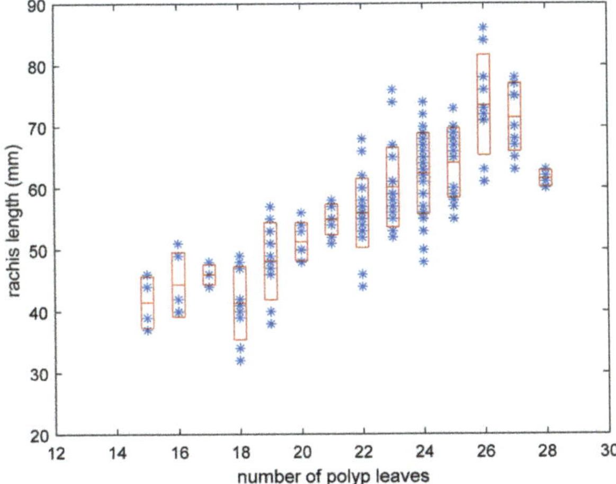

Figure 3. Distribution of rachis length in the population of *Pennatula rubra*; red boxes include all values ranging in average ± standard deviation.

Similar linear dependence was obtained by analyzing the behavior of total length estimate $\hat{l}_t$ as a function of the number of polyp leaves, reported in Equation (4), with an obtained root mean square relative error of 10.6%.

$$\hat{l}_t(n) = 5.8 \cdot \mathrm{n} + 8.1 \tag{4}$$

The major or minor variation of rachis length for each value of the number of polyp leaves can be due to the natural variability of the population and the number of colonies sampled. Despite a large number of samples analyzed (168 colonies), it is expected that the rachis length variation among the different values of the number of polyp would be more homogeneous by analyzing a larger number of colonies.

3.2. Model for Fresh Weight Estimation

The number of polyp leaves was related to the fresh weight of *P. rubra* colonies and served as a model to assess colonies' biomass. A suitable *P. rubra* envelop was considered in order to identify the best curve for data fitting. In particular, by supposing a 2D outline of *P. rubra*, it is possible to consider a second order enveloping curve expressed as a function of distance along the rachis, as shown in Figure 4. The curve intersects the rachis at distances zero and l_r.

The weight $\hat{w}_f$ of a *P. rubra* colony is assumed to be proportional to the enveloping area, and consequently can be expressed by integrating that curve between 0 and l_r, obtaining a third-degree power of l_r. Therefore, by taking into account the linear dependence between length and the number of polyp leaves, the dependence of fresh weight from n can be described by a third-degree polynomial, where all powers of n up to the third have been considered for generality (Figure 5).

Tests consisting of the use of different nonlinear models confirmed that the best fitting curve providing the minimum mean square error is expressed by the following relationship, corresponding to a root mean square relative error of 46%:

$$\hat{w}_f(n) = -0.013 \cdot n^3 + 0.076 \cdot n^2 - 1.170 \cdot n + 5.342 \tag{5}$$

The validity of the model was not assured when $n < 15$ because the weight estimated in this condition required a higher number of data from juvenile or young colonies and a more accurate weight measurement with very low values. Anyway, small colonies resulted infrequent with both trawling and visual methods, and their contribution to the biomass of the sampled population is about 0.06% of the mass.

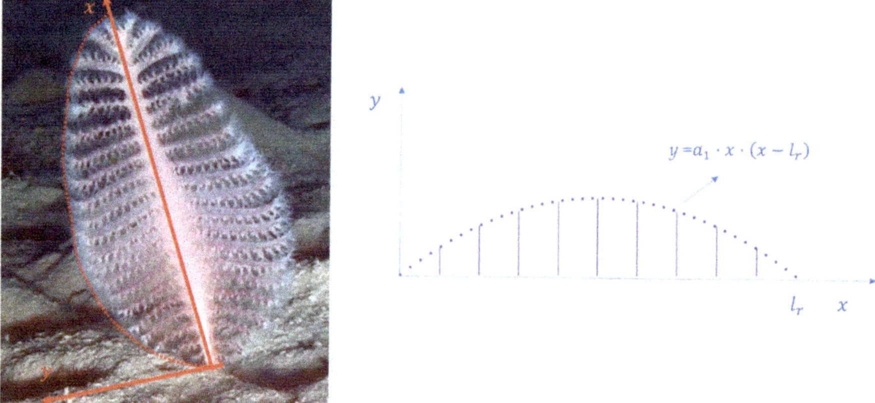

Figure 4. 2D outline of *Pennatula rubra*.

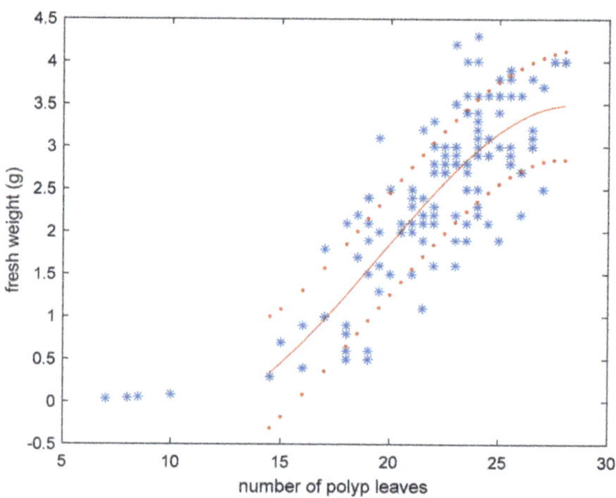

Figure 5. Fresh weight vs. the number of polyp leaves in *Pennatula rubra*; red curve is the third-degree polynomial fitting experimental data, and red dot curves include 95% of data.

A further model of the estimated weight $\hat{w}_{2f}$, taking into account also the linear dependence on rachis length, was considered with the aim to reduce the fitting error (Equation (6)). The following fitting surface was identified, as shown in Figure 6.

$$\hat{w}_{2f}(n, l_r) = 2.28 \cdot 10^{-4} \cdot n^3 - 0.023 \cdot n^2 + 0.826 \cdot n + 0.039 \cdot l_r - 9.182 \quad (6)$$

In this way, the root mean square relative error was reduced to 38.7%.

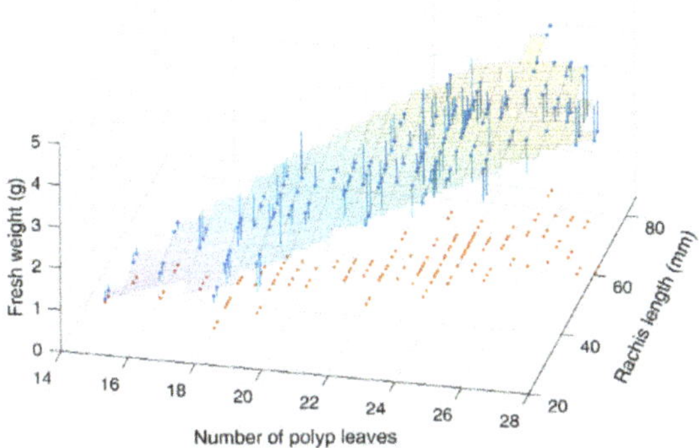

Figure 6. The behavior of fresh weight as a function of both the number of polyp leaves and rachis length in *Pennatula rubra*. Blue bars represent errors between measured weight and least squares surface fitting. Red dots are projections on the horizontal plane.

3.3. Dry Weight Estimation

The correlation between dry and fresh weight was analyzed based on measurement data on a reduced set of *P. rubra* samples represented by 54 colonies. Despite the fact that the dry weight is not frequently used for sea pens, it is considered a more reliable measure of the biomass because of the variable water content in living colonies of *P. rubra*, driving their exposure to different currents and their withdrawal as a defense strategy [31].

Figure 7 shows fresh weight behavior vs. dry weight, which can be described by means of a polynomial function of the third order expressed by means of

$$\hat{w}_d(n) = 0.062 \cdot {w_f}^3 - 0.412 \cdot {w_f}^2 + 1.134 \cdot w_f - 0.416 \tag{7}$$

where w_f is the measured fresh weight, and $\hat{w}_d$ is the dry weight estimation.

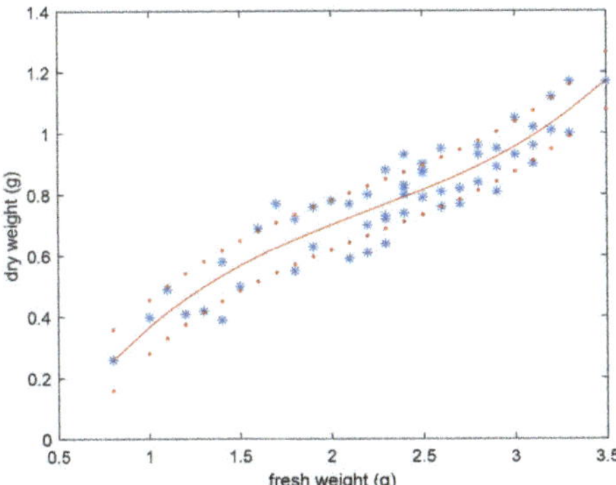

Figure 7. Fresh weight vs. dry weight in *Pennatula rubra*; the red curve is the polynomial interpolation of the third order, and red dot curves include 95% of data.

The obtained model allows to predict dry weight based on the fresh weight and shows root mean square relative error of 19.2%. This can allow the use of the model to assess dry biomass starting from both fresh weight estimation (e.g., from ROV imaging) or direct fresh weight measures (e.g., from fishery samples). In this last case, the model would help to rapidly understand the dry biomass of a sea pen field (e.g., onboard a fishing vessel) without the need of drying procedures.

3.4. Distribution of the Number of Polyp Leaves

Starting from ROV imaging data, n_i, the number of polyp leaves of the i-th colony, was counted for each of the $M = 207$ *P. rubra* colonies observed and analyzed, giving the set $\boldsymbol{n} = [n_i]_{i=0,\ldots,206}$. This set was compared with the one obtained by trawling, consisting of $M = 168$ colonies, $\boldsymbol{n} = [n_i]_{i=0,\ldots,167}$.

The distributions of the number of polyp leaves in both sets are compared in Figure 8, while descriptive statistics are reported in Table 1.

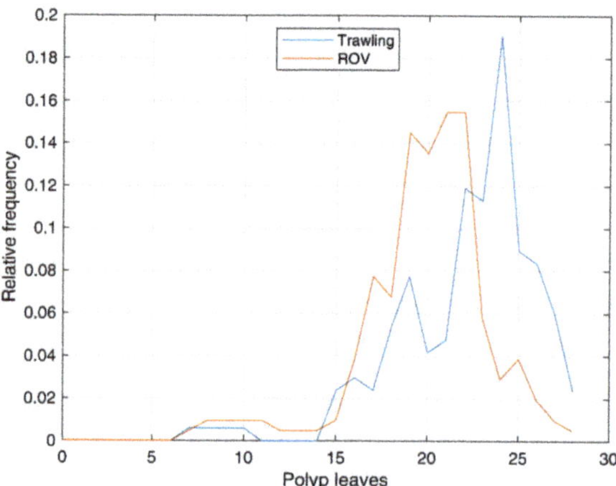

Figure 8. Comparison of the distribution of the number of polyp leaves in *Pennatula rubra*: in blue, samples obtained with trawling; in red, observations from Remotely Operated Vehicle (ROV) surveys.

Table 1. Statistics of the number of polyp leaves of *Pennatula rubra* for trawling- and Remotely Operated Vehicle (ROV)-based sampling.

Set	Mean μ	Standard Deviation σ	Skewness s	Kurtosis k
Trawling	22.0	3.7	−1.3	5.6
ROV	19.9	3.5	−1.1	5.5

Bias-corrected standard deviation, standard deviation, bias-corrected skewness, and bias-corrected kurtosis have been defined, respectively, as follows:

$$\sigma = \sqrt{\sum_{i=0}^{M-1} \frac{(n_i - \mu)^2}{M-1}} \tag{8}$$

$$\sigma' = \sqrt{\sum_{i=0}^{M-1} \frac{(n_i - \mu)^2}{M}} \tag{9}$$

$$s = \frac{\sqrt{M(M-1)}}{(M-2)} \sum_{i=0}^{M-1} \frac{\frac{1}{M}(n_i - \mu)^3}{\sigma'^3} \tag{10}$$

$$k = \frac{M-1}{(M-2)(M-3)} \left[[M+1] \sum_{i=0}^{M-1} \frac{\frac{1}{M}[n_i - \mu]^4}{\sigma'^4} - 3[M-1] \right] + 3 \tag{11}$$

Standard deviation and kurtosis are similar for trawling and ROV data, while mean values and skewness differ by about 10% and 18%, respectively (Table 1). These differences can be due to many causes, such as: a slight difference in the two nearby subpopulations studied, since trawling hauls and ROV transects do not completely overlap; a minor catch efficiency of the trawl net on smaller colonies, that can be easily passed by the net without being collected; underestimation and errors in counting the number of polyp leaves from video analysis, considering that the first and the last polyp leaves can be very small, and their identification can be less easy from images than from samples.

By using the previously identified relationship between the number of polyp leaves and fresh weight of the colonies, an estimation of the total weight can be obtained from ROV images. In particular, for each colony i, the predicted fresh weight is $\hat{w}_i = \hat{w}(n_i)$ according to the model (5). $\hat{w}_i$ has been set to 0 for small colonies ($n \leq 15$), where the model is not applicable.

Based on ROV images, the predicted mean weight of a colony is

$$\hat{w}_{mean} = \frac{1}{M}\sum_{i=0}^{M-1} \hat{w}_i = 1.86 \text{ g}$$

The total weight of a population present on an area S can then be estimated through

$$\hat{w}_{tot} = \hat{w}_{mean}\delta S$$

where δ is the density of the colonies present on the surface S.

The mean weight of trawling samples, obtained by averaging the weights measured in the laboratory, is $w_{mean} = 2.39$ g. Therefore, the relative error of weight prediction with ROV, for a given surface density and area, is −22%. This result takes into account the previously mentioned error contributions, as well as the weight modeling error.

The weight estimation based on the number of polyp leaves through ROV imaging proved to be a feasible alternative to destructive sampling. Considering the low catch efficiency of trawl nets on sea pens [10,35,36], it cannot be excluded that small colonies have less possibility to be sampled, thus justifying the 22% difference observed between the mean weight of samples with the two methods.

4. Conclusions

Zoological and ecological studies on VME indicator taxa are often based on abundance, biomass, and size structure of the population studied. This information is fundamental to assess the extent and the main features of soft-bottom coral communities, such as sea pen fields and coral gardens, in order to plan and apply proper protection initiatives. Except for abundance, considered as colonies density, whose estimation is known to be more accurate using visual methods than trawled sampling gears, the gathering of both size and biomass data has been historically carried out through destructive sampling. This study showed that the number of polyp leaves could be used for non-invasive studies of vulnerable species, such as *P. rubra*. In fact, this information can be retrieved from ad hoc ROV surveys, with a good level of accuracy and good reliability.

Despite the fact that the estimation of colonies' size and biomass using ROV imaging is a time-consuming process compared to the direct measurement of the samples, it can allow a non-destructive study and monitoring of vulnerable and protected species, for both studying and preserving the natural population. Data from different populations of *P. rubra* within the Mediterranean Sea would contribute to improve and refine the models, in order to make them applicable on a basin scale, as well as to highlight potential morphometric differences among the populations.

The use of polyp leaves to estimate other biometric parameters could be used worldwide for other species belonging to the suborder Subsessiliflorae, being characterized by the presence of polyp leaves. The identification of proper biometric relationships has been recently done for other sea pen species in eastern Canada [21] and is common for size–frequency distribution of other octocorals [37], as well as for other marine invertebrates [38]. However, this approach cannot currently be applied to rare species, such as the endemic sea pen *Crassophyllum thessalonicae* Vafidis & Koukouras, 1991 [39] and the wip-like sea pen *Protoptilum carpenteri* Kölliker, 1872 [40].

Records of benthic species collected from accidental catches (e.g., fishery) could be used to obtain useful preliminary information to identify VMEs over very large areas. The analysis of large-enough sets of samples can also represent a basis to build further ad hoc models for the future study and monitoring of these species, particularly concerning soft-bottom coral populations. This sustainable approach supports the restrictions that should take place after the finding of dense populations of vulnerable species, such as the adoption of encounter protocols and the establishment of no-fishing areas on VMEs.

Author Contributions: Conceptualization, G.C.; Data curation, G.C., A.D.N. and A.M.L.L.; Formal analysis, A.D.N. and A.M.L.L.; Funding acquisition, G.C., G.A. and A.T.; Investigation, G.C. and F.M.; Methodology, G.C., A.D.N. and A.M.L.L.; Resources, G.C., A.D.N., A.M.L.L., G.A., A.T. and F.M.; Software, A.D.N. and A.M.L.L.; Validation, G.C., A.D.N., A.M.L.L. and F.M.; Writing—original draft, G.C.; Writing—review & editing, G.C., A.D.N., A.M.L.L. and F.M.

Funding: This research was funded by Ministero delle Politiche Agricole Alimentari e Forestali (MIPAAF). Part of this research was funded by National Geographic Society, grant number EC-176R-18. The APC was funded by Polytechnic University of Bari.

Conflicts of Interest: The authors declare no conflict of interest. The funders had no role in the design of the study; in the collection, analyses, or interpretation of data; in the writing of the manuscript, or in the decision to publish the results.

References

1. Chimienti, G.; Bo, M.; Mastrototaro, F. Know the distribution to assess the changes: Mediterranean cold-water coral bioconstructions. *Rendiconti Lincei Scienze Fisiche e Naturali* **2018**, *29*, 583–588. [CrossRef]
2. Chimienti, G.; Bo, M.; Taviani, M.; Mastrototaro, F. Occurrence and Biogeography of Mediterranean Cold-Water Corals. In *Mediterranean Cold-Water Corals: Past, Present and Future*; Orejas, C., Jimennez, C., Eds.; Springer International Publishing: Berlin, Germany, 2019; Volume 9, pp. 1–64.
3. Rossi, S.; Bramanti, L.; Gori, A.; Orejas, C. An overview of the animal forests of the world. In *Marine Animal Forests*; Rossi, S., Ed.; Springer International Publishing: Berlin, Germany, 2017; pp. 1–26.
4. Mastrototaro, F.; Chimienti, G.; Acosta, J.; Blanco, J.; Garcia, S.; Rivera, J.; Aguilar, R. *Isidella elongata* (Cnidaria: Alcyonacea) facies in the western Mediterranean Sea: Visual surveys and descriptions of its ecological role. *Eur. Zool. J.* **2017**, *84*, 209–225. [CrossRef]
5. Porporato, E.M.D.; De Domenico, F.; Mangano, M.C.; Spanò, N. *Macropodia longirostris* and *Latreillia elegans* (Decapoda, Brachyura) climbing on Mediterranean Pennatulidae (Anthozoa, Octocorallia): A preliminary note. *Crustaceana* **2011**, *84*, 1777–1780.
6. Baillon, S.; Hamel, J.F.; Wareham, V.E.; Mercier, A. Deep cold-water corals as nurseries for fish larvae. *Front. Ecol. Environ.* **2012**, *10*, 351–356. [CrossRef]
7. Kenchington, E.; Murillo, F.J.; Lirette, C.; Sacau, M.; Koen-Alonso, M.; Kenny, A.; Ollerhead, N.; Wareham, V.; Beazley, L. Kernel Density Surface Modelling as a Means to Identify Significant Concentrations of Vulnerable Marine Ecosystem Indicators. *PLoS ONE* **2014**, *9*, e109365. [CrossRef] [PubMed]
8. Ruiz-Pico, S.; Serrano, A.; Punzón, A.; Altuna, A.; Fernández-Zapico, O.; Velasco, F. Sea pen (Pennatulacea) aggregations on the northern Spanish shelf: Distribution and faunal assemblages. *Sci. Mar.* **2017**, *81*, 1–11. [CrossRef]
9. FAO (Food and Agriculture Organization). *International Guidelines for the Management of Deep-Sea Fisheries in the High Seas*; FAO: Rome, Italy, 2009; pp. 1–21.
10. Chimienti, G.; Angeletti, L.; Rizzo, L.; Tursi, A.; Mastrototaro, F. ROV vs. trawling approaches in the study of benthic communities: The case of *Pennatula rubra* (Cnidaria: Cennatulacea). *J. Mar. Biol. Assoc. UK* **2018**, *98*, 1859–1869. [CrossRef]
11. Adamo, F.; Attivissimo, F.; Di Nisio, A.; Savino, M. An automated visual inspection system for the glass Industry. In Proceedings of the 16th IMEKO TC-4 International Symposium and 13th Workshop on ADC Modelling and Testing, Florence, Italy, 22–24 September 2008; pp. 442–447.

12. Adamo, F.; Attivissimo, F.; Di Nisio, A. Calibration of an inspection system for online quality control of satin glass. *IEEE Trans. Instrum. Meas.* **2010**, *59*, 1035–1046. [CrossRef]
13. Attivissimo, F.; Di Nisio, A.; Guarnieri Calò Carducci, C.; Spadavecchia, M. Fast Thermal Characterization of Thermoelectric Modules Using Infrared Camera. *IEEE Trans. Instrum. Meas.* **2017**, *66*, 305–314. [CrossRef]
14. Schindelin, J.; Rueden, C.T.; Hiner, M.C.; Eliceiri, K.W. The ImageJ ecosystem: An open platform for biomedical image analysis. *Mol. Reprod. Dev.* **2015**, *82*, 518–529. [CrossRef]
15. Sharma, N.; Aggarwal, L.M. Automated medical image segmentation techniques. *J. Med. Phys.* **2010**, *35*, 3–14. [CrossRef]
16. Adamo, F.; Attivissimo, F.; Di Nisio, A.; Spadavecchia, M. An automatic document processing system for medical data extraction. *Measurement* **2015**, *61*, 88–99. [CrossRef]
17. Andria, G.; Attivissimo, F.; Di Nisio, A.; Lanzolla, A.M.L.; Maiorana, A.; Mangiatini, M.; Spadavecchia, M. Dosimetric Characterization and Image Quality Assessment in Breast Tomosynthesis. *IEEE Trans. Instrum. Meas.* **2017**, *66*, 2535–2544. [CrossRef]
18. Gleason, A.C.R.; Reid, R.P.; Voss, K.J. Automated classification of underwater multispectral imagery for coral reef monitoring. In Proceedings of the OCEANS 2007, Vancouver, BC, Canada, 29 September–4 October 2007; pp. 1–8.
19. Phinn, S.R.; Roelfsema, C.M.; Mumby, P.J. Multi-scale, object-based image analysis for mapping geomorphic and ecological zones on coral reefs. *Int. J. Remote Sens.* **2012**, *33*, 3768–3797. [CrossRef]
20. Foglini, F.; Grande, V.; Marchese, F.; Bracchi, V.A.; Prampolini, M.; Angeletti, L.; Castellan, G.; Chimienti, G.; Hansen, I.M.; Gudmundsen, M.; et al. Underwater Hyperspectral Imaging for seafloor and benthic habitat mapping in the southern Adriatic Sea (Italy). In Proceedings of the 2018 IEEE International Workshop on Metrology for the Sea, Learning to Measure Sea Health Parameters (MetroSea), Bari, Italy, 8–10 October 2018; pp. 201–205.
21. Murillo, F.J.; MacDonald, B.W.; Kenchington, E.; Campana, S.E.; Sainte-Marie, B.; Sacau, M. Morphometry and growth of sea pen species from dense habitats in the Gulf of St. Lawrence, eastern Canada. *Mar. Biol. Res.* **2018**, *14*, 366–382. [CrossRef]
22. Ludvigsen, M.; Johnsen, G.; Sørensen, A.J.; Lågstad, P.A.; Ødegård, Ø. Scientific operations combining ROV and AUV in the Trondheim Fjord. *Mar. Technol. Soc. J.* **2014**, *48*, 59–71. [CrossRef]
23. Mastrototaro, F.; Aguilar, R.; Chimienti, G.; Gravili, C.; Boero, F. The rediscovery of *Rosalinda incrustans* (Cnidaria: Hydrozoa) in the Mediterranean Sea. *Ital. J. Zool.* **2016**, *83*, 244–247. [CrossRef]
24. Petritoli, E.; Leccese, F.; Cagnetti, M. A High Accuracy Buoyancy System Control for an Underwater Glider. In Proceedings of the 2018 IEEE International Workshop on Metrology for the Sea, Learning to Measure Sea Health Parameters (MetroSea), Bari, Italy, 8–10 October 2018; pp. 257–261.
25. Adamo, F.; Andria, G.; Di Nisio, A.; Calò Carducci, C.G.; Lay-Ekuakille, A.; Mattencini, G.; Spadavecchia, M. Designing and prototyping a sensors head for test and certification of UAV components. *Int. J. Smart Sens. Intell. Syst.* **2017**, *10*, 646–672. [CrossRef]
26. Barwise, A.; Cowie, M. Advancement in Geotechnical Site Investigation Practice Using ROV Technology. In Proceedings of the Offshore Technology Conference, Houston, TX, USA, 6–9 May 2013; pp. 1–8.
27. McLean, D.L.; Macreadie, P.; White, D.J.; Thomson, P.G.; Fowler, A.; Gates, A.R.; Benfield, M.; Horton, T.; Skropeta, D.; Bond, T.; et al. Understanding the Global Scientific Value of Industry ROV Data, to Quantify Marine Ecology and Guide Offshore Decommissioning Strategies. In Proceedings of the Offshore Technology Conference Asia, Kuala Lumpur, Malaysia, 20–23 March 2018; pp. 1–10.
28. Adamo, F.; Andria, G.; Bottiglieri, O.; Cotecchia, F.; Di Nisio, A.; Miccoli, D.; Sollecito, F.; Spadavecchia, M.; Todaro, F.; Trotta, A.; et al. GeoLab, a measurement system for the geotechnical characterization of polluted submarine sediments. *Measurement* **2018**, *127*, 335–347. [CrossRef]
29. Chimienti, G.; Tursi, A.; Mastrototaro, F. Biometric relationships in the red sea pen *Pennatula rubra* (Cnidaria: Pennatulacea). In Proceedings of the 2018 IEEE International Workshop on Metrology for the Sea, Learning to Measure Sea Health Parameters (MetroSea), Bari, Italy, 8–10 October 2018; pp. 212–216.
30. Rooper, C.N.; Wilborn, P.; Goddard, P.; Williams, K.; Towler, R.; Hoff, G.R. Validation of deep-sea coral and sponge distribution models in the Aleutian Islands, Alaska. *ICES J. Mar. Sci.* **2017**, *75*, 199–209. [CrossRef]
31. Chimienti, G.; Angeletti, L.; Mastrototaro, F. Withdrawal behaviour of the red sea pen *Pennatula rubra* (Cnidaria: Pennatulacea). *Eur. Zool. J.* **2018**, *85*, 64–70. [CrossRef]

32. Otero, M.M.; Numa, C.; Bo, M.; Orejas, C.; Garrabou, J.; Cerrano, C.; Kružić, P.; Antoniadou, C.; Aguilar, R.; Kipson, S.; et al. *Overview of the Conservation Status of Mediterranean Anthozoans*; IUCN: Málaga, Spain, 2017; pp. 1–73.
33. Bertrand, J.A.; Gil De Sola, L.; Papaconstantinou, C.; Relini, G.; Souplet, A. The general specifications of the MEDITS surveys. *Sci. Mar.* **2002**, *66*, 9–17. [CrossRef]
34. Fiorentini, L.; Dremière, P.Y.; Leonori, I.; Sala, A.; Palombo, V. Efficiency of the bottom trawl used for the Mediterranean International Trawl Survey (MEDITS). *Aquat. Living Resour.* **1999**, *12*, 187–205. [CrossRef]
35. Kenchington, E.; Murillo, F.J.; Cogswell, A.; Lirette, C. *Development of Encounter Protocols and Assessment of Significant Adverse Impact by Bottom Trawling for Sponge Grounds and Sea Pen Fields in the NAFO Regulatory Area*; NAFO Scientifc Council Report 11/75 Serial No N6005; Agris: Dartmouth, NS, Canada, 2011; Volume 53, pp. 1–51.
36. Troffe, P.M.; Levings, C.D.; Piercey, G.E.; Keong, V. Fishing gear effects and ecology of the sea whip (*Halipteris willemoesi* (Cnidaria: Octocorallia: Pennatulacea)) in British Columbia, Canada: Preliminary observations. *Aquat. Conserv.* **2005**, *15*, 523–533. [CrossRef]
37. Ambroso, S.; Gori, A.; Dominguez-Carrió, C.; Gili, J.M.; Berganzo, E.; Teixidó, N.; Greenacre, M.; Rossi, S. Spatial distribution patterns of the soft corals *Alcyonium acaule* and *Alcyonium palmatum* in coastal bottoms (Cap de Creus, northwestern Mediterranean Sea). *Mar. Biol.* **2014**, *160*, 3059–3070. [CrossRef]
38. Mastrototaro, F.; Chimienti, G.; Matarrese, A.; Gambi, M.C.; Giangrande, A. Growth and population dynamics of the non-indigenous species *Branchiomma luctuosum* Grube (Annelida, Sabellidae) in the Ionian Sea (Mediterranean Sea). *Mar. Ecol.* **2015**, *36*, 517–529. [CrossRef]
39. Fryganiotis, K.; Antoniadou, C.; Chintiroglou, C.; Vafidis, D. Redescription of the Mediterranean endemic sea-pen *Crassophyllum thessalonicae* (Octocorallia: Pteroeididae). *Mar. Biodivers. Rec.* **2011**, *4*, 1–3. [CrossRef]
40. Mastrototaro, F.; Chimienti, G.; Capezzuto, F.; Carlucci, R.; Williams, G. First record of *Protoptilum carpenteri* (Cnidaria: Octocorallia: Pennatulacea) in the Mediterranean Sea. *Ital. J. Zool.* **2015**, *82*, 61–68. [CrossRef]

Article

Application of Hyperspectral Imaging to Underwater Habitat Mapping, Southern Adriatic Sea

Federica Foglini [1], Valentina Grande [1,*], Fabio Marchese [2], Valentina A. Bracchi [2], Mariacristina Prampolini [1], Lorenzo Angeletti [1], Giorgio Castellan [1,3], Giovanni Chimienti [4], Ingrid M. Hansen [5], Magne Gudmundsen [5], Agostino N. Meroni [2], Alessandra Mercorella [1], Agostina Vertino [2,6], Fabio Badalamenti [7], Cesare Corselli [2], Ivar Erdal [5], Eleonora Martorelli [8], Alessandra Savini [2] and Marco Taviani [1,9,10]

1 National Research Council, Institute of Marine Sciences, Via Gobetti 101, 40129 Bologna, Italy; federica.foglini@bo.ismar.cnr.it (F.F.); mariacristina.prampolini@bo.ismar.cnr.it (M.P.); lorenzo.angeletti@bo.ismar.cnr.it (L.A.); giorgio.castellan@bo.ismar.cnr.it (G.C.); alessandra.mercorella@bo.ismar.cnr.it (A.M.); marco.taviani@bo.ismar.cnr.it (M.T.)

2 Department of Earth and Environmental Sciences and CoNISMa LRU, University of Milano - Bicocca, Piazza della Scienza 1, 20126 Milano, Italy; fabio.marchese1@unimib.it (F.M.); valentina.bracchi@unimib.it (V.A.B.); a.meroni9@campus.unimib.it (A.N.M.); agostina.vertino@unimib.it (A.V.); cesare.corselli@unimib.it (C.C.); alessandra.savini@unimib.it (A.S.)

3 Department for the Cultural Heritage, University of Bologna, Via degli Ariani 1, 48121 Ravenna, Italy

4 Department of Biology and CoNISMa LRU, University of Bari, Via Orabona 4, 70124 Bari, Italy; giovanni.chimienti@uniba.it

5 Ecotone AS, Pirsenteret Havnegata 9, inngang 1 – 3 etg, 7010 Trondheim, Norway; ingrid@ecotone.com (I.M.H); magne@ecotone.com (M.G.); ivar@ecotone.com (I.E.)

6 Ghent University, Campus Ufo, Rectorate, Sint-Pietersnieuwstraat 25, B-9000 Ghent, Belgium

7 National Research Council, Institute for the Study of Anthropic Impacts and Sustainability in Marine EnvironmentsVia Giovanni da Verrazzano 17, 91014 Castellammare del Golfo, Trapani, Italy; fabio.badalamenti@cnr.it

8 National Research Council, Institute of Environmental Geology and Geoengineering, Via Salaria km 29,300, 00015 Monterotondo, Rome, Italy; eleonora.martorelli@uniroma1.it

9 Biology Department, Woods Hole Oceanographic Institution, 266 Woods Hole Road, Woods Hole, MA 02543, USA

10 Stazione Zoologica Anton Dohrn, Villa Comunale, 80121 Naples, Italy

* Correspondence: valentina.grande@bo.ismar.cnr.it; Tel.: +39-051-6398875

Received: 31 March 2019; Accepted: 12 May 2019; Published: 16 May 2019

Abstract: Hyperspectral imagers enable the collection of high-resolution spectral images exploitable for the supervised classification of habitats and objects of interest (OOI). Although this is a well-established technology for the study of subaerial environments, Ecotone AS has developed an underwater hyperspectral imager (UHI) system to explore the properties of the seafloor. The aim of the project is to evaluate the potential of this instrument for mapping and monitoring benthic habitats in shallow and deep-water environments. For the first time, we tested this system at two sites in the Southern Adriatic Sea (Mediterranean Sea): the cold-water coral (CWC) habitat in the Bari Canyon and the Coralligenous habitat off Brindisi. We created a spectral library for each site, considering the different substrates and the main OOI reaching, where possible, the lower taxonomic rank. We applied the spectral angle mapper (SAM) supervised classification to map the areal extent of the Coralligenous and to recognize the major CWC habitat-formers. Despite some technical problems, the first results demonstrate the suitability of the UHI camera for habitat mapping and seabed monitoring, through the achievement of quantifiable and repeatable classifications.

Keywords: hyperspectral camera; spectral library; habitat mapping; coralligenous; cold-water coral; Adriatic Sea

1. Introduction

Traditionally, underwater habitat mapping has been carried out coupling acoustic remote sensing techniques with red/green/blue (RGB) images, videos and bottom sampling [1,2]. The analysis of video and images is performed manually by expert interpretation, or automatically when a photomosaic is available [3]. Recently, the European programs such as the EU Marine Strategy Framework Directive (MSFD: 2008/56/EC), require the monitoring of benthic habitat extent and distribution (Criteria 1.4 and 1.5 of the Descriptor 1 "Biological Diversity"). The MSFD scope is to assess the good environmental status (GES) of European water [4] with the lowest possible impact on the seafloor. This effort is translated into quantifiable operational indicators that should be measurable at different scale and repeatable in time [4]. To fulfill these requirements, there is a need for innovative approaches and tools to obtain detailed, reliable, quantifiable and repeatable maps of relevant habitats in different underwater environments [2].

During the last decade, the implementation of hyperspectral devices has become a viable alternative to regular photography. In contrast to ordinary cameras that acquire three colour bands (RGB), hyperspectral cameras record the full spectrum of reflected light, in each pixel of the acquired image. Therefore, the spectral resolution and amount of information obtainable from an image transect is vastly increased compared to traditional photography [5]. As a result, UHI can detect the subtle and otherwise unnoticeable spectral properties of a given OOI and record object-specific optical fingerprints. Optical fingerprinting increases classification accuracy for both qualitative and quantitative mapping [5].

This technology has previously been applied to airborne remote sensing, both in terrestrial and marine environments, through passive sensors requiring sunlight. However, sunlight is highly attenuated in marine waters [6,7]. As a consequence, this technique is suitable only in coastal areas and relatively shallow water (up to 50 m depth: [8–10]). Hence, works on hyperspectral imaging from satellite or airplanes are focussed on oceanographic and biological studies [11,12], mapping of ocean colour [12,13] and shallow benthic habitats [8,10] such as coral reefs [14–17], seagrasses [18–20] and kelp forests [9].

Recently, different instrument carriers for the underwater hyperspectral imager (UHI) have been used in underwater field applications, such as the customized scanning rig [21,22], remotely operated vehicle (ROV) [6,7,23–27] or autonomous underwater vehicle (AUV) [28]. The UHI has been tested and utilized for different purposes from shallow (< 6 m) [22] to abyssal depths (ca. 4200 m depth) [6,7]. Among the many applications, UHI was related to the identification of manganese nodules [6], infrastructure inspection, seafloor impact of offshore drilling [29] and marine archaeology [25,27,30]. However, the most reported UHI application is within the field of benthic habitat mapping, modelling and monitoring. Underwater hyperspectral imaging with ROV has been used to study coastal kelp forests [25], vertical rock wall habitat and soft sediments [24], red calcareous algae and associated fauna [25,26], deep-sea megafauna [7] and CWC communities [23,25]. Laboratory experiments to measure changes in the health status of CWCs exposed to hydrocarbons emissions is another application of UHI [31].

As a first application to the Mediterranean basin, we tested the UHI in the Adriatic Sea [32]. The semi-enclosed Adriatic Sea hosts a variety of benthic habitats, including the shallow oyster reefs and sponge communities in the Venice Lagoon [33], coralligenous formations on the shelf (e.g., [34] with references therein), down to the CWC habitat in deep water (> 200 m) in the south (e.g., [35–40]). At present, the Adriatic Sea is under siege by a number of stressors, such as high demographic pressure on its coastal areas, pollution, marine littering and dumping, fishing practices, ship traffic, harbour activities and industrial operations [41]. Our study targets the distribution and extent of two biogenic habitats (CWC and Coralligenous), in different geomorphological and depth contexts, considered to be of key importance in monitoring plans. In this perspective, the UHI may prove useful in habitat mapping to meet the requirements of European programs (e.g., EU MSFD).

2. Materials and Methods

2.1. Study Area

The two selected sites are located in the Southern Adriatic Sea (Figure 1): the Bari Canyon (CWC habitat) and the continental shelf off-shore Brindisi (Coralligenous habitat).

The Bari Canyon site is located ca. 40 km away from the city of Bari, on the continental margin, within a well-known Cold-Water Coral ecosystem, extending between −200 and −700 m on the southern flank of the canyon [35,36,38,42,43]. The CWC habitat is here characterised by complex megabenthic communities, mainly represented by the colonial scleractinian *Madrepora oculata,* subordinately *Desmophyllum pertusum* (*Lophelia pertusa* [44]) and the solitary *Desmophyllum dianthus*, and by large fan-shaped sponges (i.e. *Pachastrella monilifera* and *Poecillastra compressa*) [35–37,40].

The Brindisi site is placed on a flat continental shelf, about 10 km far from the coast at an average depth of 30 m. Coralligenous outcrops, mosaicking coarse biogenic sediments [34–46], dominate the seafloor. The coralligenous is a very complex habitat where crustose coralline algae (CCA) and red algae belonging to the order of Peyssonelliales are often the main bioconstructors in shallower waters, generating a new solid substrate and constituting a three-dimensional biogenic build-up [47–52]. It represents a key habitat of the Mediterranean continental shelf because of its structural and functional importance, as well as for its considerable aesthetic value [53]. In the study area, discrete coralligenous build-ups [46] characterize the seafloor, with a thickness up to 70 cm. CCA, usually growing in dim light conditions, and other algae such as Peyssonelliales primarily form these solid substrates; bryozoans and serpulids contribute to the bioconstruction [54,55]. Moreover, these hard substrates host different fauna and flora, often overgrowing the calcified red algae [51].

2.2. Underwater Hyperspectral Imager (UHI)

The underwater hyperspectral imager (UHI), developed and patented by Ecotone AS, consists of a waterproof housing containing camera system, computer and data storage. It is operated with a light source for proper illumination, and represents a new system for the identification, mapping and monitoring of OOI at the seabed [5,25].

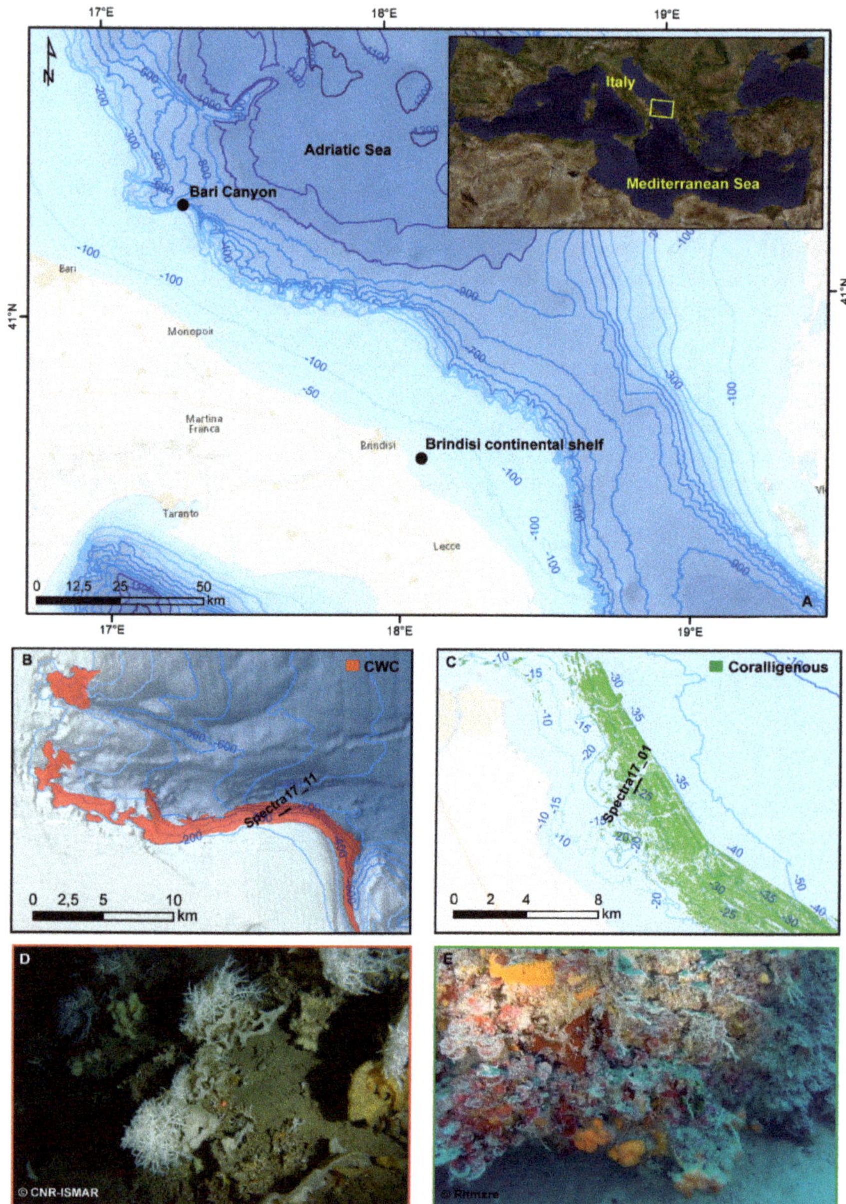

Figure 1. (**A**) Location of the two sites, inset shows the position in the Mediterranean Sea; (**B**) the extension of the Bari Canyon CWC province (from [56]) and (**C**) the extension of the coralligenous in the Brindisi area (black lines indicate the ROV surveys). Habitat maps produced by the BIOMAP project and further updated within the CoCoNet project. (**D**) Example of CWC habitat complexity showing colonies of *M. oculata* and large fan-shaped sponges (from [38]); (**E**) example of coralligenous characterized by CCA and Peyssonelliales, serpulids and orange encrusting sponges overprinting the calcified red algae.

The UHI is a push broom camera, which records one line at time. It is equipped with a spectrograph that receives light through a thin entrance slit [57]. Light entering the instrument is diffracted into separate wavelengths and projected onto the camera sensor as a contiguous spectrum. A continuous image is built up line by line and can be presented on a monitor for visualisation [26].

The UHI needs to be mounted on a moving platform (e.g., ROV) that operates at a fairly constant speed and altitude from the seafloor. Mobile platforms equipped with dynamic positioning systems permit larger areal coverage, and the possibility of re-visiting the surveyed sites based on their geolocation data [58]. External lamps provide seafloor illumination. Due to the rapid attenuation of light in the marine environment, UHI is normally confined to scanning altitudes <5 m above the OOI, depending on water quality and turbidity.

Image lines are captured perpendicular to the direction of movement. The result is a hyperspectral image, featuring detected intensities for all the wavelengths used [57]. The spatial resolution provided by hyperspectral cameras varies with altitude, exposure time of the camera and speed vessel.

2.3. Data Acquisition

In February 2017, we carried out the SPECTRA17 cruise on board of the R/V Minerva Uno, aimed at testing the ability of the UHI to acquire seafloor hyperspectral images in the Southern Adriatic Sea.

An Ecotone UHI (Model 4) was mounted vertically beneath the ROV Super Mohawk II 34 Observation Class, facing directly towards the seafloor, together with two LED lamps (3200 lumens per lamp) oriented at 90°, a 2D high resolution RGB camera with two lamps oriented at 60° and a system to correct UHI camera motion for pitch, roll and yaw. Acquisition and pre-processing of data were managed through the C++ based UHI customized software Immersion installed on the topside control unit.

The ROV maintained constant speed (about 1 knot) and heading in each dive (45° for CWC and 25° for coralligenous, respectively), at a constant altitude of 1–1.5 m. The track length was about 2.5 km for the CWC site and 785 m for the coralligenous site.

ROV navigation and position was provided by Low-Accuracy TrackLink USBL Positioning System (accuracy of <2% of the water depth), positioning data were recorded at 1 sec. The positioning system was operated using the software PDS2000. In addition to UHI, a black and white camera for ROV navigation and a high-resolution camera provided footage for manual identification of OOI.

We reoccupied previous ROV tracks, making use of high-resolution videos and photos already available for comparing the UHI results and classification.

2.4. High Resolution Camera Image Data Processing

The RGB images for coralligenous and CWC sites, collected by the high-resolution camera, were processed using ADELIE Software by IFREMER and used to identify and classify the OOI. ADELIE is a software able to synchronise video and navigation and then automatically capture georeferenced still images (or image sequences) for a chosen time interval (e.g., every 10 s). Through a specific module of ADELIE based on ArcGIS Desktop, it is possible to filter and smooth vehicle navigation, to have direct access to pictures and to localize video in real time. Following this procedure, we mapped all track lines producing benthic habitat maps and estimated the extent of habitat at ROV scale.

2.5. UHI Data Processing

Processing of hyperspectral images consisted of three main steps: (1) Radiometric processing correcting for sensor influence; (2) Georeferencing to assign geospatial information and perform image geocorrection; (3) Conversion of radiance to reflectance by correction of external influences from illumination source. Using Immersion, we performed the georeferencing and radiometric correction of the acquired images to produce non-distorted and georeferenced hyperspectral images. The spectral resolution is up to 2.2 nm, while the spatial resolution is 1 cm for CWCs and 0.5 cm for coralligenous. The resulting UHI coverage is about 1–1.2 m width, depending on height above seafloor.

The pre-processed UHI images used in this work are available as Research Object (for further details see Supplementary Materials).

2.5.1. Radiometric Processing

The UHI was calibrated in the laboratory before the oceanographic mission. A standard lamp with known spectral properties was used to find the ratio between observed the digital counts of intensity for each wave band on the sensor, to spectral radiance (W m^{-2} sr^{-1} nm^{-1}). These measurements can be applied on raw recordings from the field to correct for sensor-specific noise and dark current, as well as data acquisition parameters such as exposure time and binning. Radiance conversion was automatically performed as part of the georeferencing algorithm in Immersion.

2.5.2. Georeferencing

Geographic position and spatial correction of the hyperspectral images were provided by the georeferencing procedure through Immersion software by using: (1) USBL data for the ROV position and (2) altitude data from the Ecotone IMU (Inertial Measurement Unit). As the navigation produced by USBL track link contains frequent spikes and some metrical gaps, we statistically filtered navigation and altitude data through 20- and 5-point-wide windows, respectively, using an adjacent averaging smoothing algorithm to improve resolution [59]. In addition to navigation, motion and altitude data the following parameters for Immersion were set: (1) course-made-good option for ROV heading calculation; (2) a spectral binning of 8 resulting in 28 bands with 15 nm resolution; (3) a spatial binning of 1 and (4) a cell resolution of 1 cm for CWCs and 0.5 cm for coralligenous.

2.5.3. Reflectance Processing

Following the procedure in [1], radiance data was converted into reflectance correcting for the external influence on the spectral characteristics from LED lamps or their combination with sunlight, at the deep and shallow site, respectively. These spectral characteristics are not definable, so we approximated the illumination influence by using a reference spectrum calculated for the entire analysed segment of both sites in R software [60]. Then, we divided each image pixel by its respective reference spectrum.

2.6. *UHI Spectral Supervised Classification*

We selected a segment from the CWC track line of about 10 m and a segment from the coralligenous track line of about 7.5 m for the spectral supervised classification and further analysis. Classification was performed through the software ENVI 5.5, using the spectral angle mapper (SAM) method. The accuracy was determined generating a confusion matrix for each site.

2.6.1. Spectral Angle Mapper (SAM)

The spectral angle mapper (SAM) is a supervised classification technique that measures the similarity of image spectra to reference spectra. The reference spectra can be measured in field or laboratory, they can be taken directly from the image as region of interest (ROI) or imported from already known spectral libraries. SAM measures similarity by calculating the angle between the two spectra, treating them as vectors in n-dimensional space, with n being the number of bands [61]. The angle is the arccosine of the dot product of the two spectra. Smaller values for the angle indicate higher similarity between pixel and reference spectra. As the angle between two vectors is independent of the vector length, this method is unaffected by gain factors, such as solar illumination [62]. The SAM only compares the angle between the spectral directions of the reference and test pixels considered, there is no specific requirement for a large amount of training samples [63].

We trained the model selecting ROIs that include spectral signatures representing substrates, megaflora and megafauna (> 2 cm) present in the surveyed areas. The selected ROIs reflect the highest spectral diversity due to different pigmentation of the OOI and make up the benthic classes that will constitute the final classification.

In particular for the CWC site, we used 10 ROIs to train the SAM classification representing 4 benthic classes: (1) colonial cnidarian, (2) sponge, (3) mud, (4) bedrock (Table 1). For colonial cnidarian, bedrock and sponge we selected multiple ROIs due to the differences of the UHI RGB colours along the segment. For the coralligenous site, we created 13 ROIs identifying 5 benthic classes: (1) CCA and Peyssonelliales (P) forming the build-ups (CCA+P); (2) green algae on build-ups (in particular *Codium bursa* and *Flabellia petiolata*) and on sediment (*Caulerpa prolifera*); (3) Seagrass (*Posidonia oceanica*); (4) organisms associated with the presence of build-ups; and (5) sand (Table 2). For the CCA+P and the green algae (Green algae 1 and 2) classes, we selected more than one ROI due to the illumination unevenness caused by the slight altitude variation along the track.

After running the SAM with a maximum angle of 0.1 radiant, we applied the ENVI 'Rule classifier' post classification tool to adjust the threshold angles for each class and improve the classification results. We choose the appropriate thresholds classifying by minimum values and visualizing the histogram that shows the frequency of pixel with different angles. Finally, we used the ENVI 'Majority/Minority analysis' tool with a kernel size of 3 × 3 to clean and smooth the SAM classification.

Table 1. Selected ROIs for the benthic classes of the CWC site and relative threshold values used in the 'Rule classifier' tool.

Benthic Class	**Colonial Cnidarian**			**Sponge**				**Mud**	**Bedrock**	
ROI	**1**	**2**	**3**	**1**	**2**	**3**	**4**	**1**	**1**	**2**
Taxonomy	*M. oculata/D. pertusum*			*Hexactinellida* sp.	*Demospongiae* sp. 1	*Demospongiae* sp. 2	*Demospongiae* sp. 3			
Threshold	0.035	0.025	0.08	0.08	0.15	0.045	0.035	0.08	0.08	0.08

Table 2. Selected ROIs for the benthic classes of the coralligenous site and relative threshold values used in the 'Rule classifier' tool.

Benthic Class	**CCA+P**			**Green Algae**			**Seagrass**	**Associated Organism**					**Sand**
ROI	**1**	**2**	**3**	**1**	**2**	**3**	**1**	**Sponge 1**	**Sponge 2**	**Serpulids**	**Red Starfish**	**Red Ascidia**	**1**
Taxonomy				*C. bursa/F. petiolata*		*C. prolifera*	*P. oceanica*	*Axinella* sp. 1	*Axinella* sp. 2		*E. sepositus*	*H. papillosa*	
Threshold	0.2	0.02	0.07	0.08	0.08	0.065	0.045	0.08	0.07	0.03	0.18	0.06	0.2

2.6.2. SAM Classification Accuracy

We generated a confusion matrix for both test sites using ENVI and ArcGIS desktop to determine the accuracy of the classification results. Firstly, we produced eight random sampling points for each benthic class using the 'Generate random sample' ENVI tool with the equalized random technique to divide the population into homogeneous subgroups, while ensuring that each class sample size was the same (1 pixel). Then, we compared within ArcGIS Desktop the random sampling points predicted classes with the UHI image (ground truthing), to assign the real class for each point as defined by the expert interpretation. We generated the confusion matrix for each classification using the predicted classes and the real class values specifying the overall, producer's and user's accuracies. The overall accuracy is calculated by summing the number of correctly classified values and dividing by the total number of values. The user's accuracy (UA) is the number of correctly identified pixels in a class, divided by the total number of pixels of the class in the classified image; it shows false positives, where pixels are incorrectly classified as a known class when they should have been classified as something else. Producer's accuracy (PA) is the number of correctly identified pixels divided by the total number of pixels in the reference image; it gives a false negative, where pixels of a known class are classified as something other than that particular class.

3. Results

3.1. Spectral Library for CWC Site

The mean spectra of the 10 ROIs selected for the CWC site are shown in Figure 2. Mud and bedrock have an almost constant level of reflectance along all wavelengths. The bedrock shows a deflection point at 473 nm, with a minimum at 500 nm (Figure 2A,B). The sponge 1 has an inflection point at 500 nm and an increasing trend between 555 and 680 nm (from green to red) (Figure 2C). Sponge 2 and 3 display similar patterns with a dissimilar level of reflectance due to different illumination, an inflection point at 530 nm and highest values in the orange/red part of the spectrum between 630 nm and 670 nm. Sponge 4 has a slightly different shape and a smoothed slope, possibly caused by a minor colour difference. Despite all sponges seem to pertain to the same morphotype (belonging to the white/orange large fan-shaped *P. compressa* and/or *P. monilifera*), it is not possible to attribute the accurate species to each ROI and to each spectrum (Figure 2D). The colonial cnidarian 2 and 3 have a spectral signature with a similar pattern, constant along the entire wavelength range, with a difference in the level of reflectance, while colonial cnidarian 1 shows a slight peak in the blue part of the spectrum (about 470 nm) (Figure 2E). According to previous studies (e.g., [35,36,38,43,56]), we can assume that the most probable species belong to *M. oculata*, while the *D. pertusum* appears to be rarer in this site.

For benthic classes with multiple ROIs (colonial cnidarian, sponge and bedrock), we analyzed the spectral differences and considered the mean for spectra with the same pattern but a different reflectance intensity (Figure 2F). The three ROIs of the colonial cnidarians are recognized as a unique class, because the peak in the blue part of the spectrum for colonial cnidarian 1 is considered an artefact.

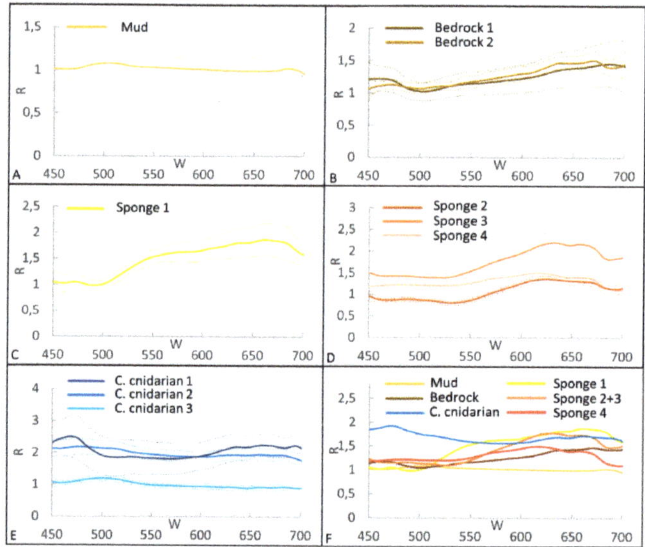

Figure 2. Mean spectra of the ROIs for the CWC site. R is the reflectance and W the wavelength. In **A**, **B**, **C**, **D** and **E**, dashed lines represent the standard deviation. **F** shows the synthesis of all spectra.

3.2. Spectral Library for the Coralligenous Site

The mean spectra of the 13 ROIs selected for the coralligenous site are shown in Figure 3. The CCA+P1, CCA+P2, CCA+P3 have the inflection point at 600 nm with a maximum between 630 and 670 nm in the red part of the spectrum. The wide range of standard deviation probably reflects the high biodiversity of this category. We can assume that the three spectra, with a similar pattern and a different level of reflectance intensity, belong to the same benthic class CCA+P.

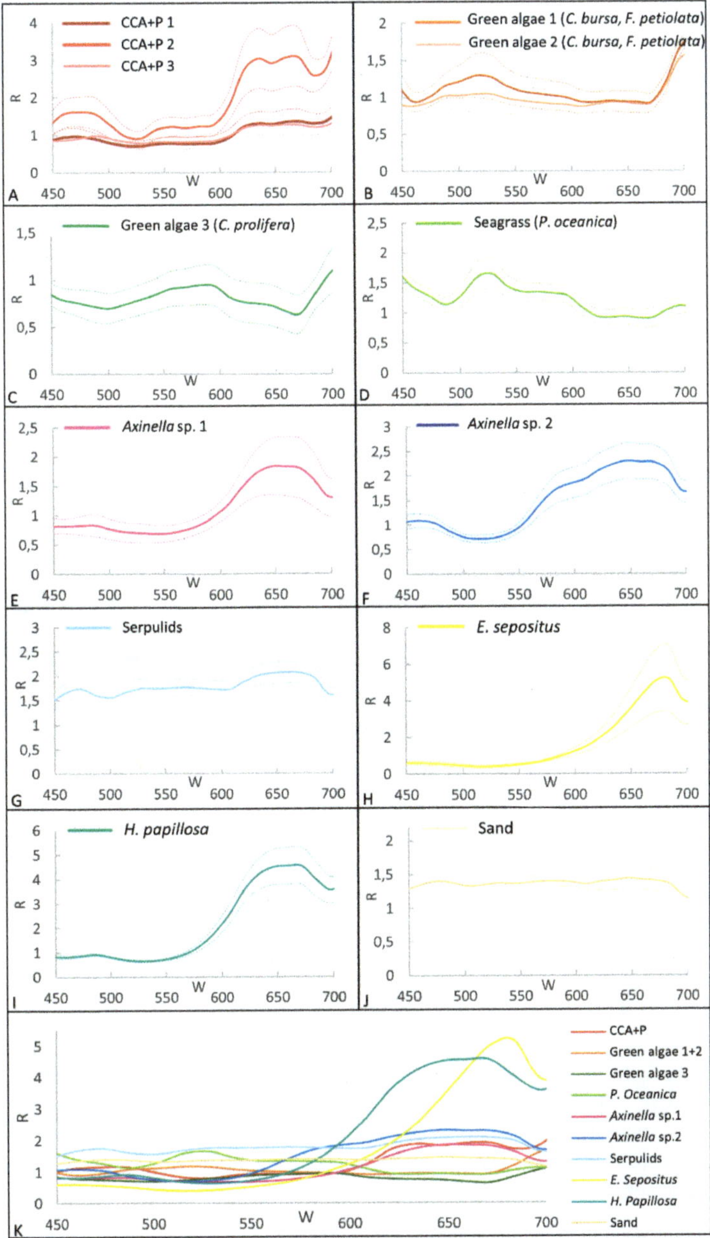

Figure 3. Mean spectra based on the ROIs relative to the coralligenous site. The dashed line in the graphs represents the standard deviation. **K** shows the synthesis of all spectra.

The green algae on build-ups (green algae 1+2), mainly represented by *C. bursa* and *F. petiolata*, have a maximum at about 515 nm (green part of the spectrum), while the green algae *C. prolifera* (Green algae 3) and the seagrass *P. oceanica* show a slight difference with a maximum of reflectance at 595 and 528 nm, respectively. The ascidian *Halocynthia papillosa* and the starfish *Echinaster sepositus* show a peak at 650 and 680 nm in the red part of the spectrum. The serpulids show an almost constant spectrum with a small inflection at 610 nm. The two species belonging to the genus *Axinella* have similar spectra with a maximum between 630 and 670 nm (red part of the spectrum). The sand has a constant value of reflectance along all wavelengths.

3.3. Supervised Classification Results for CWC Site

Based upon the previous considerations about the spectra, we can define six benthic classes for the final classification: (1) colonial cnidarian, (2) sponge 1, (3) sponge 2+3, (4) sponge 4, (5) mud and (6) bedrock (Figure 4). The mud class results were the most dominant (40%) followed by the bedrock (23.6%). The colonial cnidarian class is scattered along the track with a total coverage of 0.4%, the sponge classes (sponge 1, sponge 2+3, sponge 4) are patchy with a total coverage of 0.3%. The illumination problems, due to low lamp potential, ROV orientation and dipping of the substrate (the canyon flank in this site is up to 25°), cause a homogenous dark shading on the right (deeper) section of the image. This results in a high percentage of unclassified pixels (26.5%), also including colonial cnidarians and sponges still visible on the RGB image. However, these technical issues do not prevent the clear discrimination of the benthic classes and the estimation of percentage cover.

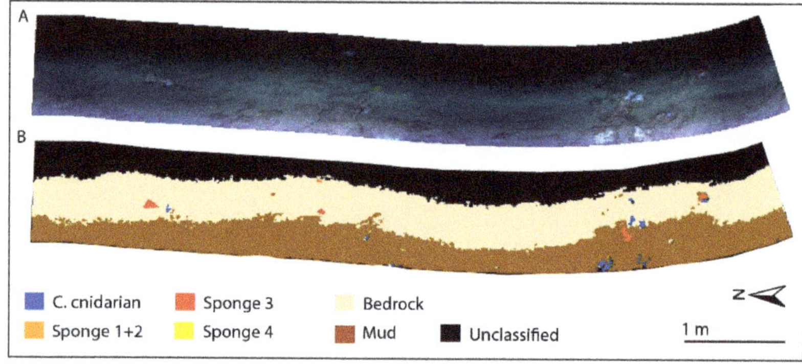

Figure 4. RGB UHI image of the CWC site in **A** and its SAM classification in **B**.

3.4. Supervised Classification Results for Coralligenous Site

For the coralligenous we defined 10 benthic classes for the final classification (Figure 5), according to the analysis of spectral signatures: (1) CCA+P, (2) green algae 1+2, (3) *C. prolifera*, (4) *P. oceanica*, (5) *Axinella* sp. 1, (6) *Axinella* sp. 2, (7) serpulids, (8) *E. sepositus*, (9) *H. papillosa*, (10) Sand. The sand class is the most dominant in the area with a coverage of 60%. The build-ups are generally well discriminated along the entire track with a coverage of 29% for CCA+P and 8% for the green algae on build-ups (Green algae 1+2). The SAM identifies well green algae *C. prolifera* (Green algae 3) and the seagrass *P. oceanica* with a coverage of 0.7% and 0.9% of the total classified area, respectively. The taxa associated with coralligenous are easily distinguishable with a total coverage of 0.13%. We highlight the low percentage of unclassified pixel (1.5% in total).

3.5. Classification accuracy for CWC and Coralligenous Sites

The CWC SAM has a high overall accuracy (84.38%) as determined from the confusion matrix comparing the predicted classes with the real classes (see Table 3). The colonial cnidarian class gives 100% UA and 87.5% PU showing the presence of a minimum number of false negative along the deepest section of the analysed segment. The sponges (sponge 1, sponge 2+3, sponge 4) show a lower value in terms of PA (87.50%) than UA (100%) due to the presence of false negative. The bedrock and mud classes show almost the same accuracy (>60% both for PA and UA) both for the presence of false negative and positive.

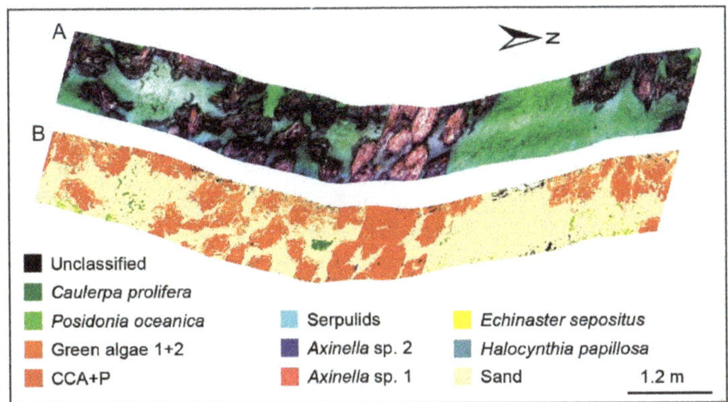

Figure 5. RGB UHI image of the coralligenous site in **A** and its SAM classification in **B**.

The overall accuracy of the SAM for the coralligenous is 72% (see Table 4). There is a high discrepancy between PA and UA for the *P. oceanica* and *C. prolifera* (Green algae 3) classes with no commission errors (100% UA), but with a level of reliability of 37.5% and 50% due to a higher omission errors. For the sand class, the UA is extremely low (25%) because of the high number of false positive. For the CCA+P both PA and UA are 100%. The green algae on build-ups (green algae 1+2) shows no false positives (100% UA) and a PA higher than the overall value (87.5%). In general, all organisms associated with the coralligenous build-ups have a high level of accuracy (> 88% both for PA and UA), with the exception of *Axinella* sp. 1 (75% PA) showing the presence of false negatives.

Table 3. Confusion Matrix for the CWC site classification.

	Overall Accuracy 84.38%					
	Mud	**Sponge**	**Bedrock**	**C. cnidarian**	**TOT**	**UA**
Mud	7	0	2	1	10	**67.78**
Sponge	0	7	0	0	7	**100.00**
Bedrock	1	1	6	0	8	**66.67**
C. cnidarian	0	0	0	7	7	**100.00**
TOT	8	8	8	8	32	
PA	**87.50**	**87.50**	**75.00**	**87.5**		

Table 4. Confusion Matrix for the coralligenous site classification.

	Overall Accuracy 72%										
	Serpulids	***E. sepositus***	***H. papillosa***	***Axinella* sp.1**	***Axinella* sp.2**	**CCA+P**	**Green Algae 3**	**Sand**	***P. oceanica***	**Green Algae 1+2**	**TOT**	**UA**
Serpulids	7	0	0	0	0	0	0	0	0	0	7	**100.0**
E. sepositus	0	8	0	0	0	0	0	0	0	0	8	**100.0**
H. papillosa	0	0	8	1	0	0	0	0	0	0	9	**88.9**
Axinella sp.1	0	0	0	6	0	0	0	0	0	0	6	**100.0**
Axinella sp.2	0	0	0	0	8	0	0	0	0	0	8	**100.0**
CCA+P	1	0	0	1	0	8	0	1	0	0	11	**61.5**
Green algae 3	0	0	0	0	0	0	3	0	0	0	3	**100.0**
Sand	0	0	0	0	0	0	5	5	4	1	15	**25.0**
P. oceanica	0	0	0	0	0	0	0	0	4	0	4	**100.0**
Green algae 1+2	0	0	0	0	0	0	0	2	0	7	9	**70.0**
TOT	8	8	8	8	8	8	8	8	8	8	80	
PA	**87.5**	**100**	**100**	**75**	**100**	**100**	**37.5**	**62.5**	**50**	**87.5**		

4. Discussion

4.1. Evaluation of the Acquisition Set-up and Suggestion of Best Practice for Data Collection

The acquisition of high-quality UHI is challenging because of the requirements to perform a satisfactory survey, such as maintaining a constant speed, heading and altitude, as well as high-resolution navigation data [5–7]. Proper illumination is mandatory to avoid the acquisition of "striped" images, characterised by shadowed and hyper-illuminated areas, leading to misclassification of the acquired images.

In our survey, the ROV met most requirements, yet seafloor illumination and navigation data were not optimal. At both sites, the illumination was uneven due to a non-optimal configuration of the lamps (two lamps for the ROV RGB camera running simultaneously with two UHI lamps) and articulated topography. This condition affected the CWC site the most, where pixels on the western side of the image, characterised by a constant darker area, could not be strictly compared to those of the eastern side. For the coralligenous site, the presence of sunlight improved the condition, giving a more homogeneous illumination of the surveyed track, which was also favoured by a rather flat topography.

Furthermore, the low accuracy of the underwater positioning system sometimes induced image distortion and a lower spatial resolution of the geocorrected images, more evident at the CWC site. Finally, the difficulty in maintaining the correct ROV altitude in areas characterized by seafloor heterogeneity, a common trait at both study sites, may have influenced the reliability of UHI acquisition and, therefore, classification [26].

Based on our experience, we can summarise some best practices needed to acquire good quality hyperspectral images using an ROV. A rigorous UHI survey functional to seafloor mapping requires:

- an ROV ensuring a constant heading and altitude above the seafloor and suitable to host the UHI and other devices (e.g., RGB camera, lamps);
- an efficient positioning system for the ROV and the UHI itself, able to provide accurate and adequately dense navigation data;
- an appropriate lamp system to illuminate the surveyed area uniformly in function of water depth and sunlight;
- an RGB camera mounted vertically alike the UHI camera, to record concomitantly the seafloor for the OOI identification;
- an advanced background knowledge of the target area.

4.2. Evaluation of Spectral Libraries for Seafloor Mapping

A spectral library permits a quick and reliable classification of benthic habitats and their individual components up to taxonomic level, if a substantial number of reference spectra has been filed [7]. Its construction represents one of the most challenging and time-consuming aspects of the automatic classification of the UHI images.

In this perspective, it appears obvious that there is a strong need to implement substantially the spectral library with respect to the deep-water scleractinians (CWCs). Living *M. oculata* appears to contain a variety of colored facies from white to pinkish hues [64]. The same holds true for *D. pertusum*, a species also present at the CWC site here considered, whose living colonies cover a chromatic spectrum from white to orange, up to reddish [65,66]. Dead skeletons of these and other CWCs are often co-occurring, further complicating the hyperspectral approach as they are characterized by whitish, yellowish and brownish colours. However, UHI is documented [5] to discriminate the optical fingerprints of white, orange and dead *D. pertusum*.

A robust sponge taxonomic classification requires the analysis of spicules and genetics, while the external morphology or colour commonly provides only an indication of 'morpho-species' or 'morpho-categories' [67]. In the specific case of the deep-water situation here considered, the large fan-shaped sponges (i.a., *P. compressa* and *P. monilifera*, often co-occurring together with other sponges) are characterized by a wide range of colours even within the same taxon [36,68–70]. Therefore, collected spectra are not unequivocally associated to species, making difficult a precise taxonomic assignment.

The resilient core of the Coralligenous habitat is hard, hosting a variety of fauna and flora, the latter often seasonal and epibiontic. At times, such seasonal overgrowth may mask, often significantly but ephemerally, the underlying substrate provided by CCA and Peyssonelliales. These considerations are relevant in defining the spectral library associated with coralligenous as a unique habitat, when tested using the UHI camera. Concerning algae, supervised UHI classification was unable to map accurately different red algae species, due to their similarity on the optical fingerprint, deciding instead for grouping [26]. We tested that the spectral fingerprint of CCA and Peyssonelliales as an opting group is conspicuous enough to be distinguished in the natural environment, where their presence prevails with seasonal and accidental signals (such as green algae bloom or megabenthos).

4.3. SAM Classification

For our scope, the SAM classification is considered ideal because it is intensity independent (LED lamps and sunlight illumination) and focuses only on identifying the spectral similarity (i.e., colour). The SAM method can eliminate the effect of the spectral brightness values (i.e., spectral vector lengths in feature space) on the classification and it is insensitive to the data variance, imparting a significant advantage for the analysis of regions with complex terrain [63]. On the other hand, this method is highly dependent on the wavelength ranges and on the thresholds selected, which are arbitrary [62]. In our study, we choose several ROIs for the same benthic class for different illumination conditions, because the illumination is not influencing only the reflectance intensity, but also colour variation (e.g., colonial cnidarian 1 in Figure 2E results blue).

At the CWC site, SAM was functional in recognizing colonial cnidarians and sponges. However, the method proved inadequate in discriminating between mud and bedrock substrates, probably characterized here by similar spectra, hampering a reliable mapping of the seafloor (Figure 4 B).

The CCA+P and associated organisms were correctly classified at the coralligenous site, despite habitat heterogeneity. Green algae as *C. bursa* and *F. petiolata* (Green algae 1+2) appear overestimated, since SAM imported artifacts such as the build-up shadows and distortions (Figure 6).

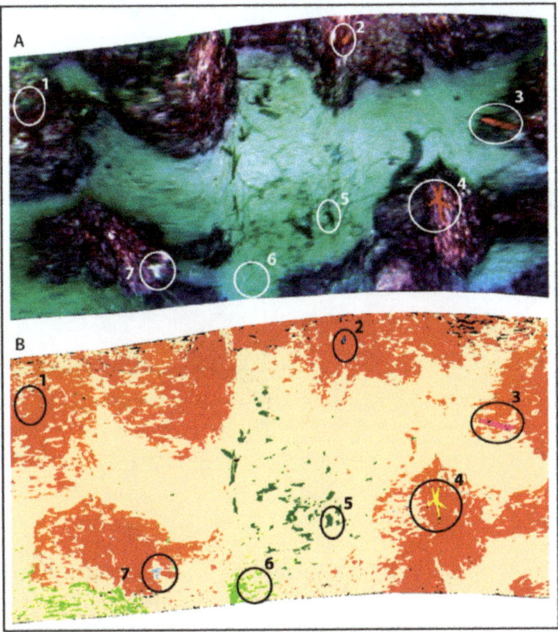

Figure 6. Zoom in the classified segment of coralligenous site, where numbers indicate the organism identified (**A**) and classified (**B**): 1. *C. bursa*, 2. *H. papillosa*, 3. *Axinella* sp. 1, 4. *E. sepositus*, 5. *C. prolifera*, 6. *P. oceanica*, 7. Serpulids. For the colour legend in B, see Figure 5.

According to confusion matrices limited to the dataset analysed in this study, the SAM classification accuracy is higher for the CWC (84.38%) than for the coralligenous (72%) site. This result could derive from differences in habitat complexity. Firm numbers on the overall classification accuracy could be obtained by increasing the number of iterations or considering a larger dataset.

4.4. Evaluation of the UHI for Seabed Monitoring

Our tests document that the UHI method is able to map the habitat extent independently of the water depth and at a high level of spatial detail. The UHI provides the effective spatial coverage of CWCs habitat-formers and coralligenous builds-ups (Figure 7), which is hard to estimate using conventional methods. This level of detail is extremely useful for monitoring purposes (e.g., MSFD) enabling a quantitative and repeatable measure of habitat extent and distribution. However, this process is heavily time-consuming compared to the conventional ROV approach, mostly due to inadequate spectral libraries, which is the major limitation to date.

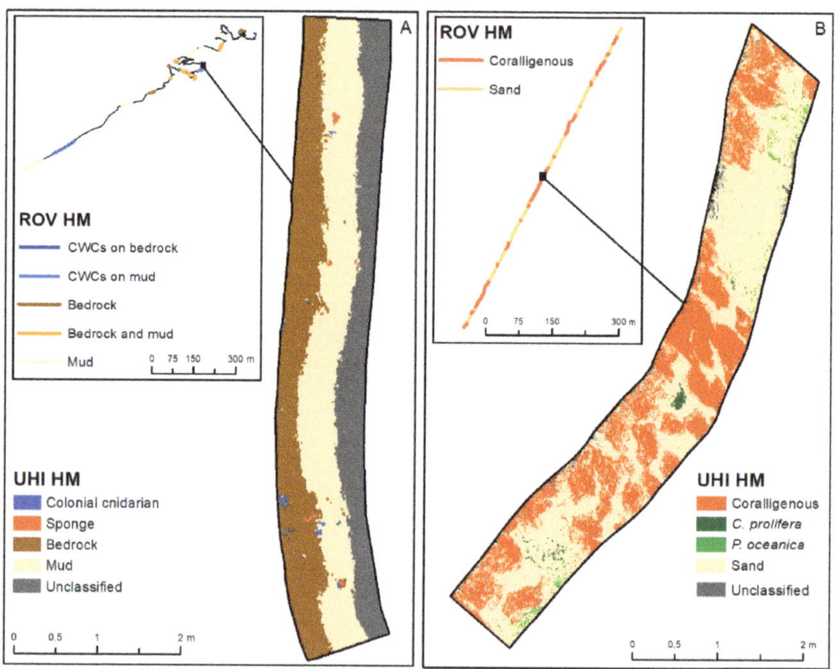

Figure 7. Comparison between the ROV transect classified with conventional methodologies (A and B insert) and the UHI classification for (**A**) CWC and (**B**) coralligenous sites. The image shows the higher level of detailed obtained by the UHI camera, in contrast with the larger amount of data analyzed by the ROV video.

5. Conclusions

This first application of the UHI camera in the Mediterranean Sea (Southern Adriatic Sea) confirmed its potential for underwater habitat mapping in shallow and deep water.

We tested the UHI camera in two geomorphological contexts containing charismatic marine benthic habitats. We noticed that the quality of the positioning system, the illumination settings and the complexity of the seafloor affected the UHI performance and the hyperspectral image analysis. We created a preliminary spectral library for each site enabling a supervised classification (SAM), which discriminated between substrates, megafauna and megaflora in a satisfactory manner.

Given substantially implemented spectral libraries, the UHI camera will likely represent a valid aid for habitat mapping and monitoring, in the perspective of quantifiable and repeatable classifications and European MSFD indicators.

Supplementary Materials: The UHI raw data analyzed in this work and relative documentation are stored as Data Research Object (RO) in ROHUB at the following link: http://www.rohub.org/rodetails/UHI_Data/overview.

Author Contributions: Conceptualization, Federica Foglini; Data curation, Valentina Grande, Fabio Marchese, Valentina Alice Bracchi, Mariacristina Prampolini, Ingrid Myrnes Hansen, Magne Gudmundsen and Agostino Niyonkuru Meroni; Formal analysis, Federica Foglini, Valentina Grande, Fabio Marchese, Valentina Alice Bracchi, Mariacristina Prampolini and Giorgio Castellan; Funding acquisition, Federica Foglini, Fabio Badalamenti, Cesare Corselli, Eleonora Martorelli and Alessandra Savini; Investigation, Federica Foglini, Valentina Grande, Fabio Marchese, Mariacristina Prampolini, Lorenzo Angeletti, Giovanni Chimienti, Agostino Niyonkuru Meroni, Alessandra Mercorella, Agostina Vertino and Marco Taviani; Resources, Ivar Erdal ; Validation, Federica Foglini; Writing—original draft, Federica Foglini, Valentina Grande, Fabio Marchese, Valentina Alice Bracchi, Mariacristina Prampolini, Lorenzo Angeletti, Giovanni Chimienti, Ingrid Myrnes Hansen, Fabio Badalamenti, Eleonora Martorelli and Marco Taviani.

Funding: Flagship Project RITMARE (La Ricerca Italiana per il Mare) and EVER-EST projects (ID: 674907).

Acknowledgments: Captain, crew and scientific staff on-board of the R/V Minerva Uno and three ROV pilots are acknowledged for their support during the operations at sea. The research is part of the National Flagship Project RITMARE (La Ricerca Italiana per il Mare), and a contribution to EU EVER-EST, EU IDEM and MIUR GLIDE projects. ISMAR-CNR Bologna scientific contribution n. 1996. This paper is a scientific contribution of Project MIUR—Dipartimenti di Eccellenza 2018–2022.

Conflicts of Interest: The authors declare no conflict of interest.

References

1. Brown, C.J.; Smith, S.J.; Lawton, P.; Andersond, J.T. Benthic habitat mapping: A review of progress towards improved understanding of the spatial ecology of the seafloor using acoustic techniques. *Estuar. Coast. Shelf Sci.* **2011**, *92*, 502–520. [CrossRef]
2. Angeletti, L.; Bargain, A.; Foglini, F.; Grande, V.; Prampolini, M.; Taviani, M. Cold-water coral multiscale habitat mapping: Methodologies and perspectives. In *Mediterranean Cold-Water Corals: Past, Present and Future, Coral Reefs of the World*; Orejas, C., Jiménez, C., Eds.; Springer International Publishing: Berlin, Germany, 2019; Volume 9.
3. Lim, A.; Wheeler, A.J.; Arnaubec, A. High-resolution facies zonation within a cold-water coral mound: The case of the Piddington Mound, Porcupine Seabight, NE Atlantic. *Mar. Geol.* **2017**, *390*, 120–130. [CrossRef]
4. Berg, T.; Fürhaupter, K.; Teixeira, H.; Uusitalo, L.; Zampoukas, N. The Marine Strategy Framework Directive and the ecosystem-based approach–pitfalls and solutions. *Mar. Pollut. Bull.* **2015**, *96*, 18–28. [CrossRef]
5. Johnsen, G.; Volent, Z.; Dierssen, H.; Pettersen, R.; Ardelan, M.V.; Søreide, F.; Fearns, P.; Ludvigsen, M.; Moline, M. Underwater hyperspectral imagery to create biogeochemical maps of seafloor properties. In *Subsea Optics and Imaging*, 1st ed.; Watson, J., Zielinski, O., Eds.; Elsevier: Amsterdam, the Netherlands, 2013; pp. 508–540.
6. Dumke, I.; Nornes, S.M.; Purser, A.; Marcon, Y.; Ludvigsen, M.; Ellefmo, S.L.; Johnsen, G.; Søreide, F. First hyperspectral imaging survey of the deep seafloor: High-resolution mapping of manganese nodules. *Remote Sens. Environ.* **2018**, *209*, 19–30. [CrossRef]
7. Dumke, I.; Purser, A.; Marcon, Y.; Nornes, S.M.; Johnsen, G.; Ludvigsen, M.; Søreide, F. Underwater hyperspectral imaging as an in situ taxonomic tool for deep-sea megafauna. *Sci. Rep.* **2018**, *8*, 12860. [CrossRef] [PubMed]
8. Klonowski, W.M.; Fearns, P.R.C.S.; Lynch, M.J. Retrieving key benthic cover types and bathymetry from hyperspectral imagery. *JARS* **2007**, *1*, 011505. [CrossRef]
9. Volent, Z.; Johnsen, G.; Sigernes, F. Kelp forest mapping by use of airborne hyperspectral imager. *J. Appl. Remote Sens.* **2007**, *1*, 011505. [CrossRef]
10. Fearns, P.R.C.; Klonowski, W.; Babcock, R.C.; England, P.; Phillips, J. Shallow water substrate mapping using hyperspectral remote sensing. *Cont. Shelf Res.* **2011**, *31*, 1249–1259. [CrossRef]
11. Chang, C.; Member, S.; Du, Q. Estimation of Number of Spectrally Distinct Signal Sources in Hyperspectral Imagery. *IEEE Trans. Geosci. Remote Sens.* **2004**, *42*, 608–619. [CrossRef]

12. Dickey, T.; Lewis, M.; Chang, G. Optical oceanography: Recent advances and future directions using global remote sensing and in situ observations. *Rev. Geophys.* **2006**, *44*, 1–39. [CrossRef]
13. Dierssen, H.M.; Randolph, K. Remote Sensing of Ocean Color. In *Earth System Monitoring*; Orcutt, J., Ed.; Springer: New York, NY, USA, 2013; pp. 439–472. ISBN 978-1-4614-5683-4.
14. Hochberg, E.J.; Atkinson, M.J. Spectral discrimination of coral reef benthic communities. *Coral Reefs* **2000**, *19*, 164–171. [CrossRef]
15. Hochberg, E.J.; Atkinson, M.J.; Andréfouët, S. Spectral reflectance of coral reef bottom-types worldwide and implications for coral reef remote sensing. *Remote Sens. Environ.* **2003**, *85*, 159–173. [CrossRef]
16. Kutser, T.; Miller, I.; Jupp, D.L.B. Mapping coral reef benthic substrates using hyperspectral space-borne images and spectral libraries. *Estuar. Coast. Shelf Sci.* **2006**, *70*, 449–460. [CrossRef]
17. Petit, T.; Bajjouk, T.; Mouquet, P.; Rochette, S.; Vozel, B.; Delacourt, C. Hyperspectral remote sensing of coral reefs by semi-analytical model inversion—Comparison of different inversion setups. *Remote Sens. Environ.* **2017**, *190*, 348–365. [CrossRef]
18. Phinn, S.; Roelfsema, C.; Dekker, A.; Brando, V.; Anstee, J. Mapping seagrass species, cover and biomass in shallow waters: An assessment of satellite multi-spectral and airborne hyper-spectral imaging systems in Moreton Bay (Australia). *Remote Sens. Environ.* **2008**, *112*, 3413–3425. [CrossRef]
19. Dierssen, H.M. Overview of hyperspectral remote sensing for mapping marine benthic habitats from airborne and underwater sensors. In *Imaging Spectrometry XVIII, Proceedings of SPIE- International Society for Optics and Photonics, San Diego, CA, USA, 26–28 August 2013*; Mouroulis, P., Ed.; 2013; Volume 8870.
20. Dierssen, H.M.; Chlus, A.; Russell, B. Hyperspectral discrimination of floating mats of seagrass wrack and the macroalgae *Sargassum* in coastal waters of Greater Florida Bay using airborne remote sensing. *Remote Sens. Environ.* **2015**, *167*, 247–258. [CrossRef]
21. Chennu, A.; Färber, P.; Volkenborn, N.; Al-Najjar, M.A.A.; Janssen, F.; de Beer, D.; Polerecky, L. Hyperspectral imaging of the microscale distribution and dynamics of microphytobenthos in intertidal sediments. *Limnol. Oceanogr. Methods* **2013**, *11*, 511–528. [CrossRef]
22. Pettersen, R.; Johnsen, G.; Bruheim, P.; Andreassen, T. Development of hyperspectral imaging as a bio-optical taxonomic tool for pigmented marine organisms. *Org. Divers. Evol.* **2014**, *14*, 237–246. [CrossRef]
23. Ludvigsen, M.; GJohnsen, G.; Sørensen, A.J.; Lågstad, P.A.; Ødegård, Ø. Scientific operations combining ROV and AUV in the Trondheim Fjord. *Mar. Technol. Soc. J.* **2014**, *48*, 59–71. [CrossRef]
24. Tegdan, J.; Ekehaug, S.; Hansen, I.M.; Sandvik Aas, L.M.; Steen, K.J.; Pettersen, R.; Beuchel, F.; Camus, L. Underwater hyperspectral imaging for environmental mapping and monitoring of seabed habitats. In Proceedings of the OCEANS 2015, Genova, Italy, 18–21 May 2015; pp. 1–6.
25. Johnsen, G.; Ludvigsen, M.; Sørensen, A.; Sandvik Aas, L.M. The use of underwater hyperspectral imaging deployed on remotely operated vehicles—Methods and applications. *IFAC* **2016**, *49*, 476–481. [CrossRef]
26. Mogstad, A.A.; Johnsen, G. Spectral characteristics of coralline algae: A multi-instrumental approach, with emphasis on underwater hyperspectral imaging. *Appl. Opt.* **2017**, *56*, 9957. [CrossRef]
27. Ødegård, Ø.; Mogstad, A.A.; Johnsen, G.; Sørensen, A.J.; Ludvigsen, M. Underwater hyperspectral imaging: A new tool for marine archaeology. *Appl. Opt.* **2018**, *57*, 3214. [CrossRef]
28. Sture, Ø.; Ludvigsen, M.; Søreide, F.; Sandvik Aas, L.M. Autonomous underwater vehicles as a platform for underwater hyperspectral imaging. In Proceedings of the OCEANS 2017, Aberdeen, UK, 19–22 June 2017; pp. 1–8.
29. Cochrane, S.K.J.; Ekehaug, S.; Refit, E.C.; Hansen, I.M.; Sandvik Aas, L.M. Detection of deposited drill cuttings on the sea floor—A comparison between underwater hyperspectral imagery and the human eye. *Mar. Pollut. Bull.* **2019**. in review.
30. Ødegård, Ø.; Sørensen, A.J.; Hansen, R.E.; Ludvigsen, M. A new method for underwater archaeological surveing using sensors and unmanned platforms. *IFAC-PapersOnLine* **2016**, *49*, 486–493.
31. Letnes, P.A.; Hansen, I.M.; Sandvik Aas, L.M.; Eide, I.; Pettersen, R.; Tassara, L. Underwater hyperspectral classification of deep sea corals exposed to 2-methylnaphthalene. *PLoS ONE* **2019**, *14*, e0209960. [CrossRef]
32. Foglini, F.; Angeletti, L.; Bracchi, V.A.; Chimienti, G.; Grande, V.; Hansen, I.M.; Meroni, A.N.; Marchese, F.; Mercorella, A.; Prampolini, M.; et al. Underwater Hyperspectral Imaging for seafloor and benthic habitat mapping. In Proceedings of the 2018 IEEE International Workshop on Metrology for the sea (MetroSea 2018), Bari, Italy, 8–10 October 2018; pp. 201–205.

33. Gavazzi, G.M.; Madricardo, F.; Janowski, L.; Kruss, A.; Blondel, P.; Sigovini, M.; Foglini, F. Evaluation of seabed mapping methods for fine-scale classification of extremely shallow benthic habitats—Application to the Venice Lagoon, Italy. *Estuar. Coast. Shelf Sci.* **2016**, *170*, 45–60. [CrossRef]
34. Ingrosso, G.; Abbiati, M.; Badalamenti, F.; Bavestrello, G.; Belmonte, G.; Cannas, R.; Benedetti-Cecchi, L.; Bertolino, M.; Bevilacqua, S.; Bianchi, C.N.; et al. Mediterranean Bioconstructions Along the Italian Coast. *Adv. Mar. Biol.* **2018**, *79*, 61–136.
35. Freiwald, A.; Beuck, L.; Rüggeberg, A.; Taviani, M.; Hebbeln, D.; R/V Meteor Cruise M70-1 Participants. The white coral community in the Central Mediterranean Sea Revealed by ROV Surveys. *Oceanography* **2009**, *22*, 58–74. [CrossRef]
36. Angeletti, L.; Taviani, M.; Canese, S.; Foglini, F.; Mastrototaro, F.; Argnani, A.; Trincardi, F.; Bakran-Petricioli, T.; Ceregato, A.; Chimienti, G.; et al. New deep-water cnidarian sites in the southern Adriatic Sea. *Mediterr. Mar. Sci.* **2014**, *15*, 263–273. [CrossRef]
37. D'Onghia, G.; Capezzuto, F.; Cardone, F.; Carlucci, R.; Carluccio, A.; Chimienti, G.; Corriero, G.; Longo, C.; Maiorano, P.; Mastrototaro, F.; et al. Macro-and megafauna recorded in the submarine Bari Canyon (southern Adriatic, Mediterranean Sea) using different tools. *Mediterr. Mar. Sci.* **2015**, *16*, 180–196. [CrossRef]
38. Taviani, M.; Angeletti, L.; Beuck, L.; Campiani, E.; Canese, S.; Foglini, F.; Freiwald, A.; Montagna, P.; Trincardi, F. Reprint of 'On and off the beaten track: Megafaunal sessile life and Adriatic cascading processes'. *Mar. Geol.* **2016**, *375*, 146–160. [CrossRef]
39. Taviani, M.; Angeletti, L.; Cardone, F.; Montagna, P.; Danovaro, R. A unique and threatened deep water coral-bivalve biotope new to the Mediterranean Sea offshore the Naples megalopolis. *Sci. Rep.* **2019**, *9*, 3411. [CrossRef] [PubMed]
40. Chimienti, G.; Bo, M.; Taviani, M.; Mastrototaro, F. Occurrence and Biogeography of Mediterranean Cold-Water Corals. In *Mediterranean Cold-Water Corals: Past, Present and Future, Coral Reefs of the World*; Orejas, C., Jiménez, C., Eds.; Springer International Publishing: Berlin, Germany, 2019; Volume 9. (in press)
41. Foglini, F.; Angeletti, L.; Campiani, E.; Correggiari, A.; Grande, V.; Leidi, E.; Madricardo, F.; Mercorella, A.; Remia, R.; Taviani, M. Habitat mapping in the Adriatic (Mediterranean Sea) from coastal areas to deep sea: Approaches and methodologies for assessing seafloor habitat for sustainable and integrated sea management strategy. In Proceedings of the GeoHab 2015, Salvador, Brazil, 3–8 May 2015.
42. Sanfilippo, R.; Vertino, A.; Rosso, A.; Beuck, L.; Freiwald, A.; Taviani, M. *Serpula* aggregates and their role in deep-sea coral communities in the southern Adriatic Sea. *Facies* **2013**, *59*, 663–677. [CrossRef]
43. Bargain, A.; Foglini, F.; Pairaud, I.; Bonaldo, D.; Carniel, S.; Angeletti, L.; Taviani, M.; Rochette, S.; Fabri, M.C. Predictive habitat modeling in two Mediterranean canyons including hydrodynamic variables. *Prog. Oceanogr.* **2018**, *169*, 151–168. [CrossRef]
44. Addamo, A.M.; Vertino, A.; Stolarski, J.; García-Jiménez, R.; Taviani, M.; Machordom, A. Merging scleractinian genera: The overwhelming genetic similarity between solitary *Desmophyllum* and colonial *Lophelia*. *BMC Evol. Biol.* **2016**, *16*, 108. [CrossRef]
45. Bracchi, V.; Savini, A.; Marchese, F.; Palamara, S.; Basso, D.; Corselli, C. Coralligenous habitat in the Mediterranean Sea: A geomorphological description from remote data. *Ital. J. Geosci.* **2015**, *134*, 32–40. [CrossRef]
46. Bracchi, V.A.; Basso, D.; Marchese, F.; Corselli, C.; Savini, A. Coralligenous morphotypes on subhorizontal substrate: A new categorization. *Cont. Shelf Res.* **2017**, *144*, 10–20. [CrossRef]
47. Laborel, J. Le concrétionnement algal 'coralligène' et son importance geomorphologique en Méditerranée. *Recueil des travaux Station Marine d'Endoume* **1961**, *23*, 37–60.
48. Pérès, J.M.; Picard, J. Nouveau manuel de bionomie benthique de la mer Méditerranée. *Recent Trav. De La Stn. Mar. De Endoume* **1964**, *31*, 1–137.
49. Bellan-Santini, D.; Lacaze, J.C.; Poizat, C. Les biocénoses marines et littorales de Méditerranée, synthèse, menaces et perspectives. *Collection Patrimoines Naturels. Muséum National d'Histoire Naturelle* **1994**, *19*, 1–246.
50. Bressan, G.; Babbini, I.; Ghirardelli, L.; Basso, D. Bio-costruzione e bio-distruzione di corallinales nel Mar Mediterraneo. *Biol. Mar. Mediterr.* **2001**, *8*, 131–174.
51. Ballesteros, E. Mediterranean coralligenous assemblages: A synthesis of present knowledge. *Oceanogr. Mar. Biol. Annu. Rev.* **2006**, *44*, 123–195.
52. Piazzi, L.; Gennaro, P.; Balata, D. Threats to macroalgal coralligenous assemblages in the Mediterranean Sea. *Mar. Poll. Bull.* **2012**, *64*, 2623–2629. [CrossRef]

53. Chimienti, G.; Stithou, M.; Mura, I.D.; Mastrototaro, F.; D'Onghia, G.; Tursi, A.; Izzi, C.; Fraschetti, S. An explorative assessment of the importance of Mediterranean Coralligenous habitat to local economy: The case of recreational diving. *J. Environ. Account. Manag.* **2017**, *5*, 315–325. [CrossRef]
54. Sarà, M. *Research on Benthic Fauna of Southern Adriatic Italian Coast: Final Scientific Report*; Office of Naval Research: Washington, DC, USA, 1968; pp. 1–53.
55. Sarà, M. Un biotopo da proteggere: Il coralligeno pugliese. In Proceedings of the Atti del I Simposio Nazionale sulla Conservazione della Natura, Bari, Italy, 21–25 April 1971; pp. 145–151.
56. Angeletti, L.; Prampolini, M.; Foglini, F.; Grande, V.; Taviani, M. Cold-water coral habitat in the Bari Canyon System, Southern Adriatic Sea (Mediterranean Sea). In *Seafloor Geomorphology as Benthic Habitat*, 2nd ed.; Harris, P., Baker, E., Eds.; Elsevier: Amsterdam, the Netherlands, 2019. (in press)
57. Johnsen, G.; Volent, Z.; Sakshaug, E.; Sigernes, F.; Pettersson, L. Remote sensing in the Barents Sea. In *Ecosystem Barents Sea*; Sakshaug, E., Johnsen, G.H., Kovacs, K.M., Eds.; Fagbokforlaget: Bergen, Norway, 2009; Volume 2, pp. 139–166.
58. Dukan, F.; Ludvigsen, M.; Sorensen, A.J. Dynamic positioning system for a small size ROV with experimental results. In Proceedings of the IEEE OCEANS, Santander, Spain, 6–9 June 2011; pp. 1–10. [CrossRef]
59. Williams, D.J.; Shah, M. A fast algorithm for active contours and curvature estimation. *CVGIP Image Under.* **1992**, *55*, 14–26. [CrossRef]
60. R Core Team. *R: A Language and Environment for Statistical Computing*; R Foundation for Statistical Computing: Vienna, Austria, 2013; Available online: http://www.R-project.org/ (accessed on 18 May 2019).
61. Kruse, F.A.; Heidebrecht, K.B.; Shapiro, A.T.; Barloon, P.J.; Goetz, A.F.H. The Spectral Image Processing System (SIPS) Interactive Visualization and Analysis of Imaging Spectrometer Data. *Remote Sens. Environ.* **1993**, *44*, 145–163. [CrossRef]
62. Crósta, A.P.; Sabine, C.; Taranik, J.V. Hydrothermal Alteration Mapping at Bodie, California, Using AVIRIS Hyperspectral Data. *Remote Sens. Environ.* **1998**, *64*, 309–319. [CrossRef]
63. Liu, Y.; Lu, S.; Lu, X.; Wang, Z.; Chen, C.; He, H. Classification of Urban Hyperspectral Remote Sensing Imagery Based on Optimized Spectral Angle Mapping. *J. Indian Soc. Remote Sens.* **2019**, *47*, 289–294. [CrossRef]
64. Boland, G.S.; Etnoyer, P.J.; Fisher, C.R.; Hickerson, E.L. State of deep-sea coral and sponge ecosystems of the Gulf of Mexico Region: Texas to the Florida Straits. In *The State of Deep-Sea Coral and Sponge Ecosystems of the United States*; Hourigan, T.F., Etnoyer, P.J., Cairns, S.D., Eds.; NOAA Tech. Memo. NMFS-OHC-4; Silver Spring: Silver, MD, USA, 2017; pp. 321–378.
65. Freiwald, A. Geobiology of *Lophelia pertusa* (Scleractinia) Reefs in the North Atlantic. Unpublished Habilitation Thesis, Bremen University, Bremen, Germany, 1998.
66. Freiwald, A.; Fosså, J.H.; Grehan, A.; Koslow, T.; Roberts, J.M. Cold-water coral reefs—Out of sight—No longer out of mind. In *UNEP-WCMC Biodiversity*; UNEP Coral Reef Unit: Cambridge, UK, 2004; Volume 22, pp. 1–84.
67. Santín, A.; Grynó, J.; Ambroso, S.; Uriz, M.J.; Dominguez-Carrió, C.; Gili, J.M. Distribution patterns and demographic trends of demosponges at the Menorca Channel (Northwestern Mediterranean Sea). *Prog. Oceanogr.* **2019**, *173*, 9–25. [CrossRef]
68. Bell, J.J.; Barnes, D.K.A. Sponge morphological diversity: A qualitative predictor of species diversity? *Aquat. Conserv.* **2001**, *11*, 109–121. [CrossRef]
69. Calcinai, B.; Moratti, V.; Martinelli, M.; Bavestrello, G.; Taviani, M. Uncommon sponges associated with deep coral bank and maerl habitats in the Strait of Sicily (Mediterranean Sea). *Ital. J. Zool.* **2013**, *80*, 412–423. [CrossRef]
70. Rueda, J.L.; Urra, J.; Aguilar, R.; Angeletti, L.; Bo, M.; García-Ruiz, C.; González-Duarte, M.; López, E.; Madurell, T.; Maldonado, M.; et al. Cold-water coral associated fauna in the Mediterranean Sea and adjacent areas. In *Mediterranean Cold-Water Corals: Past, Present and Future, Coral Reefs of the World*; Orejas, C., Jiménez, C., Eds.; Springer International Publishing: Berlin, Germany, 2019; Volume 9.

MDPI
St. Alban-Anlage 66
4052 Basel
Switzerland
Tel. +41 61 683 77 34
Fax +41 61 302 89 18
www.mdpi.com

Sensors Editorial Office
E-mail: sensors@mdpi.com
www.mdpi.com/journal/sensors

www.ingramcontent.com/pod-product-compliance
Lightning Source LLC
LaVergne TN
LVHW080504180726
843515LV00016B/1389

* 9 7 8 3 0 3 9 2 8 4 0 6 1 *